Springer Proceedings in Mathematics & Statistics

Volume 275

Springer Proceedings in Mathematics & Statistics

This book series features volumes composed of selected contributions from workshops and conferences in all areas of current research in mathematics and statistics, including operation research and optimization. In addition to an overall evaluation of the interest, scientific quality, and timeliness of each proposal at the hands of the publisher, individual contributions are all refereed to the high quality standards of leading journals in the field. Thus, this series provides the research community with well-edited, authoritative reports on developments in the most exciting areas of mathematical and statistical research today.

More information about this series at http://www.springer.com/series/10533

Julio Delgado · Michael Ruzhansky
Editors

Analysis and Partial Differential Equations: Perspectives from Developing Countries

Imperial College London, UK, 2016

Springer

Editors
Julio Delgado
School of Mathematical Sciences
Queen Mary University of London
London, UK

Michael Ruzhansky
Department of Mathematics: Analysis, Logic and Discrete Mathematics
Ghent University
Ghent, Belgium

School of Mathematical Sciences
Queen Mary University of London
London, UK

ISSN 2194-1009 ISSN 2194-1017 (electronic)
Springer Proceedings in Mathematics & Statistics
ISBN 978-3-030-05656-8 ISBN 978-3-030-05657-5 (eBook)
https://doi.org/10.1007/978-3-030-05657-5

Library of Congress Control Number: 2018964685

Mathematics Subject Classification (2010): 47Fxx, 47F05, 47G10, 47G30, 34K30, 42C05, 65M70, 49K20, 45B05, 35S05, 01-XX, 00A99

This Springer imprint is published by the registered company Springer Nature Switzerland AG
The registered company address is: Gewerbestrasse 11, 6330 Cham, Switzerland

This volume is dedicated to the memory of Professor Francis Kofi Allotey who passed away on the 2nd November 2017. Professor Allotey was a prominent Ghanian mathematical physicist and a remarkable promotor of science in Africa. Professor Allotey was visiting us at Imperial College London in February 2017 as part of the project 'Analysis in Developing Countries' which was the starting point of this volume. The volume contains his biography as well as his last paper 'State of Mathematics in Africa and the way forward' that he has kindly written for our volume following his visit.

Preface

The present volume is an attempt to collect some of the research in analysis that is being currently conducted in the developing countries. Certainly, it does not pretend to make a complete overview but to give a glimpse into various research topics and directions in the framework of developing countries activities.

The starting point of these activities has been the project 'Analysis in Developing Countries' that the editors of this volume have been organising at Imperial College London in 2016–2017. Part of it was realised as a conference 'Analysis and Partial Differential Equations' that took place at Imperial College London during 26–30 September 2016, followed by a series of smaller intensive meetings. During these activities, mathematicians from many developing countries had a chance of visiting Imperial College London for presenting their work and for the subsequent productive exchange of ideas, supported by the EPSRC GCRF grant administered by Imperial College London.

More specifically, mathematicians from *Burkina Faso, Ethiopia, Ghana, Jamaica, Cuba, Costa Rica, Uzbekistan, Kyrgyzstan, Armenia, Azerbaijan, India, Vietnam, Brazil, Colombia, Kazakhstan, Serbia and China* took part in these events, complemented by a remote participation to various extents by further colleagues from *Benin, Senegal, Morocco, Iran, Chile, Tajikistan and Romania.*

Part of this took place also during the conference 'Noncommutative Analysis and Partial Differential Equations' that we organised at Imperial College London during 11–15 April 2016.

Altogether, this volume is one of the results of these activities. In preparing it, we have decided to combine the papers by participants from the developing countries with those by colleagues from more developed ones, to put the presented research in a broader context, emphasising their interplays and mutual interactions.

As such, the purpose of this volume is to exhibit some of the research activities related to the mathematical analysis in a broad sense, taking place in developing countries, to facilitate further exchanges and cooperations between mathematicians in different countries.

Among the mathematical papers, the volume also contains the biography of the late Prof. Francis Allotey and his paper 'State of Mathematics in Africa and the way forward'. He has written it following our request to write about his historical perspective and his vision for the development of mathematics in Africa. This request was made during his visit to Imperial College London in February 2017, and of course, we had no idea that this could become his last paper: we are privileged to include it in our volume.

We would like to thank all the participants and contributors as well as EPSRC and Imperial College London for providing funds for supporting the participants, enabling their travels to these meetings.

June 2018

Julio Delgado
Michael Ruzhansky

Contents

Contributors

Elmira Abdyldaeva Department of Mathematics, Kyrgyz-Turkish University 'Manas', Bishkek, Kyrgyzstan

Francis K. A. Allotey University of Ghana, Accra, Ghana

Rakesh Balhara Centre for Applicable Mathematics (CAM), Tata Institute of Fundamental Research, Bangalore, India

Edgardo Samuel Barraza Department of Computer Science and Artificial Intelligence, Universidad de Sevilla, Seville, Spain

Divyang G. Bhimani Centre for Applicable Mathematics (CAM), Tata Institute of Fundamental Research, Bangalore, India

Andrea Bonfiglioli Dipartimento di Matematica, Università degli Studi di Bologna, Bologna, Italy

Duván Cardona Mathematics Department, Pontificia Universidad Javeriana, Bogotá, Colombia

Victor Chabu Universidade de Sao Paulo, IF-USP, DFMA, CP 66.318, Sao Paulo, SP, Brazil

Giani Egaña Fernández Facultad de Matemática y Computación, San Lázaro y L, Vedado, Universidad de La Habana, Havana, Cuba

Emmanuel K. Essel Department of Mathematics and Statistics, University of Cape Coast, Cape Coast, Ghana

Clotilde Fermanian-Kammerer LAMA, UMR CNRS 8050, Université Paris Est, Créteil Cedex, France

Martin G. Grigoryan Yerevan State University, Yerevan, Armenia

Tigran M. Grigoryan Yerevan State University, Yerevan, Armenia

Qi Huang School of Mathematics and System Science, Beihang University, Beijing, China

Akylbek Kerimbekov Department of Applied Mathematics and Informatics, Kyrgyz-Russian Slavic University, Bishkek, Kyrgyzstan

Fabricio Macià Universidad Politécnica de Madrid, DCAIN, ETSI Navales, Madrid, Spain

Marius Măntoiu Name, Facultad de Ciencias, Departamento de Matematicas, Universidad de Chile, Santiago, Chile

Bo Meng School of Energy and Power Enigineering, Beihang University, Beijing, China

Juan Carlos Muñoz Grajales Departamento de Matemáticas, Universidad del Valle, Cali, Colombia

Isaac Z. Pesenson Department of Mathematics, Temple University, Philadelphia, 19122, USA

Nusrat Rajabov Research Institute of the Tajik National University, Dushanbe, Tajikistan

Mariano Rodríguez Ricard Facultad de Matemática y Computación, San Lázaro y L, Vedado, Universidad de La Habana, Havana, Cuba

Maximiliano Sandoval Facultad de Ciencias, Universidad de Chile, Santiago, Chile

J. Sarría González Facultad 4, Universidad de Ciencias Informáticas, Havana, Cuba

L. S. Simonyan Yerevan State University, Yerevan, Armenia

Sundaram Thangavelu Centre for Applicable Mathematics (CAM), Tata Institute of Fundamental Research, Bangalore, India

Hamidou Toure Unité de Recherche et de Formation en Sciences Exactes et Appliquées, Département de Mathématiques, Université de Ouaga 1 Pr Joseph Ki-Zerbo, Ouagadougou 03, Burkina Faso

Luisa Fernanda Vargas Departamento de Ciencias Naturales y Matemáticas, Universidad Javeriana, Cali, Colombia

Haihui Wang School of Mathematics and System Science, Beihang University, Beijing, China

Issa Zabsonre Unité de Recherche et de Formation en Sciences Exactes et Appliquées, Département de Mathématiques, Université de Ouaga 1 Pr Joseph Ki-Zerbo, Ouagadougou 03, Burkina Faso

Analysis in Developing Countries

Francis K. A. Allotey (Saltpond, Ghana 9 August 1932 – Accra, Ghana 2 November 2017)

Emmanuel K. Essel(✉)

Department of Mathematics and Statistics, University of Cape Coast, Cape Coast, Ghana
ekessel04@yahoo.co.uk

Abstract. This is a brief Biography of Professor Francis K. Allotey.

Keywords: History and biography · History of mathematics · Mathematicians

1 Francis Kofi Ampenyin Allotey

Professor Francis Kofi Ampenyin Allotey, who filled the scientific and academic world with its fragrance was a Professor of Mathematics, Scholar, Nuclear Physicist and a Consultant in Informatics for Development. Prof Allotey, a prodigy, born in August 1932 to a humble parentage at Saltpond in the Central Region of Ghana, received his early education at the Ghana National College among the 1952 pioneer group. He pursued further studies at the University Tutorial College, London Borough Polytechnic and London Imperial College of Science and Technology where he obtained the then coveted Diploma of Imperial College (London) in 1960.

He returned in the same year to take up a lectureship position in the Department of Mathematics at Kwame Nkrumah University of Science and Technology (KNUST). He had a distinguished academic career at the KNUST rising from the position of a lecturer to a full Professor of Mathematics and Head of Department within seven years. This was after his return in 1966 from study leave at Princeton University where he obtained his Master's and Doctorate degrees.

Founder and First Director of the KNUST Computer Centre, he was the first to introduce computer education into Ghana. The Faculty Board between 1971 and 1980 elected him four times as the Dean of the Faculty of Science. During this same period, he was elected five times as a Representative of the Academic Staff. He was ultimately appointed as the Pro-Vice Chancellor of the University in 1978. He became a world authority and an instant fame with his work on Soft X-Ray Spectroscopy which established the principle widely known as the

J. Delgado and M. Ruzhansky (Eds.): *Analysis and Partial Differential Equations: Perspectives from Developing Countries*, PROMS 275, pp. 3–6, 2018.
https://doi.org/10.1007/978-3-030-05657-5_1

"Allotey Formalism" for which he received the Prince Philip Gold Medal Award in 1973.

At the national level he held the positions of Chairman, Ghana Atomic Energy Commission, Chairman, Council for Scientific and Industrial Research, Chairman of the Management Board, Soil Research Institute, Chairman, Ghana Technical Committee on Nuclear Energy, Vice-President, Ghana Academy of Arts and Sciences, President, Ghana Institute of Physics, National Vice-President Ghana Science Association, President, Mathematical Association of Ghana, Vice-Chairman of the Science and Technology Sector, National Development Commission,Consultant, National Committee of Curriculum Development, Founder and National Co-ordinator, Ghana Energy Research Group, President, Ghana Institute of Physics and Member National Energy Commission, just to mention a few.

At the international level he gave several seminars in Universities in USA, Europe, Asia and Africa on his Research Work and participated and contributed in many International Conferences and Symposia by invitation. He published many papers in international journals. He acted as external examiner and also as an external assessor for professorial appointments for Universities inside and outside Ghana.

He was a Consultant to several International Institutions, including United Nations Organization, United Nations Educational Scientific and Cultural Organization (UNESCO), International Atomic Energy Agency, Intergovernmental Bureau for Information, and United Nations Industrial and Development Organization.

He also visited many nuclear installations in Russia, Poland, East Germany, Iraq, USA, India, West Germany, United Kingdom, Sweden, Yugoslavia etc.

He was involved locally and internationally on policies and issues related to science and technology for development. Thus he was a focal person for UN Advanced Technology Alert System since its formation. He was a Co-author of the Book "Comprehensive study of Nuclear Weapons", a UN Secretary General's Report.

He held numerous appointments, including the following: Visiting Scientist, International Centre for Theoretical Physics (ICTP), President, Society of African Physicists and Mathematicians,

Member, International panel of authors who wrote Entebbe Mathematics Series for Africa, Visiting Scientist, Chalmers University of Technology, Gothenburg, Sweden, Appointed to contribute a chapter on a book "Band Spectroscopy of Metals and Alloys" in honor of the 70th birthday of Sir Neville Mott, a Nobel Prize Winner.

Member of the organizing panel for the 3rd Congress held in Karlsruhe for the International Commission on Mathematical Instructions, Director, Summer College of Teaching of Physics at the tertiary level, International Centre for Theoretical Physics, Trieste, Italy,
Member of the Editorial Board of the Ethiopian Journal of science, Vice President, African Union of Physicists,

Member, Advisory Planning Committee of the 1984 International Conference on the Physics of super lattices, super structure interfaces and micro devices, USA, President, Society of African Physicists and Mathematicians, Co-Founder and joint organizer of the Annual Abidjan Symposium of Mathematics Series,
Organizer and Chairman, Workshop on curriculum development and design in Physics, Mathematics and Computer Science for Universities in Africa, Organizer and Chairman, State of Physics and Mathematics in Africa,
Member of International experts who did feasibility studies for publishing a journal of science for Africa, Innovation and Discovery,
Member of Executive, Africa Mathematical Union, Organizer and Chairman, First Addis Ababa School of Applicability of Environmental Physics and Meteorology in Africa,
Member, Steering Committee for the Energy Agency beyond the year 2000",
Member, UNESCO Physics Action Council,
Member, Scientific Council of the International Centre for Theoretical Physics,
Member, Technical Advisory Committee, Commission on Science and Technology for Sustainable Development in the South, Governor, International Atomic Energy Agency,
Member of Preparatory Committee and Vice-President of the Conference on the Non-Proliferation of Nuclear Weapons in Geneva, Switzerland, Appointed by the Secretary-General of United Nations to be among a group of 12 experts commissioned to advise United Nations in Nuclear Weapons,
Member, Sub-Committee on Technology transfer of the UN Conference on New and renewable Energy Sources, Vice-President of the Preparatory Committee of the UN Conference for promotion of international peaceful uses of Nuclear Energy, Organizer and Chairman,UNESCO/ANSTI International Conference on Renewable Energy, Vice president, 7th General Assembly of Intergovernmental Bureau of Informatics (IBI), Organizer and Chairman of International UNESCO and IBI Conference on Informatics for Development, Chairman, Panel Discussion on Financial and quantitative aspects of Computer Education in developing countries in France-1975 Organiser and Chairman, "IFIP CONGRESS80" section on Computer Education in developing Countries in Australia-1980
Member of International Panel on Requirements of developing Countries Regarding Informatics Education in Switzerland. Invited by IBM International to join a Group of European Professors touring Computer Science Centers in Universities in USA and Canada for four weeks Chairman, International Working Party for the Analysis of Economics and Commercial Impact of Transferred Data Flow in Italy.

Member of Advisory Panel, Transnational Data Report in Amsterdam, Holland. Rapporteur, Scientific and Technological Communication, a USA sponsored Regional Seminar for Africa.
Member of Executive Secretariat Responsible for Sensitizing African Leaders on Modern Technology for development.
Member of Editorial Board, Journal of Information Technology for development.
Member of International Consultative Committee on Transborder Data Flow (TDF) Consultant to IBI Centers in Africa. Chairman, Information technology for development, Williamsburg Conference on International Information Economy, USA.
Member, UNCSTTD Editorial Workshop in Rome for Advanced Information Technology Trend and Global Implications. A discussant and a Member of panel on Information Economy Risks and Opportunities, Atwater Institute, Quebec, Canada.

He has been a phenomenal inspirer and influence in the study of Physics and Mathematics in Ghanaian Schools, Colleges and Universities and he has become a living legend. He played no mean part in the establishment of the Laser Research Unit in the Physics Department of the University of Cape Coast, projected to be a Centre of excellence to serve the whole of the West Africa Sub-Region.

He established "African institute for Mathematical Sciences" (AIMS) in Ghana in 2012 and ensured its acceptance by UNESCO as a UNESCO Centre of Excellence in 2017 before his death on 2nd November 2017.

State of Mathematics in Africa and the Way Forward

Francis K. A. Allotey(✉)

University of Ghana, Accra, Ghana
fkallotey@gmail.com

Abstract. The paper discusses some of the factors influencing students poor performance in mathematics in Africa and suggests ways to improve it. Examples of international centres and networks for capacity building in mathematical sciences are given.

Keywords: Mathematics education · Higher education · Educational policy and systems

1 Introduction

Africa is a vast continent with a total population of 1.2 billion (2016) and an area of 30.2 million sq km. It consists of 54 sovereign states exhibiting variability in size, population level, wealth and culture. It has three major colonial languages: English, French, and Portuguese. Africa also has over 2,000 indigenous languages; Ghana and Nigeria alone have over 87 and 527 languages respectively. These variations are also seen in educational development including Mathematics development. While African countries differ from one another in many features, they are broadly similar in key areas that concern institutional and national conditions that help or hinder mathematical development. These conditions are virtually the same throughout the continent. Despite the increasing importance of mathematics to Africa socio-economic development, the study of mathematics in Africa is declining. Some of the reasons for this state of affairs are:

- Low numbers of teachers with degrees in mathematics teaching in secondary schools and mathematicians at the masters and Ph.D. levels. With few professors to train the next generation mathematicians. African countries cannot meet the growing demand for mathematicians with advanced up-to-date training.
- Professional and geographical isolation, which limits opportunities to advance professionally.
- Low salaries, a poor public image, and shortage of mentors
- Record numbers of students clamoring for a college education

J. Delgado and M. Ruzhansky (Eds.): *Analysis and Partial Differential Equations: Perspectives from Developing Countries*, PROMS 275, pp. 7–13, 2018.
https://doi.org/10.1007/978-3-030-05657-5_2

- Deficient and outdated infrastructure, instrumentation, and teaching materials
- Brain-drain
- Most governments (with exception of South Africa) do not award research funds, so there is little or no incentive for doing research in mathematics. Thus students hoping for advanced study in mathematics must search for funding from foundations or foreign government which is scarce and difficult to secure.

2 Levels of Research

Africa needs its own research and innovation solutions to tackle its development challenges. However the researcher-to-population ratio is very low in African countries. Burkina Faso, for example has 45 researchers and development (R&D specialists) per million people and Nigeria has 38, in comparison to an average of 481 in Latin America and 1,714 in East Asia (World Bank, Africa Centre of Excellence News Bulletin (Washington, April 2014). In Mathematical Sciences, the number of Ph.D. holders for the whole of Sub-Saharan Africa, according to 2009 estimates was less than 2,000. Mathematical publications was also very low, while North Africas share is highest, about 0.87% of the world output, the shares for the rest of Africa were extremely low, for Southern Africa 0.39%, for West Africa 0.08%, for Central Africa 0.03% and for East African 0.01% world output. (Mathematics in Africa: challenges and opportunities, International Mathematical Union Report 2014). North African Countries are relatively advanced in mathematics due at least in part to their government broad commitment to research and education at all levels and to nearby European countries.

3 The Gender Question

- Why are there so few women lecturers in mathematical sciences in our universities in Africa?
- What are the main contributory factors?
- What should we do to address the situation?

4 Capacity Building

The most important and urgent needs in Africa are capacity building in Mathematics both (human and infrastructure) and research collaborations with mathematicians inside and outside Africa. Human capacity, building programme in mathematics in Africa must have the following objectives:

- Promote investment into human resource development,
- Promote regional and international cooperation in research and training
- Stimulate the growth in the utilization of newer and emerging technologies

- Increase the participation of women and other disadvantaged groups
- Seek more sponsorship or assistance from international agencies and from private sector, recognizing Africa's inadequate local resources to help narrow the gap between Africa's educational system and that of developed world,
- Raise the level of awareness of the value of mathematics among the general population and among those who influence educational policies
- Strengthen, the cooperation between local and external experts to ensure full of socio-economic and cultural problems
- Increase the participation of African mathematicians in international scientific bodies
- Establish regional centres of excellence
- Establish regional summer school programmes to expose students to different environments and explore varying real-life problems in different regional settings.

5 Academic Exchange Programmes

- Exchange of Experts or Professionals,
- Collaborative training and research in mathematical sciences,
- Access of African scientists to international facilities, intra and extra African students, for doctoral and faculty exchanges,
- University-industry links: internships, joint research programmes, sandwich Ph.D. degree programmes and
- Sharing of common facilities and infrastructure.

6 Benefits of Research Cooperation

- Aggregation of financial supports
- Benefitting from wider expertise
- Innovation arising from cultural diversity
- Cooperation and Competition
- Advantages/Disadvantages using different funding systems
- Research as a pilot project for political cooperation endeavors

7 Some Centres of Excellence and Networks in Mathematical Sciences in Africa

(1) Abdus Salam International Centre for Theoretical Physics (ICTP)
I CTP was founded by the late Professor Abdus Salam in Trieste, Italy in 1964. Though it is not located in Africa, it has greatly impacted positively in the development and promotion of Mathematical Sciences in Africa. It has established affiliated centres and networks in Africa. It organizes topical conferences, summer schools, and workshops in Africa. It runs Advanced Postgraduate Diploma Programme in Mathematical Sciences which are attended by

many African students. It organizes Masters and Ph.D. Sandwich Programmes for African Students. The Society of African Physicists and Mathematicians (SAPAM) and African Academy of Sciences were founded at ICTP. In 1986, with support from ICTP SAPAM organized a Pan African workshop in Nairobi, Kenya on harmonization of curriculum mathematics at tertiary level education in Africa and it was attended by over 100 mathematicians from all over Africa. It was the first of its kind.

(2) African Institute for Mathematical Sciences, Next Einstein Initiative

(AIMS-NEI) is a pan-African network of centres of excellence. AIMS core mission is to provide a high quality Masters education to talented African women and men and to enable Africas brightest students to flourish as independent thinkers, problems solvers and innovators capable of propelling Africas future scientific, educational and economic self-sufficient. It was founded by Prof. Neil Turok, a South African and a professor of Mathematical Physics at Cambridge University. Currently, Director of Perimeter Institute, Waterloo, Canada. AIMS-NEI has the following objectives:

- World-class post-graduate training
- Cutting-edge research for innovation and solutions in Africa
- Industrys leading academic partner in Africa
- Impactful public engagement
- Communication that drives results internally and externally
- Organizational excellence and sustainability

Key features:

- Outstanding African and international lecturers teach with the support of resident tutors
- Broad variety of theoretical and applied courses in mathematical sciences
- Emphasis on problem solving and computational methods.
- A high interactive teaching styles that encourages questioning, creativity and discovery.
- Students and lecturers share learning environment with 24/7 access to the library, computer lab and internet.
- Pan-African student body, of which 30% or more are women.
- Graduates progress to advanced-degree programmes at leading institutions in Africa and worldwide. They also take up careers in academia and industry, working in financial, agriculture, health, ICT, mining etc.
- AIMS Senegal, along with its regular programme, offers a 17-month Co-op stream, which integrate learning with work placements preparing students for careers in industry.

The ultimate goal of AIMS-NEI is to create 15 AIMS Centres across Africa with the main objectives to address the impact of the critical shortage of mathematical and scientific expertise in Africa. Currently, there are six AIMS-centers, South Africa, Senegal, Ghana, Cameroon, Tanzania and Rwanda. Outstanding lecturers from all over the world teach at AIMS-Centres leveraging the expertise and good will of the international mathematics community to deliver the highest quality training at low cost. 2016 saw the graduation of AIMS 1000th student across the network, 237 students including 78 women graduated from their respective centres. This brings the total number of AIMS alumni to 1210 with 32% of women from 42 African countries.

Children and Youth: Each AIMS centre implements public outreach programmes to support local schools and educators building scientific literacy and numeracy from the ground up. AIMS South Africa already runs a highly successful Enrichment Centre, offering training courses to over a hundred teachers annually.

3. Institute of Mathematical Sciences (IMS) Ghana

IMS was established in 1995. It is a research centre in Mathematics and Physics for three Universities in Ghana: university of Ghana, Kwame Nkrumah University of Science and Technology and University of Cape Coast. Among the functions of IMS are;

- Bring together the best international lecturers to train and mentor young African graduates in the mathematical sciences, especially in fields which promote the economic growth and prosperity and sustainable development in Africa,
- Build a community of highly skilled young Africans who will be part of the dynamic labour force in public and private sectors
- Provide young Africans from across the continent with opportunities to share their solving network as well as building their leadership and team skills
- Provide African mathematicians and scientists with excellent conditions for research in Ghana and
- Provide African researchers with opportunities to establish strong links and partnership with the international scientific community IMS organizes sandwich degree programmes leading to M.Sc., MPhil and Ph.D. degrees. The main objectives of the sandwich Ph.D. programmes are:
- To pool resources to offer programmes at postgraduate levels.
- To help curb the deplorable state of mathematical sciences in Ghana
- To establish a stable and vital mathematical sciences community.

IMS also attracts students from neighboring countries such as Nigeria, Togo, Burkina Faso, Ivory Coast, Senegal, Liberia and Cameroon for courses, conferences and workshops. Annually, about 110 participants from Ghana, neighboring countries, USA and Europe participate in some of its activities. The objectives of the conferences, postgraduate courses, workshops and schools are to provide;

- Advanced training of young mathematicians and scientists for teaching positions in institutions of higher learning and research
- To update the knowledge of African mathematicians
- To provide a forum for interaction of African mathematicians
- Advanced school/workshop/conferences

Among the annual workshops organized in Ghana since 1995 are: a. Abdus Salam and Edward Bouchet African Regional College on functional Analysis, Differential Equations and Applications. b. Abdus Salam and Edward Bouchet African Regional Workshop on Modelling, Simulation and Optimization College.

(4) African Mathematics Millennium Science Initiative (AMMSI)

The African Mathematics Millennium Science Initiative (AMMSI), established in 2004, is a distribution network of mathematics research, training and promotion. There are six Region Offices, each run by a Regional Coordinator, located in Botswana, Cameroon, Morocco, Nigeria, Senegal and Tanzania. The Programme Office, is at the University of Nairobi, and oversees overall coordination. The mission of the AMMSI is to nurture the next general of African mathematicians and mathematical leaders. Its activities include:

- Research in mathematics and its applications
- Training and education in mathematics and its applications
- Linkages and networking to enhance capacity
- Outreach and public education, to raise general awareness in mathematics
- Enhancing the availability and use of ICT in teaching and learning mathematics

(5) International Programme in Mathematical Sciences (IPMS)

IPMS is a project of Uppsala University with the funding from Swedish International Development Agency (SIDA). It supports several projects in Africa among them are:

- Eastern Africa Universities Mathematics Programme and
- Partial Differential Equations Modelling and Contract Projects in West African

8 The Way forward

1. Governments should channel more resources and prioritize the learning of mathematics at the basic education level.
2. Governments should motivate good students with attractive incentives to enroll in teacher training colleges in order to churn out high grade mathematics teachers for our basic schools.
3. Local governments and philanthropists must be encouraged to support financially the training of brilliant local citizens in their community in teacher training colleges and employ them as mathematics teachers in their localities. Their salaries must be topped up by the local authorities.
4. Local governments must take special interest in the output of teachers in their jurisdiction.
5. There should be a continuous professional education development for primary and secondary school teachers.
6. Existing and new projects aimed at encouraging girls to study mathematics and science should be vigorously pursued.

7. Strengthen and expand training and research activities, especially regional networks of people and institutions. University teachers should be encouraged by sponsoring them to participate in local, regional and international seminars, conferences, workshops, etc.

The Optimal Vector Control for the Elastic Oscillations Described by Fredholm Integral-Differential Equations

Elmira Abdyldaeva[1](✉) and Akylbek Kerimbekov[2]

[1] Department of Mathematics, Kyrgyz-Turkish "Manas" University, Bishkek, Kyrgyzstan
efa_69@mail.ru

[2] Department of Applied Mathematics and Informatics, Kyrgyz-Russian Slavic University, Bishkek, Kyrgyzstan
akl7@rambler.ru,applmath@krsu.edu.kg

Abstract. In this paper, we investigate the nonlinear problem of the optimal vector control for oscillation processes described by Fredholm integro-differential equations in partial derivatives when function of external sources nonlinearly depend on control parameters. It was found that the system of nonlinear integral equations,which obtained relatively to the components of the optimal vector control, have the property of equal relations. This fact lets us to simplify the procedure of the constructing the solution of the nonlinear optimization problem. We have developed algorithm for constructing the solution of the nonlinear optimization problem.

Keywords: Optimal control · Fredholm integral equations · Boundary value problems

1 Introduction

Many application problems are investigated by the integro-differential equations [1–3]. With the formation of the theory of control systems with distributed parameters, the fundamentals of which were laid by A.I.Egorov, A.G.Butkovskii, T.K.Sirazetdinov etc. [4–8], the optimal control problem for processes described by integro-differential [9–15] and functional equations has become possible to investigate. However, despite the large number of investigations, owing to their complexity, such problems of optimal control theory are little studied. In this paper we investigate the problem of optimal vector control for the elastic oscillations described by Fredholm integral-differential equations in partial derivatives, when the external influence are described by the nonlinear function with respect to the control parameters. The research was performed using the notion generalized solution to the boundary value problem for the controlled process and

J. Delgado and M. Ruzhansky (Eds.): *Analysis and Partial Differential Equations: Perspectives from Developing Countries*, PROMS 275, pp. 14–30, 2018.
https://doi.org/10.1007/978-3-030-05657-5_3

it was found that its Fourier coefficients are determined as the solution to the countable number of linear Fredholm integral equations of the second kind. In the optimal control problem where it is required to minimize quadratic integral functional on the set of solutions to the boundary value problem, the optimality conditions, as a consequence of the maximum principle [4–8,16], are obtained in the form of system of the nonlinear integral equations and system of inequalities. These systems are provide the positive definiteness of Hessian matrix of the Pontryagin type function. Note that the optimization problem has such a property that the system of nonlinear integral equations satisfies the conditions of equal relations. This fact lets us to simplify the procedure constructing of the optimal vector control. Comparisons with the results of [17,18] shows that this property is characteristic for all the optimal vector control problem, i. e. it does not depend on the nature of the controlled process. Sufficient conditions for the existence the unique solution of the nonlinear optimization problem are found. We have developed algorithm for constructing of the solution to the nonlinear optimization problem.

2 Boundary Value Problem of the Controlled Process

Let the state of the oscillation process is described by a scalar function $V(t,x)$, which in the region $Q_T = Q\times(0,T]$, where Q is a region of the space R^n bounded by the piecewise smooth curve γ, satisfies the integro-differential equation

$$V_{tt}-AV = \lambda\int_0^T K(t,\tau)V(\tau,x)d\tau+g(t,x)f[t,u_1(t),...,u_m(t)],\ x\in Q,\ 0<t\leq T, \tag{1}$$

on the boundary of Q satisfies the initial conditions

$$V(0,x)=\psi_1(x),\ V_t(0,x)=\psi_2(x),\ x\in Q, \tag{2}$$

and the boundary conditions

$$\Gamma V(t,x)\equiv\sum_{i,j=1}^{n} a_{ij}(x)V_{x_j}cos(\nu,x_i)+a(x)V=0,\ x\in\gamma,\ 0<t<T. \tag{3}$$

Here A is the elliptic operator defined by the formula

$$AV(t,x)=\sum_{i,j=1}^{n}(a_{ij}(x)V_{x_j}(t,x))_{x_i}-c(x)V(t,x),$$

and $a(x)\geq 0,\ c(x)\geq 0$ are known functions; ν is a normal vector, emanating from the point $x\in\gamma$; $K(t,\tau)$ is a given function with region $D=\{0\leq t\leq T,\ 0\leq\tau\leq T\}$ and satisfying the condition

$$\int_0^T\int_0^T K^2(t,\tau)d\tau dt = K_0<\infty, \tag{4}$$

i.e. $K(t,s) \in H(D)$; $\psi_1(x) \in H_1(Q)$, $\psi_2(x) \in H(Q)$, $f[t, u_1(t), ..., u_m(t)] \in H(0,T)$ are given functions. The function of external effects $f[t, u_1(t), ..., u_m(t)] \in H(0,T)$ depends nonlinearly on the control functions $u_1(t), ..., u_m(t), u_i(t) \in H(0,T)$, $i = 1, 2, 3, ..., m$; and satisfying the conditions

$$f_{u_i}[t, u_1(t), ..., u_m(t)] \neq 0, \ \forall t \in (0,T), \ i = 1, 2, 3, ..., m; \tag{5}$$

T is a fixed moment of time, $\alpha > 0$ is a constant; $H(Y)$ is the Hilbert space of functions determined on the Y. $H_1(Q)$ is the Sobolev space of the first kind;

It is known that, under the conditions imposed on the initial function, boundary value problem (1)–(3) cannot have a classical solution [19]. Analogically to [20] we use the notion of weak generalized solution of the problem (1)–(3).

Definition 1. A generalized solution of boundary value problem (1)–(3) is a function $V(t,x) \in H(Q_T)$ that satisfies the integral identity

$$\begin{aligned}\int_Q [(V_t\Phi) - (V\Phi_t)]_{t_1}^{t_2}\, dx \equiv \int_{t_1}^{t_2} \int_Q \left[-V\Phi_{tt} - \Sigma_{i,j=1}^{n} a_{ij}(x) V_{x_j} \Phi_{x_i} - CV\Phi\right] dxdt + \\ + \int_{t_1}^{t_2} \int_\gamma (-a(x)V)\Phi(t,x) dxdt + \\ + \int_{t_1}^{t_2} \int_Q \left(\lambda \int_0^T K(t,\tau)V(\tau,x)d\tau + g(t,x) f[t, u_1(t), ..., u_m(t)]\right) \Phi(t,x) dxdt\end{aligned} \tag{6}$$

for any t_1 and t_2, $0 < t_1 \leq< t \leq t_2 \leq T$, and for any function $\Phi(t,x) \in C^{2,1}(Q_T)$, and the initial and boundary conditions in a weak sense, i.e. for any functions $\phi_0(x) \in H(Q)$, $\phi_1(x) \in H(Q)$ the following relations hold

$$\begin{aligned}\lim_{t \to +0} \int_Q V(t,x)\phi_0(x)dx = \int_Q \psi_1(x)\phi_0(x)dx, \\ \lim_{t \to +0} \int_Q V_t(t,x)\phi_1(x)dx = \int_Q \psi_2(x)\phi_1(x)dx,\end{aligned} \tag{7}$$

where $C^{2,1}(Q_T)$ is the space of functions, which have second derivatives with respect to t and first derivatives with respect to x.

We look for a solution of problem (1)–(2) in the form:

$$V(t,x) = \sum_{n=1}^{\infty} V_n(t) z_n(x), \tag{8}$$

where $V_n(t) =< V(t,x), z_n(x) >$ are Fourier coefficients, the symbol $< , >$ is used for the scalar product in the Hilbert space $H(Q)$, $z_n(x)$, are eigen functions of the boundary value problem

$$D_n(\Phi, z_n) \equiv \int_Q \left(\sum_{i,j=1}^{n} a_{ij}(x)\Phi_{x_j} z_{nx_i} + C(x)z_n(x)\Phi(t,x) \right) dx + \int_\gamma a(x)z_n(x)\Phi(t,x)dx =$$

$$= \lambda_n^2 \int_Q z_n(x)\Phi(t,x)dx;$$

$$\Gamma z_n(x) = 0, \ x \in \gamma, \ 0 < t < T, n = 1, 2, ...$$

These eigenfunctions form a complete orthonormal system in the Hilbert space $H(Q)$, and the corresponding eigen values λ_n satisfy the following conditions $\lambda_n \leq \lambda_{n+1}, \forall n = 1, 2, 3, ..., \lim\limits_{n\to\infty} = \infty$.

The formal solution of the boundary problem (1)–(3) is found by using the integral identity (6). By the arbitrariness of function $\phi(t,x)$ in the integral identity (6) we assume that $\phi(t,x) \equiv z_n(x)$ and we obtain the following equation

$$\int_{t_1}^{t_2} \frac{\partial}{\partial t} < V_t, z_n > dt \equiv -\lambda_n^2 \int_{t_1}^{t_2} < V_t, z_n > dt +$$

$$+\lambda \int_0^T K(t,\tau) < V, z_n > dt + \int_{t_1}^{t_2} \left(g_n(t) f[t, u_1(t), ..., u_m(t)]\right) dt, \tag{9}$$

where

$$g(t,x) = \sum_{n=1}^{\infty} g_n(t)z_n(x), \ g_n(t) =< g(t,x), z_n(x) >= \int_Q g(t,x)z_n(x)dx,$$

$$\psi_1(x) = \sum_{n=1}^{\infty} \psi_{1n}z_n(x), \ \psi_{1n} =< \psi_1(x), z_n(x) >= \int_Q \psi_1(x)z_n(x)dx,$$

$$\psi_2(x) = \sum_{n=1}^{\infty} \psi_{2n}z_n(x), \ \psi_{2n} =< \psi_2(x), z_n(x) >= \int_Q \psi_2(x)z_n(x)dx. \tag{10}$$

In this identity by supposing $t_2 = t$ and differentiating with respect to t, we obtain the integro-differential equation

$$\frac{\delta}{\delta t} < V_t, z_n > +\lambda_n^2 < V, z_n >= \lambda \int_0^T K(t,\tau) < V, z_n > d\tau + g_n(t) f[t, u_1(t), ..., u_m(t)], \tag{11}$$

which we solve with the initial condition

$$< V, z_n > \big|_{t=t_1} =< V(t_1,x), z_n(x) >, < V_t, z_n > \big|_{t=t_1} =< V_t(t_1,x), z_n(x) >, \tag{12}$$

for each fixed $n = 1, 2, 3, ...$ Considering the right side of the Eq. (11) as absolute term, we solve the Cauchy problem (11)–(12) by the formula

$$V_n(t) =< V(t_1,x), z_n(x) > cos\lambda_n t + \frac{1}{\lambda_n} < V_t(t_1,x), z_n(x) > sin\lambda_n t +$$

$$+\frac{1}{\lambda_n} \int_{t_1}^{t} sin\lambda_n(t-\tau) \left[\lambda \int_0^T K(\tau,\eta)V_n(\eta)d\eta + q_n(\tau) f[\tau, u_1(\tau), ..., u_m(\tau)] \right] d\tau;$$

Tending t_1 to zero and taking account of (7), (10) we obtain the fredholm integral equations of the second kind

$$V_n(t) = \lambda \int_0^T \left(\frac{1}{\lambda_n} \int_0^t sin\lambda_n(t-\tau)K(\tau,s)d\tau \right) V_n(s)ds + \psi_{1n}cos\lambda_n t + \frac{\psi_{2n}}{\lambda_n} sin\lambda_n t +$$

$$+ \frac{1}{\lambda_n} \int_0^t sin\lambda_n(t-\tau) \left[q_n(\tau) f[\tau, u_1(\tau), ..., u_m(\tau)] \right] d\tau; \quad (13)$$

We can rewrite this equation as

$$V_n(t) = \lambda \int_0^T K_n(t,s)V_n(s)ds + a_n(t), \quad (14)$$

where

$$K_n(t,s) = \frac{1}{\lambda_n} \int_0^t sin\lambda_n(t-\tau)K(\tau,s)d\tau; \quad (15)$$

$$a_n(t) = \psi_{1n}cos\lambda_n t + \frac{\psi_{2n}}{\lambda_n} sin\lambda_n t + \frac{1}{\lambda_n} \int_0^t sin\lambda_n(t-\tau) \left[q_n(\tau) f[\tau, u_1(\tau), ..., u_m(\tau)] \right] d\tau; \quad (16)$$

We solve the integral equation (14) using the following formula [21,22]:

$$V_n(t) = \lambda \int_0^T R_n(t,s,\lambda)a_n(s)ds + a_n(t); \quad (17)$$

where

$$R_n(t,s,\lambda) = \sum_{i=1}^{\infty} \lambda^{i-1} K_{n,i}(t,s), \; n = 1,2,3,..., \quad (18)$$

is the resolvent of the kernel $K_n(t,s)$, the iterated kernels $K_{n,i}(t,s)$ are defined by the formula

$$K_{n,i+1}(t,s) = \int_0^T K_n(t,\eta)K_{n,i}(\eta,s)d\eta, \; i = 1,2,3,..., \quad (19)$$

for each $n = 1,2,3,....$

Now we investigate the convergence of Neumann series (18). According to (15) and (19) by direct computation the following estimates hold

$$|K_{n,i}(t,s)|^2 \le \left(\frac{T}{\lambda_n^2} \right)^i (K_0 T)^{i-1} \int_0^T K^2(y,s)dy; \; \forall t \in (0,T);$$

$$\int_0^T K_{n,i}^2(t,s)ds \le \left(\frac{T}{\lambda_n^2} \right)^i (K_0 T)^{i-1} \int_0^T \int_0^T K^2(y,s)dyds \le \left(\frac{TK}{\lambda_n^2} \right)^i (T)^{i-1}; \quad (20)$$

Convergence of the Neumann series (18) follows from the inequality

$$R_n(t,s,\lambda) = \sum_{i=1}^{\infty} |\lambda|^{i-1} |K_{n,i}(t,s)| \le \sum_{i=1}^{\infty} |\lambda|^{i-1} \sqrt{\left(\frac{T}{\lambda_n^2} \right)^i (K_0 T)^{i-1} \int_0^T K^2(y,s)dy} \le$$

$$\leq \sqrt{\int_0^T K^2(y,s)dy} \sum_{i=1}^{\infty} |\lambda|^{i-1} \sqrt{\frac{T}{\lambda_n^2}} \sqrt{\left(\frac{K_0T^2}{\lambda_n^2}\right)^{i-1}} \leq$$

$$\sqrt{\int_0^T K^2(y,s)dy} \sqrt{\frac{T}{\lambda_n^2}} \sum_{i=1}^{\infty} \left(|\lambda| \sqrt{\frac{K_0T^2}{\lambda_n^2}}\right)^{i-1} \leq \sqrt{\frac{T}{\lambda_n^2}} \sqrt{\int_0^T K^2(y,s)dy} \frac{1}{1-|\lambda|\frac{\sqrt{K_0T^2}}{\lambda_n^2}} =$$

$$= \frac{\sqrt{T}}{\lambda_n} \frac{\lambda_n}{\lambda_n - |\lambda|\sqrt{K_0T^2}} \sqrt{\int_0^T K^2(y,s)dy} \leq \frac{\sqrt{T}\sqrt{\int_0^T K^2(y,s)dy}}{\lambda_n - |\lambda|\sqrt{K_0T^2}}.$$

This inequality converges for the values of the parameter λ that satisfy the inequality

$$|\lambda| \frac{T}{\lambda_n} \sqrt{K_0} < 1.$$

Note that the Neumann series for values of the parameter λ satisfying

$$|\lambda| < \frac{\lambda_n}{T\sqrt{K_0}} \to_{n\to\infty} \infty \tag{21}$$

converges absolutely for each $n = 1, 2, 3, ...$,i.e. the radius of convergence increases when n is growing. However, the Neumann series, for the parameter λ that satisfy the condition

$$|\lambda| < \frac{\lambda_1}{T\sqrt{K_0}}.$$

converges absolutely for any $n = 1, 2, 3, ...$. In this case the resolvent $R_n(t, s, \lambda)$ as the sum of an absolutely convergent series, is the continuous function and satisfies the following estimates

$$|R_n(t,s,\lambda)| \leq \frac{\sqrt{T}\sqrt{\int_0^T K^2(y,s)dy}}{\lambda_n - |\lambda|\sqrt{K_0T^2}}. \tag{22}$$

By direct calculation the following inequality hold

$$\int_0^T R_n^2(t,s,\lambda)ds = \frac{T}{(\lambda_n - |\lambda|\sqrt{K_0T^2})^2} \int_0^T \int_0^T K^2(y,s)dyds = \frac{K_0T}{(\lambda_n - |\lambda|\sqrt{K_0T^2})^2}.$$

Thus, we find the solution of problem (1)–(3) by formula

$$V(t,x) = \sum_{n=1}^{\infty} V_n(t)z_n(x) = \sum_{n=1}^{\infty} \left(\lambda \int_0^T R_n(t,s,\lambda)a_n(s)ds + a_n(t)\right) z_n(x), \tag{23}$$

where $V_n(t)$ is defined by (17) as the unique solution of integral equation (14). Taking into account the (15) and (17) the function (23) is rewrited to the following form

$$V(t,x) = \sum_{n=1}^{\infty} \left\{ \psi_n(t,\lambda) + \frac{1}{\lambda_n} \int_0^T \varepsilon_n(t,\eta,\lambda) g_n(\eta) f(\eta, u_1(\eta), ..., u_m(\eta)) d\eta \right\} z_n(x); \tag{24}$$

where

$$\psi_n(t,\lambda) = \psi_{1n}\left[cos\lambda_n t + \lambda\int_0^T R_n(t,s,\lambda)cos\lambda_n sds\right] +$$
$$+\frac{\psi_{2n}}{\lambda_n}\left[sin\lambda_n t + \lambda\int_0^T R_n(t,s,\lambda)sin\lambda_n sds\right]; \tag{25}$$

$$\varepsilon_n(t,\eta,\lambda) = \begin{cases} sin\lambda_n(t-\eta) + \lambda\int_\eta^T R_n(t,s,\lambda)sin\lambda_n(s-\eta)ds, & 0\le\eta\le t, \\ \lambda\int_\eta^T R_n(t,s,\lambda)sin\lambda_n(s-\eta)ds, & t\le\eta\le T, \end{cases} \tag{26}$$

Lemma 1. Generalized solution of problem (1)–(3) which defined by (24) is an element of the Hilbert space.

Proof. Taking into account (26) we have the inequality

$$\int_0^T\int_Q V^2(t,x)dxdt =$$
$$= \int_0^T\sum_{n=1}^{\infty}\left\{\psi(t,\lambda) + \left(\frac{1}{\lambda_n}\int_0^T \varepsilon_n(t,\eta,\lambda)g_n(\eta)f(\eta,u(\eta))d\eta\right)\right\}^2 dt \le$$
$$\le 2\int_0^T\sum_{n=1}^{\infty}\left\{\psi^2(t,\lambda) + \left(\frac{1}{\lambda_n}\int_0^T \varepsilon_n(t,\eta,\lambda)g_n(\eta)f(\eta,u(\eta))d\eta\right)^2\right\} dt.$$

Then, taking into account the inequalities

$$1)\ \sum_{n=1}^{\infty}\psi_n^2(t,\lambda) \le 2\sum_{n=1}^{\infty}\left\{2\psi_{1n}^2\left[1+\lambda^2\frac{K_0T}{(\lambda_n-|\lambda|\sqrt{K_0T})^2}\right] +\right.$$
$$\left.+\frac{2\psi_{2n}^2}{\lambda_n^2}\left[1+\lambda^2\frac{K_0T^2}{(\lambda_n-|\lambda|\sqrt{K_0T})^2}\right]\right\} = 4\left[1+\lambda^2\frac{K_0T^2}{(\lambda_1-|\lambda|\sqrt{K_0T})^2}\right]\times$$
$$\times\left\{\|\psi_1(x)\|_H^2 + \frac{1}{\lambda_1^2\|\psi_2(x)\|_H^2}\right\} < \infty;$$

$$2)\ \sum_{n=1}^{\infty}\frac{1}{\lambda_n^2}\int_0^T \varepsilon_n^2(t,\eta,\lambda)g_n^2(\eta)d\eta\int_0^T f^2(\eta,u_1(\eta),...,u_m(\eta))d\eta \le$$
$$\le\frac{2}{\lambda_1^2}\left[1+\lambda^2\frac{K_0T^2}{(\lambda_1-|\lambda|\sqrt{K_0T})^2}\right]\|g(t,x)\|_{H(Q_T)}^2\|f[t,u_1(t),...,u_m(t)]\|_{H(Q_T)}^2 < \infty;$$

we obtain that $V(t,x)\in H(Q_T)$. □

Thus, the solution of the boundary problem (1)–(3) be found by formula (8), where $V_n(t)$ is determined by the formula (17) as unique solution of the integral equation (14). It is easy to verify that this solution satisfies the initial conditions (2).

Analogically, it is not difficult to verify, that $V_t(t,x)\in H(Q_T)$.

3 Formulation of Optimal Control Problem and Conditions Optimality

We consider the optimization problem in which it is required to minimize the quadratic integral functional

$$J[u_1(t),...,u_m(t)] = \int_Q \Big\{[V(T,x) - \xi_1(x)]^2 +$$

$$+[V_t(T,x) - \xi_2(x)]^2\Big\}dx + \beta\int_0^T \sum_{k=1}^{m} u_k^2(t)dt,\ \beta > 0, \tag{27}$$

on the set of solutions of problem (1)–(3), i.e. we need to found the optimal vector control $(u_1^0(t),...,u_m^0(t)) \in H^m(0,T)$, ($H^m(0,T)$ is Cartesian product of m spaces $H(0,T)$), which together with the corresponding solution $V^0(t,x)$ of boundary value problem (1)–(3) gives the least possible value of functional (27). In this case $(u_1^0(t),...,u_m^0(t))$ is called the optimal vector control, and $V^0(t,x)$ is the optimal process.

Since each vector control $u_1(t),...,u_m(t)$ uniquely defines the controlled process $V(t,x)$, the solution of boundary value problem (1)–(3) in the form $V(t,x)+\delta V(t,x)$ correspond to the control $(u_1(t) + \delta u_1(t),...,u_m(t) + \delta u_m(t))$. Here $\delta V(t,x)$ is the increment corresponding to the increment $(\delta u_1(t),...,\delta u_m(t))$. According to the maximum principle [3–8,16], the increment of functional (27) can be written as

$$\Delta\bar{J}(u_1(t),...,u_m(t)) = \bar{J}(u_1 + \Delta u_1,...,u_m + \delta u_m) - J(u_1,...,u_m) =$$

$$\int_Q \{\Delta V^2(T,x) + \Delta V_t(T,x)\}dx + \beta\int_0^T \sum_{k=1}^{m}\left[(u_k(t) + \delta u_k(t))^2 - (u_k(t))^2\right]dt-$$

$$-\int_0^T \left(\int_Q g(t,x)\omega(t,x)dx\{f[t,u_1 + \Delta u_1,...,u_m + \Delta u_m] - f[t,u_1,...,u_m]\}\right)dt =$$

$$= -\int_0^T \int_Q g(t,x)\omega(t,x)dx\{f[t,u_1(t) + \Delta u_1(t),...,u_m(t)$$

$$+\Delta u_m(t)] - f[t,u_1(t),...,u_m(t)]\}dt+$$

$$+\beta\int_0^T \sum_{k=1}^{m}\left[2u_k(t)\Delta u_k(t) + (\Delta u_k(t))^2\right]dt + \int_Q \left\{\Delta V^2(T,x) + \Delta V_t^2(t,x)\right\}dx. \tag{28}$$

$$\Delta\bar{J}(u_1(t),...,u_m(t)) = -\int_0^T \Delta\Pi(t,V(t,x),\omega(t,x),u_1(t),...,u_m(t))dt+$$

$$+\int_Q \{V^2(T,x) + \Delta V_t^2(T,x)\}dx;$$

where

$$\Pi[t,x,V(t,x),\omega(t,x),u_1(t),...,u_m(t)]=$$
$$=\int_Q g(t,x)\omega(t,x)dx\cdot f(t,u_1(t),...,u_m(t))-2\beta\sum_{k=1}^{m}u_k^2(t), \tag{29}$$

function $\omega(t,x)$ is the solution of the following conjugate boundary value problem

$$\omega_{tt}-A\omega=\lambda\int_0^T K(\tau,t)\omega(\tau,x)d\tau,\ x\in Q,\ 0\le t<T,$$

$$\omega(T,x)+2[V_t(T,x)-\xi_2(x)]=0,\ \omega_t(T,x)-2[V(T,x)-\xi_1(x)]=0,\ x\in Q,$$

$$\Gamma\omega(t,x)=\sum_{i,j=1}^{n}\left(a_{ij}(x)\omega_{x_j}(t,x)\right)_{x_i}-c(x)\omega(t,x)=0,\ x\in\gamma,0\le t<T, \tag{30}$$

As the consequence of the maximum principle for systems with distributed parameters, the optimal control is determined by the relations

$$\int_Q g(t,x)\omega(t,x)dxf_{u_i}-2\beta u_i(t)=0,\ i=1,2,3,...,m, \tag{31}$$

$$\prod_{j=1}^{k}f_{u_j}\begin{vmatrix}\left(\frac{u_1}{f_{u_1}}\right)_{u_1} & \cdots & \left(\frac{u_1}{f_{u_1}}\right)_{u_k}\\ \cdots & \cdots & \cdots\\ \left(\frac{u_k}{f_{u_k}}\right)_{u_1} & \cdots & \left(\frac{u_k}{f_{u_k}}\right)_{u_k}\end{vmatrix}>0,\ k=1,2,3,...,m, \tag{32}$$

These relations are called *optimality conditions.*

We note that the optimality condition (31) can be written as

$$2\beta\frac{u_i(t)}{f_{u_i}}=\int_Q g(t,x)\omega(t,x)dx,\ i=1,2,3,...,m, \tag{33}$$

i.e. the prosperity of equal relations is hold.

Condition (32) obtained by means of the Sylvester's criterion for the Hessian matrix of the function (29).

4 Solution of the Conjugate Boundary Value Problem

We look for a solution to conjugate boundary value problem (30) in the form of the series

$$\omega(t,x)=\sum_{n=1}^{\infty}\omega_n(t)z_n(x). \tag{34}$$

Its not difficult to verify that the Fourier coefficients $\omega_n(t)$ satisfy the following linear inhomogeneous Fredholm integral equation of the second kind

$$\omega_n(t)=\lambda\int_0^T B_n(s,t)\omega_n(s)ds+q_n(t), \tag{35}$$

where

$$B_n(s,t) = \frac{1}{\lambda_n}\int_t^T sin\lambda_n(\tau - t)K(s,\tau)d\tau,$$

$$q_n(t) = -2\left\{[V_n'(T) - \xi_{2n}]cos\lambda_n(T-t) + \frac{1}{\lambda_n}[V_n(T) - \xi_{1n}]sin\lambda_n(T-t)\right\} \quad (36)$$

We find the solution of Eq. (35) by the formula [21,22]

$$\omega_n(t) = \lambda\int_0^T P_n(s,t,\lambda)q_n(s)ds + q_n(t), \quad (37)$$

where the resolvent $P_n(s,t,\lambda)$ of the kernel $B_n(s,t)$ is given by

$$P_n(s,t,\lambda) = \sum_{i=1}^{\infty}\lambda^{i-1}B_{n,i}(s,t),\ B_{n,i+1}(s,t) = \int_0^T B_n(\eta,t)B_{n,i}(s,\eta)d\eta,\ i = 1,2,3,...,$$

and according to condition (21), these resolvents are continuous functions and satisfy the inequality

$$|P_n(s,t,\lambda)| \le \frac{\sqrt{\int_0^T K^2(s,\eta)d\eta}}{\lambda_n - |\lambda|\sqrt{K_0T^2}},\ \int_0^T P_n^2(s,t,\lambda)ds \le \frac{TK_0}{(\lambda_n - |\lambda|\sqrt{K_0T^2})^2}. \quad (38)$$

In further, taking into account (34) and (37) the solution to the conjugate boundary value problem can be written as

$$\omega(t,x) = -2\sum_{n=1}^{\infty}\Big\{ - E_n^*(T-t)h_n +$$

$$+\int_0^T E_n^*(T-t)G_n(T-\tau)g_n(\tau)f[\tau,u_1(\tau),...,u_m(\tau)]d\tau\Big\}z_n(x), \quad (39)$$

where the symbol $*$ is a sign of transposition;

$$h_n = (h_{1n}, h_{2n}), \quad (40)$$

$$h_{1n} = \xi_{2n} - \psi_{1n}[-\lambda_n sin\lambda_n T] + \lambda\int_0^T R_{nt}'(T,s,\lambda)cos\lambda_n sds-$$

$$-\psi_{2n}[cos\lambda_n T] + \frac{\lambda}{\lambda_n}\int_0^T R_{nt}'(T,s,\lambda)sin\lambda_n sds;$$

$$h_{2n} = \xi_{1n} - \psi_{1n}\left[cos\lambda_n T + \lambda\int_0^T R_n(T,s,\lambda)cos\lambda_n sds\right] -$$

$$-\frac{\psi_{2n}}{\lambda_n}\left[sin\lambda_n T + \lambda\int_0^T R_n(T,s,\lambda)sin\lambda_n sds\right];$$

$$G_n[T-t] = (G_{n1}[T-t], G_{n2}[T-t]), \tag{41}$$

$$G_{n1}[T-t,\lambda] = cos\lambda_n(T-t) + \frac{\lambda}{\lambda_n}\int_\tau^T R'_{nt}(T,s,\lambda)sin\lambda_n(s-\tau)ds;$$

$$G_{n2}[T-t,\lambda] = \frac{1}{\lambda_n}\left[sin\lambda_n(T-t) + \lambda\int_\tau^T R_n(T,s,\lambda)sin\lambda_n(s-\tau)ds\right];$$

$$E_n[T-t,\lambda] = (\varepsilon_{n1}[T-t], \varepsilon_{n2}[T-t]), \tag{42}$$

$$E_{n1}[T-t,\lambda] = cos\lambda_n(T-t) + \lambda\int_0^T P_n(s,t,\lambda)cos\lambda_n(T-t)ds,$$

$$E_{n2}[T-t,\lambda] = \frac{1}{\lambda_n}\left[sin\lambda_n(T-t) + \lambda\int_0^T P_n(s,t,\lambda)sin\lambda_n(T-t)ds\right].$$

Lemma 2. Function $h(t,x) = \sum_{n=1}^{\infty} E_n^*(T-t,\lambda)h_n z_n(x)$ is the element of the Hilbert space $H(Q_T)$.

Proof. By straightforward calculation we obtain the inequality

$$\int_0^T\int_Q h^2(t,x)dxdt = \int_0^T\int_Q\left(\sum_{n=1}^{\infty} E_n^*(T-t,\lambda)h_n z_n(x)\right)^2 dxdt =$$

$$\int_0^T\sum_{n=1}^{\infty}|< E_n(T-t,\lambda), h_n >_{R^2}|^2\,dt \leq \int_0^T\sum_{n=1}^{\infty}\|E_n(T-t,\lambda)\|_{R^2}^2\cdot\|h_n\|_{R^2}^2 dt \leq$$

$$\leq \int_0^T\sum_{n=1}^{\infty} 2\left(1+\frac{1}{\lambda_n^2}\right)\left(1+\frac{\lambda^2 K_0 T^2}{(\lambda_n - |\lambda|\sqrt{K_0 T^2})^2}\right)\cdot 3\Big\{\xi_{1n}^2 + \xi_{2n}^2 +$$

$$+2\left(1+\frac{\lambda^2 K_0 T^2}{(\lambda_n - |\lambda|\sqrt{K_0 T^2})^2}\right)\left[(1+\lambda_n^2)\psi_{1n}^2 + \left(1+\frac{1}{\lambda_n^2}\right)\psi_{2n}^2\right]\Big\} \leq$$

$$\leq 3TE_0^2\Big\{\|\xi_1(x)\|_{H(Q)}^2 + \|\xi_2(x)\|_{H(Q)}^2 + E_0^2\left[\|\psi_1(x)\|_{H_1(Q)}^2 + \left(1+\frac{1}{\lambda_1^2}\right)\|\psi_2(x)\|_{H(Q)}^2\right]\Big\} < \infty,$$

where $E_0^2 = 2\left(1+\frac{1}{\lambda_1^2}\right)\left(1+\frac{\lambda^2 K_0 T^2}{(\lambda_1 - |\lambda|\sqrt{K_0 T^2})^2}\right)$, $< \cdot,\cdot >_{R^2}$ is a symbol of the scalar product of vectors in the Euclidean space , from this inequality follows that $h(t,x) \in H(Q_T)$.

Lemma 3. Function

$$E(t,x,\lambda) = \sum_{n=1}^{\infty} E_n^*(T-t,\lambda)\int_0^T G_n(T-\tau,\lambda)g_n(\tau)f(\tau,u_1(\tau),...,u_m(\tau))d\tau z_n(x)$$

is an element of Hilbert space $H(Q_T)$.

Proof. By straightforward calculation we obtain the inequality

$$\int_0^T \int_Q E^2(t,x,\lambda)dxdt =$$

$$= \int_0^T \sum_{n=1}^{\infty} \left| \langle E_n(T-t,\lambda), \int_0^T G_n(T-\tau,\lambda)[g_n(\tau)f(\tau,u_1(\tau),...,u_m(\tau))]d\tau \rangle_{R^2} \right|^2 dt \leq$$

$$\int_0^T \sum_{n=1}^{\infty} \|E_n(T-t,\lambda)\|_{R^2}^2 \left(\int_0^T \|G_n(T-\tau,\lambda)\|_{R^2}\, g_n(\tau)f[\tau,u_1(\tau),...,u_m(\tau)]d\tau \right)^2 dt \leq$$

$$\int_0^T \sum_{n=1}^{\infty} \|E_n(T-t,\lambda)\|_{R^2}^2 \int_0^T \|G_n(T-\tau,\lambda)\|_{R^2}\, g_n^2(\tau)d\tau \int_0^T f^2(\tau,u_1(\tau),...,u_m(\tau))d\tau dt \leq$$

$$\int_0^T \sum_{n=1}^{\infty} E_0^2 \int_0^T E_0^2 g_n^2(\tau)d\tau \int_0^T f^2(\tau,u_1(\tau),...,u_m(\tau))d\tau dt =$$

$$= TE_0^4 \|g(t,x)\|_{H(Q)}^2 \|f(t,u_1(t),...,u_m(t))\|_{H(Q_T)}^2 < \infty,$$

from which the statement of the *Lemma* 3 follows, i.e. $E(t,x,\lambda) \in H(Q_T)$. Based on the *Lemma* 2 from $\omega(t,x) = -2\{-h(t,x) + E(t,x,\lambda)\}$ it follows that solution $\omega(t,x)$ to conjugate problem (13) is an element of Hilbert space $H(Q_T)$.

5 System of Nonlinear Integral Equations of the Optimal Vector Control

We find the optimal control according to optimality conditions (31) and (32). We substitute in (31) the solution to conjugate boundary-value problem (30) defined by (34) and (37). Taking into account (33) we have the following system of inequalities

$$2\beta \frac{u_1(t)}{f_{u_1}} = ... = 2\beta \frac{u_m(t)}{f_{u_m}} = -2\sum_{n=1}^{\infty} g_n(t)E_n^*(T-t,\lambda) \times$$

$$\times \int_0^T G_n(T-\tau,\lambda)g)n(\tau)f(\tau,u_1(\tau),...,u_m(\tau))d\tau +$$

$$+2\sum_{n=1}^{\infty} g_n(t)E_n^*(T-t,\lambda)h_n. \tag{43}$$

Thus, the optimal vector control is defined as the solution of the system of the nonlinear integral equations (43), where function of the external source $f[t,u_1(t),...,u_m(t)]$ should satisfy the conditions (5) and (32). These conditions restrict the class of functions of external influences $f[t,u_1(t),...,u_m(t)]$. Further, we assume that the function $f[t,u_1(t),...,u_m(t)]$ satisfies conditions (5) and (32) for any controls $u_i(t) \in H(0,T),\ i = 1,2,3,...,m$.

As system (43) has the property of equal relations, according to the method in the work [15,17,18] we use the denotation

$$\beta \frac{u_1(t)}{f_{u_1}} = ... = \beta \frac{u_m(t)}{f_{u_m}} = p(t). \tag{44}$$

According to (32), the functions $u_1(t), ..., u_m(t)$ are uniquely determined from (44), i.e. the following relations hold

$$u_i(t) = \varphi_i(t, p(t), \beta), \;\; i = 1, 2, 3, ..., m. \tag{45}$$

Taking into account (44) and (45) from (43) we have the following scalar nonlinear integral equation

$$p(t) + \sum_{n=1}^{\infty} g_n(t) E_n^*(T-t, \lambda) \int_0^T G_n(T-\tau, \lambda) g_n(\tau) \times$$

$$\times f(\tau, \phi_1(\tau, p(\tau), \beta), ..., \phi_m(\tau, p(\tau), \beta)) d\tau = \sum_{n=1}^{\infty} g_n(t) E_n^*(T-t, \lambda) h_n. \tag{46}$$

By using the following denotations

$$G[p] = -\sum_{n=1}^{\infty} g_n(t) E_n^*(T-t, \lambda) \times$$

$$\times \int_0^T G_n(T-\tau, \lambda) g_n(\tau) f(\tau, \varphi_1(\tau, p(\tau), \beta), ..., \varphi_m(\tau, p(\tau), \beta)) d\tau; \tag{47}$$

$$h(t) = \sum_{n=1}^{\infty} g_n(t) E_n^*(T-t, \lambda) h_n \tag{48}$$

we rewrite the system of equations (46) in the operator form

$$p = G[p] + h \tag{49}$$

and we investigate its solvability.

Lemma 4. The function $h(t)$ is an element of the Hilbert space $H(0, T)$.

Proof. By straightforward calculation we obtain the inequality

$$\int_0^T \|h(t)\|_{R^2}^2 = dt \int_0^T \left(h_1^2(t) + h_2^2(t)\right) dt =$$

$$= \int_0^T \left(\sum_{n=1}^{\infty} g_n(t) \langle E_n(T-t, \lambda), h_n \rangle_{R^2}\right)^2 dt \leq \int_0^T \sum_{n=1}^{\infty} g_n^2(t) \sum_{n=1}^{\infty} |\langle E_n(T-t, \lambda), h_n \rangle_{R^2}|^2 \, dt \leq$$

$$\leq \int_0^T \sum_{n=1}^{\infty} g_n^2(t) \sum_{n=1}^{\infty} \|E_n(T-t, \lambda)\|_{R^2}^2 \, \|h_n\|_{R^2}^2 dt \leq E_0^2 \|g(t, x)\|_{H(Q_T)}^2 \cdot \sum_{n=1}^{\infty} \|h_n\|_{R^2}^2 < \infty$$

from which proof of Lemma follows.

Lemma 5. The operator G maps the space $H(0, T)$ into itself.

Proof. By straightforward calculation we obtain the inequality

$$G[p]=\sum_{n=1}^{\infty}g_n(t)E_n^*(T-t,\lambda)\int_0^T G_n(T-\tau)g_n(\tau)f(\tau,\varphi_1(\tau,p(\tau),\beta),...,\varphi_m(\tau,p(\tau),\beta))d\tau;$$

$$\int_0^T G^2[p]dt=\int_0^T\left(\sum_{n=1}^{\infty}g_n(t)E_n^*(T-t,\lambda)\times\right.$$

$$\left.\times\int_0^T G_n(T-t,\lambda)g_n(\tau)f(\tau,\varphi_1(\tau,p(\tau),\beta),...,\varphi_m(\tau,p(\tau),\beta))d\tau\right)^2dt\le$$

$$\le\int_0^T\left(\sum_{n=1}^{\infty}\left\langle g_n(t)E_n(T-t,\lambda),\right.\right.$$

$$\left.\left.\int_0^T G_n(T-t,\lambda)g_n(\tau)f(\tau,\varphi_1(\tau,p(\tau),\beta),...,\varphi_m(\tau,p(\tau),\beta))d\tau\right\rangle_{E^2}\right)^2dt\le$$

$$\le\int_0^T\left(\sum_{n=1}^{\infty}\|g_n(t)E_n(T-t,\lambda)\|_{R^2}\left\|\int_0^T G_n(T-t,\lambda)g_n(\tau)\times\right.\right.$$

$$\left.\left.\times f(\tau,\varphi_1(\tau,p(\tau),\beta),...,\varphi_m(\tau,p(\tau),\beta))d\tau\right\|_{E^2}\right)^2dt\le\int_0^T\sum_{n=1}^{\infty}\|g_n(t)E_n(T-t,\lambda)\|_{R^2}^2$$

$$\left\|\int_0^T G_n(T-t,\lambda)g_n(\tau)f(\tau,\varphi_1(\tau,p(\tau),\beta),...,\varphi_m(\tau,p(\tau),\beta))d\tau\right\|_{E^2}^2dt\le$$

$$\le\int_0^T\sum_{n=1}^{\infty}g_n^2(t)\,\|E_n(T-t,\lambda)\|_{R^2}^2\times$$

$$\times\sum_{n=1}^{\infty}\left(\int_0^T\|G_n(T-t,\lambda)g_n(\tau)f(\tau,\varphi_1(\tau,p(\tau),\beta),...,\varphi_m(\tau,p(\tau),\beta))\|_{E^2}d\tau\right)^2dt\le$$

$$\le\int_0^T\sum_{n=1}^{\infty}g_n^2(t)\,\|E_n(T-t,\lambda)\|_{R^2}^2\times$$

$$\times\sum_{n=1}^{\infty}\int_0^T\|G_n(T-t,\lambda)\|_{E^2}^2\,g_n^2(\tau)d\tau\int_0^T f^2(\tau,\varphi_1(\tau,p(\tau),\beta),...,\varphi_m(\tau,p(\tau),\beta))d\tau dt\le$$

$$\le E_0^4\left(\|g(t,x)\|_{H(Q_T)}^2\right)\|f(\tau,\varphi_1(\tau,p(\tau),\beta),...,\varphi_m(\tau,p(\tau),\beta))\|_{H(0,T)}^2<\infty,$$

from which the proof of Lemma follows

Lemma 6. Suppose that the conditions

$$\|f[t,u_1(t),...,u_m(t)]-f[t,\bar{u}_1(t),...,\bar{u}_m(t)]\|_{H(0,T)}\le\sqrt{m}f_0\left(\sum_{k=1}^{m}\|u_k(t)-\bar{u}_k(t)\|_{H(0,T)}^2\right),\tag{50}$$

$$f_0=\left(\sum_{k=1}^{\infty}f_{0k}^2\right)^{1/2},$$

$$\|\varphi_i[t,p(t),\beta]-\varphi_i[t,\bar{p}(t),\beta]\|_{H(0,T)}^2\le\varphi_{0i}^2(\beta)\|p(t)-\bar{p}(t)\|_{H(0,T)}^2,\ i=1,2,3,...,m,\tag{51}$$

are satisfied. Then if the condition

$$0 < \sigma = \|g(t,x)\|^2_{H(Q_T)} E_0^2 \sqrt{m} f_0 \varphi_0(\beta) < 1, \; \varphi_0(\beta) = \left(\sum_{k=1}^{m} \varphi_{0k}^2(\beta) \right)^{1/2}, \quad (52)$$

is met, the operator G is contractive in space $H[0,T]$.

Proof. The proof of this Lemma follows from the proof of *Lemma* 5 by replacing $f[t,u_1(t),...,u_m(t)]$ the with difference $f[t,u_1(t),...,u_m(t)] - f[t,\bar{u}_1(t),...,\bar{u}_m(t)]$.

Theorem 1. *Suppose that conditions (15), (32), (50)–(52) are satisfied. Then operator Eq. (49) has a unique solution in the space $H(0,T)$.*

Proof. According to *Lemmas* 5 and 6, operator Eq. (43) can be considered in the space $H(0,T)$. According to *Lemma 4.3* operator G is a contractive. Since the Hilbert space $H(0,T)$ is a complete metric space, by the theorem on contraction mappings the operator G has a unique fixed point, i.e. operator Eq. (43) has a unique [22,23].

The solution of operator Eq. (43) can be found by the method of successive approximations, i.e. nth approximation of the solution is found by the formula

$$p_n(t) = G[p_{n-1}(t)] + h(t), \; n = 1,2,3,...$$

Here $p_0(t)$ is an arbitrary element of the space $H(0,T)$. The exact solution $p_0(t)$ can be found as the limit of the approximate solutions, i.e.

$$p^0(t) = \lim_{n\to\infty} p_n(t),$$

nd the following estimate holds

$$\|p^0(t) - p_n(t)\|_{H(0,T)} \leq \frac{\delta^n}{1-\delta} \|G[p_0(t)] + h(t) - p_0(t)\|_{H(0,T)}.$$

Substituting the solution $p^0(t)$ of nonlinear integral equation (43) in (45) we find the required optimal vector control

$$u_i^0(t) = \varphi_i[t, p^0(t), \beta], \; i = 1,2,3,...,m.$$

According to (8) and (24) we find the optimal process $V^0(t,x)$ or the solution of boundary control problem (1)–(3) corresponding to the optimal vector control $(u_1^0(t),...,u_m^0(t))$, by the formula

$$V^0(t,x) = \sum_{n=1}^{\infty} \left\{ \psi_n(t,\lambda) + \frac{1}{\lambda_n} \int_0^T \varepsilon_n(t,\eta,\lambda) g_n(\tau) f[\tau, u_1^0(\eta),...,u_m^0(\eta)] d\eta \right\} z_n(x).$$

The minimum value of the functional (27) is calculated by the formula

$$J[u_1^0(t),...,u_m^0(t)] = \int_Q \left\{ [V^0(T,x) - xi_1(x)]^2 + [V_t^0(T,x) - xi_2(x)]^2 \right\} dx +$$

$$+\beta \sum_{k=1}^{m} \left(u_k^0(t)\right)^2 dt, \ \beta > 0.$$

The found triplet $\{\left(u_1^0(t), ..., u_m^0(t)\right), V^0(t,x), J\left(u_1^0(t), ..., u_m^0(t)\right)\}$ is a complete solution to the nonlinear optimization problem.

6 Conclusion

In the process of studying the problem of optimal vector control for the oscillation processes described by Fredholm integro-differential equations in partial derivatives we obtained the following results:

- Fourier coefficients of generalized solution of the boundary value problem are defined as solution of the linear inhomogeneous Fredholm integral equation of the second kind. Convergence radius of the Neumann series with respect to the parameter λ , where the resolvent is continuous function, exist for any Fourier coefficient;
- Class of external source functions are defined, for which the second optimality condition are satisfied and the given optimization problem will be solvable;
- System of nonlinear integral equations of optimal controls possess the property of equal relations which allows simplifying the procedure of constructing the solution of the system. This property is universal, i.e. it undepend on the nature of the controlled process;
- Sufficient condition for the unique solvability of the optimal control problem were found and algorithm for constructing the complete solution of the optimization problem in the form of optimal vector control, optimal process and the minimum value of the functional were developed.

The obtained theoretical results are new in the optimal control theory of systems with distributed parameters and can be used to develop new methods of qualitative researches and of the construction of algorithms for solving the problem of technological processes of a more complex nature, as well as in practice.

References

1. Volterra, V.: Theory of Functionals and of Integral and Integro-Differential Equations. New York, USA (2005)
2. Vladimirov, V.S.: Mathematical problems of speed transport theory of particles, pp. 3–158. Trudy MIAN, T. 61 (1961) (In Russian)
3. Egorov, A.I.: Optimal control of thermal and diffusion processes. M: Nauka (1978) (In Russian)
4. Egorov, A.I.: On optimal control for processes in distributed objects. Appl. Math. Mech. **VII**(4), 688–696 (1963) (In Russian)
5. Egorov, A.I.: On optimal control for processes in some systems with distributed parameters. In: Avtomatika and Telemechanika, vol. XXV, No. 5 (1964) (In Russian)

6. Sirazetdinov, T.K.: To the theory of optimal processes with distributed parameters. In: Avtomatika and Telemechanika, vol. XXV, No. 4 (1964) (In Russian)
7. Egorov, A.I.: Optimal processes and invariance theory. SIAM J. Control **4**(4), 601–661 (1966)
8. Butkovskii, A.G., Egorov, A.I., Lurie, A.K.: Optimal control of distributed systems. SIAM J. Control **6**(13), 437–476 (1968)
9. Wanfang, Sh., Danping, Y., Wenbin, L.: Optimal control problem governed by a linear hyperbolic integro-differential equation and its finite element analysis. Bound. Value Probl. **2014**(173) (2014)
10. Kowalewski, A.: Optimal control of an infinite order hyperbolic system with multiple time-varying lags. Automatyka, Tom **15**, 53–65 (2011)
11. Asatur, zh. K.:On optimal boundary and distributed control of partial integro-differential equations. Arch. Control. Sci. **24(LX)**(1), 5–25 (2014)
12. Appel, J.M., Kalitvin, A.S., Zabrejko, P.P.: Partial Integral Operators and Integro-Differential Equations, IM. Dekkar, New York (2000)
13. Sachs, E.W., Strauss, A.K.: Efficient solution of partial integrodifferential equation in finance. Appl. Numer. Math. **58**(11), 1687–1703 (2008)
14. Thorwe, J., Bhalekar, S.: Solving partial integrodifferential equations using Laplace transform method. Am. J. Comput. Appl. Math. **2**(3), 101–104 (2012)
15. Kerimbekov, A.: On the solvability of a nonlinear optimal control problem for the thermal processes described by fredholm integro-differential equations. In: Proceedings of the 9th ISAAC Congress Current Trends in Analysis and its Applications (Krakov 2013). A series of trends in mathematics, vol. XVI, pp. 803–811. Springer International Publishing, Switzerland (2015)
16. Komkov, V.: Optimal Control Theory for the Damping of Vibrations of Simple Elastic Systems, pp. 160. Springer (1972)
17. Kerimbekov, A., Baetov, A.K.: On solvability of a nonlinear optimization problem for oscillation processes. In: Proceedings of International Scientific conference Actual Problems of Applied Mathematics and Information Technologies-al-Horezmi. Tashkent (Uzbekistan), pp. 301–303 (2012) (In Russian)
18. Kerimbekov, A., Kabaeva, Z.C.: Solution of the optimization problem for thermal processes in nonlinear vector control. In: Proceedings of the VI International Scientific Conference. Part I, Aktobe Regional University K.Zhubanova. Aktobe (Kazakhstan), 14-17 Oct 2012, pp. 100–104 (2012) (In Russian)
19. Tikhonov, A.N., Samarskiy, A.A.: Equations of Mathematical Physics. M. Nauka, Moscow (1972) (In Russian)
20. Plotnikov, V.I.: The energy inequality and over-determination property of systems of eigen function. In: The News of Academy of Science of USSR, vol. 32, No. 4, pp. 743–755 (Math. Ser.) (1968)
21. Tricomi, I.F.: Integral Equations. Intersciense Publishers, New York (1957)
22. Krasnov, M.V.: Integral Equations, pp. 303. Nauka, Moscow (1975) (In Russian)
23. Lusternik, L.A., Sobolev, V.I.: Elements of Functional Analysis, pp. 520. Nauka, Moscow (1965) (In Russian)

On Nuclear L^p-Multipliers Associated to the Harmonic Oscillator

Edgardo Samuel Barraza[1] and Duván Cardona[2(✉)]

[1] Department of Computer Science and Artificial Intelligence, Universidad de Sevilla, Seville, Spain
edgbarver@alum.us.es
[2] Mathematics Department, Pontificia Universidad Javeriana, Bogotá, Colombia
cardonaduvan@javeriana.edu.co

Abstract. In this paper we study multipliers associated to the harmonic oscillator (also called Hermite multipliers) belonging to the ideal of r-nuclear operators on Lebesgue spaces. We also study the nuclear trace and the spectral trace of these operators.

Keywords: Nuclearity · Harmonic oscillators · Spectra theory · Fourier multiplier · Hermite multiplier · Traces

1 Introduction

In this paper, we are interested in the r-nuclearity of multipliers associated to the harmonic oscillator (also called Hermite multipliers) on $L^p(\mathbb{R}^n)$-spaces. In quantum mechanics the harmonic oscillator is the unbounded operator defined by $H = -\Delta_x + |x|^2$, where Δ_x is the Laplacian. The operator H extends to an unbounded self-adjoint operator on $L^2(\mathbb{R}^n)$, and its spectrum consists of the discrete set $\lambda_\nu := 2|\nu| + n$, $\nu \in \mathbb{N}_0^n$, with real eigenfunctions ϕ_ν, $\nu \in \mathbb{N}_0^n$, called Hermite functions. Every Hermite function ϕ_ν on $\mathbb{R}^n$ has the form

$$\phi_\nu = \Pi_{j=1}^n \phi_{\nu_j}, \ \ \phi_{\nu_j}(x_j) = (2^{\nu_j}\nu_j!\sqrt{\pi})^{-\frac{1}{2}} H_{\nu_j}(x_j) e^{-\frac{1}{2}x_j^2} \tag{1}$$

where $x = (x_1, \cdots, x_n) \in \mathbb{R}^n$, $\nu = (\nu_1, \cdots, \nu_n) \in \mathbb{N}_0^n$, and $H_{\nu_j}(x_j)$ denotes the Hermite polynomial of order ν_j. Furthermore, by the spectral theorem, for every $f \in \mathscr{D}(\mathbb{R}^n)$ we have

$$Hf(x) = \sum_{\nu \in \mathbb{N}_0^n} \lambda_\nu \widehat{f}(\phi_\nu)\phi_\nu(x), \ \ \widehat{f}(\phi_\nu) := \langle f, \phi_\nu \rangle_{L^2(\mathbb{R}^n)} = \int_{\mathbb{R}^n} f(x)\phi_\nu(x)\,dx, \tag{2}$$

where $\widehat{f}(\phi_v)$ is the Hermite-Fourier transform of f at ν. In addition, every function m on $\mathbb{N}_0^n$ has associated a multiplier (or Hermite multiplier) which is a linear

J. Delgado and M. Ruzhansky (Eds.): *Analysis and Partial Differential Equations: Perspectives from Developing Countries*, PROMS 275, pp. 31–41, 2018.
https://doi.org/10.1007/978-3-030-05657-5_4

operator T_m of the form:

$$T_m f(x) = \sum_{\nu \in \mathbb{N}_0^n} m(\nu) \widehat{f}(\phi_\nu) \phi_\nu(x), \quad f \in D(T_m). \tag{3}$$

The discrete function m is called the symbol of the operator T_m. In particular, if m is a measurable function, the symbol of the spectral multiplier $m(H)$, defined by the functional calculus, is given by $m(\nu) := m(\lambda_\nu)$, thus, spectral multipliers are natural examples of multipliers associated to the harmonic oscillator.

In order that the operator $T_m : L^{p_1}(\mathbb{R}^n) \to L^{p_2}(\mathbb{R}^n)$ extends to a r-nuclear operator, we provide sufficient conditions on the symbol m. We recall the notion of r-nuclearity as follows. By following Grothendieck [18], we can recall that a linear operator $T : E \to F$ (E and F Banach spaces) is r-nuclear, if there exist sequences $(e_n')_{n\in\mathbb{N}_0}$ in E' (the dual space of E) and $(y_n)_{n\in\mathbb{N}_0}$ in F such that

$$Tf = \sum_{n\in\mathbb{N}_0} e_n'(f) y_n, \quad \text{and} \quad \sum_{n\in\mathbb{N}_0} \|e_n'\|_{E'}^r \|y_n\|_F^r < \infty. \tag{4}$$

The class of $r-$nuclear operators is usually endowed with the quasi-norm

$$n_r(T) := \inf \left\{ \left\{ \sum_n \|e_n'\|_{E'}^r \|y_n\|_F^r \right\}^{\frac{1}{r}} : T = \sum_n e_n' \otimes y_n \right\} \tag{5}$$

and, if $r = 1$, $n_1(\cdot)$ is a norm and we obtain the ideal of nuclear operators. In addition, when $E = F$ is a Hilbert space and $r = 1$ the definition above agrees with the concept of trace class operators. For the case of Hilbert spaces H, the set of r-nuclear operators agrees with the Schatten-von Neumann class of order r (see Pietsch [21,22]).

In order to study the r-nuclearity and the spectral trace of Hermite multipliers, we will use results from Delgado [5], on the characterization of nuclear integral operators on $L^p(X,\mu)$ spaces, which in this case can be applied to Lebesgue spaces on $\mathbb{R}^n$. Indeed, we will prove that under certain conditions, a r-nuclear operator $T_m : L^p(\mathbb{R}^n) \to L^p(\mathbb{R}^n)$ has a nuclear trace given by

$$\text{Tr}(T_m) = \sum_{\nu\in\mathbb{N}_0^n} m(\nu). \tag{6}$$

Furthermore, by using the version of the Grothendieck-Lidskii formula proved in Reinov and Latif [23] we will show that the nuclear trace of these operators coincides with the spectral trace for $\frac{1}{r} = 1 + |\frac{1}{p} - \frac{1}{2}|$. An important tool in the formulation of our results will be the asymptotic behavior of L^p-norms for Hermite functions (see [29]). There are complications to estimate norms of Hermite functions on different functions spaces (see e.g. [19]), thus, our results can be not extended immediately to function spaces like Lebesgue spaces with exponent variable.

Now, we present some references on the subject. Sufficient conditions for the r-nuclearity of spectral multipliers associated to the harmonic oscillator, but, in modulation spaces and Wiener amalgam spaces have been considered by Delgado et al. [7,8]. The Properties of these multipliers in L^p-spaces have been investigated in the references Bagchi and Thangavelu [1], Epperson [15], Stempak and Torrea [26–28], Thangavelu [29,30] and references therein. Hermite expansions for distributions can be found in Simon [25]. The r-nuclearity and Grothendieck-Lidskii formulae for multipliers and other types of integral operators can be found in [6,8]. Sufficient conditions for the nuclearity of pseudo-differential operators on the torus can be found in [4,16]. The references [9–12,14] include a complete study on the r-nuclearity of multipliers (and pseudo-differential operators) on compact Lie groups and more generally on compact manifolds. On Hilbert spaces the class of r-nuclear operators agrees with the Schatten-von Neumann class $S_r(H)$; in this context operators with integral kernel on Lebesgue spaces and, in particular, operators with kernel acting of a special way with anharmonic oscillators of the form $E_a = -\Delta_x + |x|^a$, $a > 0$, has been considered on Schatten classes on $L^2(\mathbb{R}^n)$ in Delgado and Ruzhansky [13].

Our main results will be presented in the next section. We end this paper by applying our trace formula to the the Hermite semigroup (or semigroup associated to the harmonic oscillator) e^{-tH}, $t > 0$. We would like to thanks César del Corral from Universidad de los Andes, Bogotá-Colombia, for discussions and his remarks about our work.

2 r-Nuclear Multipliers Associated to the Harmonic Oscillator

2.1 Preliminaries and Notations: Nuclear Operators on Lebesgue Spaces

In this section we study r-nuclear multipliers T_m on Lebesgue spaces. Our criteria will be formulated in terms of the symbols m. First, let us observe that every multiplier T_m is an operator with kernel $K_m(x,y)$. In fact, straightforward computation show that

$$T_m f(x) = \int_{\mathbb{R}^n} K_m(x,y) f(y) dy, \ K_m(x,y) := \sum_{\nu \in \mathbb{N}_0^n} m(\nu)\phi_\nu(x)\phi_\nu(y)$$

for every $f \in \mathscr{D}(\mathbb{R}^n)$. In order to analyze the r-nuclearity of T_m we study its kernel K_m by using the following theorem (see Delgado [3,5]).

Theorem 1. *Let us consider* $1 \leq p_1, p_2 < \infty$, $0 < r \leq 1$ *and let* p_1' *be such that* $\frac{1}{p_1} + \frac{1}{p_1'} = 1$. *An operator* $T : L^p(\mu_1) \to L^p(\mu_2)$ *is* r*-nuclear if and only if there exist sequences* $(g_n)_n$ *in* $L^{p_2}(\mu_2)$, *and* (h_n) *in* $L^{q_1}(\mu_1)$, *such that*

$$\sum_n \|g_n\|^r_{L^{p_2}} \|h_n\|^r_{L^{q_1}} < \infty, \text{ and } Tf(x) = \int (\sum_n g_n(x) h_n(y)) f(y) d\mu_1(y), \text{ a.e.w. } x, \tag{7}$$

for every $f \in L^{p_1}(\mu_1)$. In this case, if $p_1 = p_2$ (see Sect. 3 of [3]) the nuclear trace of T is given by

$$\operatorname{Tr}(T) := \int \sum_n g_n(x)h_n(x)d\mu_1(x). \tag{8}$$

If $1 \leq p_1, p_2 < \infty$, by Theorem 1, we have that a multiplier $T_m : L^{p_1}(\mathbb{R}^n) \to L^{p_2}(\mathbb{R}^n)$ is r-nuclear if

$$s_r(m, p_1, p_2) := \sum_{\nu \in \mathbb{N}_0^n} |m(\nu)|^r \|\phi_\nu\|^r_{L^{p_2}(\mathbb{R}^n)} \|\phi_\nu\|^r_{L^{p_1'}(\mathbb{R}^n)} < \infty, \tag{9}$$

consequently, the nuclear trace of T_m is given by

$$\operatorname{Tr}(T_m) = \int_{\mathbb{R}^n} \sum_{\nu \in \mathbb{N}^n} m(\nu)\phi_\nu(x)^2 dx = \sum_{\nu \in \mathbb{N}^n} m(\nu) \int_{\mathbb{R}^n} \phi_\nu(x)^2 dx = \sum_{\nu \in \mathbb{N}^n} m(\nu), \tag{10}$$

where we have used that the L^2-norm of every ϕ_ν is normalized.

Although the estimate (9) is a sufficient condition on m in order to guarantee the r-nuclearity of T_m, we want to study explicit conditions on m, p_1, and p_2 in order that the series (9) converges. In order to find such conditions, we need to estimate the L^p-norms of Hermite functions which are eigenfunctions of the harmonic oscillator. This situation is analogue to one in the case of compact Lie groups where, in order to establish the r-nuclearity of multipliers, it is important to estimate L^p-norms of the eigenfunctions of the Laplacian (see Cardona [2] and Delgado and Ruzhansky [9]).

Our starting point is the following lemma on the asymptotic properties of L^p-norms for Hermite functions on $\mathbb{R}$ (see Lemma 4.5.2 of [29]).

Lemma 1. *Let us denote by ϕ_ν, $\nu \in \mathbb{N}_0$, the one-dimensional Hermite functions. As $\nu \to \infty$, these functions satisfy the estimates*

- $\|\phi_\nu\|_{L^p(\mathbb{R})} \asymp \nu^{\frac{1}{2p} - \frac{1}{4}}$, $1 \leq p \leq 2$.
- $\|\phi_\nu\|_{L^p(\mathbb{R})} \asymp \nu^{-\frac{1}{8}} \ln(\nu)$, $p = 4$,
- $\|\phi_\nu\|_{L^p(\mathbb{R})} \asymp \nu^{-\frac{1}{6p} - \frac{1}{12}}$, $4 < p \leq \infty$.

We need to extend the previous lemma to the case of n-dimensional Hermite functions, we explain the process of extension in the following remark.

Remark 1. By the previous lemma, there exists $k \in \mathbb{N}$ large enough, such that

- $\|\phi_\nu\|_{L^p(\mathbb{R})} \asymp \nu^{\frac{1}{2p} - \frac{1}{4}}$, $1 \leq p < 4$,
- $\|\phi_\nu\|_{L^p(\mathbb{R})} \asymp \nu^{-\frac{1}{8}} \ln(\nu)$, $p = 4$,
- $\|\phi_\nu\|_{L^p(\mathbb{R})} \asymp \nu^{-\frac{1}{6p} - \frac{1}{12}}$, $4 < p \leq \infty$,

if $\nu \geq k$, and for all $\nu \leq k$, $\|\phi_\nu\|_{L^p(\mathbb{R})} \asymp \rho_k := \|\phi_k\|_{L^p(\mathbb{R})}$. Now, let us define the following $n+1$ subsets of $\mathbb{N}_0^n$:

- $I_0 := \{\nu \in \mathbb{N}_0^n : \nu_j > k \text{ for all } 1 \leq j \leq n\}$,
- I_s, for $s = 1, 2, \cdots, n$, consists of all $\nu \in \mathbb{N}_0^n$ such that for some $j_1, \cdots, j_s$, $\nu_{j_1}, \cdots \nu_{j_s} \leq k$ but, for $j \neq j_1, j_2, \cdots j_s$, $\nu_j > k$. So, I_s is the set of n-tuples of non negative integers with exactly s entries less than or equal to k.

The sequence of subsets $I_0, I_1, \cdots I_n$, provides a partition of $\mathbb{N}_0^n$ which will be useful in order to estimate the series (9). We end this remark by observing that for all $\nu \in I_s$,

$$\|\phi_\nu\|_{L^p(\mathbb{R}^n)} = \prod_{v_j \leq k} \|\phi_{\nu_j}\|_{L^p(\mathbb{R})} \prod_{v_j > k} \|\phi_{\nu_j}\|_{L^p(\mathbb{R})} \asymp \|\phi_k\|^s_{L^p(\mathbb{R}^n)} \prod_{v_j > k} \|\phi_{\nu_j}\|_{L^p(\mathbb{R})}.$$

With the notation of the preceding remark and by using Lemma 1 we have immediately the following proposition.

Proposition 1. *Let $s \in \mathbb{N}_0$, and $\nu \in I_s \subset \mathbb{N}_0^n$. The n-dimensional Hermite function ϕ_ν satisfies the estimates*

- $\|\phi_\nu\|_{L^p(\mathbb{R}^n)} \asymp k^{s(\frac{1}{2p}-\frac{1}{4})} \cdot (\prod_{\nu_j > k} \nu_j)^{\frac{1}{2p}-\frac{1}{4}}$, $1 \leq p < 4$.
- $\|\phi_\nu\|_{L^p(\mathbb{R}^n)} \asymp k^{-\frac{s}{8}} (\ln(k))^s \cdot \prod_{\nu_j > k} [\nu_j^{-\frac{1}{8}} \ln(\nu_j)]$, $p = 4$,
- $\|\phi_\nu\|_{L^p(\mathbb{R}^n)} \asymp k^{s(-\frac{1}{6p}-\frac{1}{12})} \cdot (\prod_{\nu_j > k} \nu_j)^{-\frac{1}{6p}-\frac{1}{12}}$, $4 < p \leq \infty$.

Extensions of Lemma 1 have been obtained in [19], however, for our purposes will be sufficient such lemma and the preceding proposition.

2.2 Nuclear Multipliers of the Harmonic Oscillator

By using the previous proposition, we study separately the r-nuclearity of multipliers by dividing our classification in the following three cases: $1 \leq p_2 < 4$ and $1 < p_1 < \infty$, $p_2 = 4$ and $1 < p_1 < \infty$, and $4 < p_2 \leq \infty$, $1 < p_1 < \infty$. We start with the following result.

Theorem 2. *Let $1 \leq p_2 < 4$ and $1 < p_1 < \infty$. Let T_m be a multiplier associated to the harmonic oscillator and m its symbol. Then $T_m : L^{p_1}(\mathbb{R}^n) \to L^{p_2}(\mathbb{R}^n)$ extends to a r-nuclear operator if one of the following conditions holds:*

- $\frac{4}{3} < p_1 < \infty$ *and*

$$\varkappa(m, p_1, p_2) := \sum_{s=0}^{n} \sum_{\nu \in I_s} k^{\frac{sr}{2}(\frac{1}{p_2}-\frac{1}{p_1})} (\prod_{\nu_j > k} \nu_j)^{\frac{r}{2}(\frac{1}{p_2}-\frac{1}{p_1})} |m(\nu)|^r < \infty. \tag{11}$$

- $p_1 = \frac{4}{3}$ *and*

$$\varkappa(m, p_1, p_2) := \sum_{s=0}^{n} \sum_{\nu \in I_s} k^{\frac{sr}{2}(\frac{1}{p_2}-\frac{3}{4})} (\ln k)^{sr} \cdot \prod_{\nu_j > k} [\nu_j^{\frac{r}{2}(\frac{1}{p_2}-\frac{3}{4})} (\ln(\nu_j))^r] |m(\nu)|^r < \infty. \tag{12}$$

- $1 < p_1 < \frac{4}{3}$ *and*

$$\varkappa(m,p_1,p_2) := \sum_{s=0}^{n}\sum_{\nu\in I_s} k^{\frac{sr}{2}(\frac{1}{p_2}+\frac{1}{3p_1}-1)}\cdot(\prod_{\nu_j>k}\nu_j)^{\frac{r}{2}(\frac{1}{p_2}+\frac{1}{3p_1}-1)}|m(\nu)|^r < \infty. \tag{13}$$

In every case we have $s_r(m,p_1,p_2) \asymp \varkappa(m,p_1,p_2)$.

Proof. In view of Theorem 1 we only need to proof (9). So, let us note that for $1 \le p_2 < 4$ and $\frac{4}{3} < p_1 < \infty$ (which imply $1 < p_1' < 4$) we have the following estimate for (9),

$$\begin{aligned}\sum_{\nu\in\mathbb{N}_0^n}|m(\nu)|^r\|\phi_\nu\|^r_{L^{p_2}(\mathbb{R}^n)}\|\phi_\nu\|^r_{L^{p_1'}(\mathbb{R}^n)} &= \sum_{s=0}^n\sum_{\nu\in I_s}|m(\nu)|^r\|\phi_\nu\|^r_{L^{p_2}(\mathbb{R}^n)}\|\phi_\nu\|^r_{L^{p_1'}(\mathbb{R}^n)}\\ &\asymp \sum_{s=0}^n\sum_{\nu\in I_s}k^{sr(\frac{1}{2p_2}-\frac{1}{4})}\cdot(\prod_{\nu_j>k}\nu_j)^{r(\frac{1}{2p_2}-\frac{1}{4})}\cdot k^{sr(\frac{1}{2p_1'}-\frac{1}{4})}\cdot(\prod_{\nu_j>k}\nu_j)^{r(\frac{1}{2p_1'}-\frac{1}{4})}\\ &= \sum_{s=0}^n\sum_{\nu\in I_s}k^{\frac{sr}{2}(\frac{1}{p_2}-\frac{1}{p_1})}(\prod_{\nu_j>k}\nu_j)^{\frac{r}{2}(\frac{1}{p_2}-\frac{1}{p_1})}|m(\nu)|^r < \infty. \end{aligned}\tag{14}$$

On the other hand, if we consider $1 \le p_2 < 4$, $p_1 = \frac{4}{3}$ (i.e $p_1' = 4$) we obtain the following asymptotic expressions for (9),

$$\begin{aligned}\sum_{\nu\in\mathbb{N}_0^n}|m(\nu)|^r\|\phi_\nu\|^r_{L^{p_2}(\mathbb{R}^n)}\|\phi_\nu\|^r_{L^4(\mathbb{R}^n)} &= \sum_{s=0}^n\sum_{\nu\in I_s}|m(\nu)|^r\|\phi_\nu\|^r_{L^{p_2}(\mathbb{R}^n)}\|\phi_\nu\|^r_{L^4(\mathbb{R}^n)}\\ &\asymp \sum_{s=0}^n\sum_{\nu\in I_s}|m(\nu)|^r k^{sr(\frac{1}{2p_2}-\frac{1}{4})}\\ &\quad\times(\prod_{\nu_j>k}\nu_j)^{r(\frac{1}{2p_2}-\frac{1}{4})}\cdot k^{-\frac{sr}{8}}(\ln(k))^{sr}\cdot\prod_{\nu_j>k}[\nu_j^{-\frac{r}{8}}(\ln(\nu_j))^r]\\ &= \sum_{s=0}^n\sum_{\nu\in I_s}|m(\nu)|^r k^{\frac{sr}{2}(\frac{1}{p_2}-\frac{3}{4})}(\ln k)^{sr}\cdot\prod_{\nu_j>k}[\nu_j^{\frac{r}{2}(\frac{1}{p_2}-\frac{3}{4})}(\ln(\nu_j))^r] < \infty.\end{aligned}\tag{15}$$

For the case where $1 \le p_2 < 4$, $1 < p_1 < \frac{4}{3}$ (now, $4 < p_1' < \infty$) we have

$$\begin{aligned}\sum_{\nu\in\mathbb{N}_0^n}|m(\nu)|^r\|\phi_\nu\|^r_{L^{p_2}(\mathbb{R}^n)}\|\phi_\nu\|^r_{L^{p_1'}(\mathbb{R}^n)} &= \sum_{s=0}^n\sum_{\nu\in I_s}|m(\nu)|^r\|\phi_\nu\|^r_{L^{p_2}(\mathbb{R}^n)}\|\phi_\nu\|^r_{L^{p_1'}(\mathbb{R}^n)}\\ &\asymp \sum_{s=0}^n\sum_{\nu\in I_s}|m(\nu)|^r k^{sr(\frac{1}{2p_2}-\frac{1}{4})}\\ &\quad\times(\prod_{\nu_j>k}\nu_j)^{r(\frac{1}{2p_2}-\frac{1}{4})}\cdot k^{sr(-\frac{1}{6p_1'}-\frac{1}{12})}\cdot(\prod_{\nu_j>k}\nu_j)^{-\frac{r}{6p_1'}-\frac{1}{12}}\\ &= \sum_{s=0}^n\sum_{\nu\in I_s}k^{\frac{sr}{2}(\frac{1}{p_2}+\frac{1}{3p_1}-1)}\\ &\quad\times(\prod_{\nu_j>k}\nu_j)^{\frac{r}{2}(\frac{1}{p_2}+\frac{1}{3p_1}-1)}|m(\nu)|^r < \infty.\end{aligned}$$

thus, we conclude the proof.

Now, we classify the r-nuclearity of multipliers when $p_2 = 4$.

Theorem 3. *Let* $1 < p_1 < \infty$. *Let* T_m *be a multiplier associated to the harmonic oscillator and* m *its symbol. Then* $T_m : L^{p_1}(\mathbb{R}^n) \to L^4(\mathbb{R}^n)$ *extends to a* r*-nuclear operator if one of the following conditions holds:*

- $\frac{4}{3} < p_1 < \infty$ *and*

$$\varkappa(m,p_1,p_2) := \sum_{s=0}^{n}\sum_{\nu\in I_s} k^{\frac{sr}{2}(\frac{1}{4}-\frac{1}{p_1})}(\ln(k))^{sr}\prod_{\nu_j>k}[(\ln(\nu_j))^r\nu_j^{\frac{r}{2}(\frac{1}{4}-\frac{1}{p_1})}]|m(\nu)|^r < \infty. \tag{16}$$

- $p_1 = \frac{4}{3}$ *and*

$$\varkappa(m,p_1,p_2) := \sum_{s=0}^{n}\sum_{\nu\in I_s} k^{-\frac{sr}{4}}(\ln k)^{2sr}\prod_{\nu_j>k}[\nu_j^{-\frac{r}{4}}(\ln\nu_j)^{2r}]\cdot|m(\nu)|^r < \infty. \tag{17}$$

- $1 < p_1 < \frac{4}{3}$ *and*

$$\varkappa(m,p_1,p_2) := \sum_{s=0}^{n}\sum_{\nu\in I_s} k^{\frac{sr}{6}(\frac{1}{p_1}-\frac{9}{4})}(\ln(k))^{sr}\prod_{\nu_j>k}[\nu_j^{\frac{r}{6}(\frac{1}{p_1}-\frac{9}{4})}\ln(\nu_j)^r]\cdot|m(\nu)|^r < \infty. \tag{18}$$

In every case we have $s_r(m,p_1,p_2) \asymp \varkappa(m,p_1,p_2)$.

Proof. First, we consider the case $\frac{4}{3} < p_1 < \infty$, $(1 < p_1' < 4.)$ In fact, we have

$$\begin{aligned}\sum_{\nu\in\mathbb{N}_0^n}|m(\nu)|^r\|\phi_\nu\|^r_{L^4(\mathbb{R}^n)}\|\phi_\nu\|^r_{L^{p_1'}(\mathbb{R}^n)} &= \sum_{s=0}^{n}\sum_{\nu\in I_s}|m(\nu)|^r\|\phi_\nu\|^r_{L^4(\mathbb{R}^n)}\|\phi_\nu\|^r_{L^{p_1'}(\mathbb{R}^n)}\\ &\asymp \sum_{s=0}^{n}\sum_{\nu\in I_s}|m(\nu)|^r k^{-\frac{sr}{8}}(\ln(k))^{sr}\cdot\prod_{\nu_j>k}[\nu_j^{-\frac{r}{8}}(\ln(\nu_j))^r]\cdot k^{sr(\frac{1}{2p_1'}-\frac{1}{4})}\cdot(\prod_{\nu_j>k}\nu_j)^{r(\frac{1}{2p_1'}-\frac{1}{4})}\\ &= \sum_{s=0}^{n}\sum_{\nu\in I_s}k^{\frac{sr}{2}(\frac{1}{4}-\frac{1}{p_1})}(\ln(k))^{sr}\prod_{\nu_j>k}[(\ln(\nu_j))^r\nu_j^{\frac{r}{2}(\frac{1}{4}-\frac{1}{p_1})}]|m(\nu)|^r < \infty.\end{aligned}$$

For the case where $p_1 = \frac{4}{3}$, (i.e. $p_1' = 4$), we have

$$\begin{aligned}\sum_{\nu\in\mathbb{N}_0^n}|m(\nu)|^r\|\phi_\nu\|^r_{L^4(\mathbb{R}^n)}\|\phi_\nu\|^r_{L^{p_1'}(\mathbb{R}^n)} &= \sum_{s=0}^{n}\sum_{\nu\in I_s}|m(\nu)|^r\|\phi_\nu\|^{2r}_{L^4(\mathbb{R}^n)}\\ &\asymp \sum_{s=0}^{n}\sum_{\nu\in I_s}k^{-\frac{sr}{4}}(\ln(k))^{2sr}\cdot\prod_{\nu_j>k}[\nu_j^{-\frac{r}{4}}(\ln(\nu_j))^{2r}]\cdot|m(\nu)|^r < \infty.\end{aligned}$$

Finally, if we consider $1 < p_1 < \frac{4}{3}$, then $4 < p_1' < \infty$, and we estimate (9) as follows

$$\begin{aligned}\sum_{\nu\in\mathbb{N}_0^n}|m(\nu)|^r\|\phi_\nu\|^r_{L^4(\mathbb{R}^n)}\|\phi_\nu\|^r_{L^{p_1'}(\mathbb{R}^n)} &= \sum_{s=0}^{n}\sum_{\nu\in I_s}|m(\nu)|^r\|\phi_\nu\|^r_{L^4(\mathbb{R}^n)}\|\phi_\nu\|^r_{L^{p_1'}(\mathbb{R}^n)}\\ &\asymp \sum_{s=0}^{n}\sum_{\nu\in I_s}|m(\nu)|^r k^{-\frac{sr}{8}}(\ln(k))^{sr}\cdot\prod_{\nu_j>k}[\nu_j^{-\frac{r}{8}}(\ln(\nu_j))^r]\cdot k^{sr(-\frac{1}{6p_1'}-\frac{1}{12})}\\ &\qquad\times(\prod_{\nu_j>k}\nu_j)^{-\frac{r}{6p_1'}-\frac{1}{12}}\\ &\asymp \sum_{s=0}^{n}\sum_{\nu\in I_s}k^{\frac{sr}{6}(\frac{1}{p_1}-\frac{9}{4})}(\ln(k))^{sr}\prod_{\nu_j>k}[\nu_j^{\frac{r}{6}(\frac{1}{p_1}-\frac{9}{4})}\ln(\nu_j)^r]\cdot|m(\nu)|^r < \infty.\end{aligned}$$

thus, we end the proof.

We end this chapter with some conditions for the r-nuclearity of multipliers when $4 < p_2 \leq \infty$. We omit the proof because we only need to repeat the arguments above on asymptotic properties of Hermite functions.

Theorem 4. *Let $4 < p_2 \leq \infty$ and $1 < p_1 < \infty$. Let T_m be a multiplier associated to the harmonic oscillator and m its symbol. Then $T_m : L^{p_1}(\mathbb{R}^n) \to L^{p_2}(\mathbb{R}^n)$ extends to a r-nuclear operator if one of the following conditions holds:*

- $\frac{4}{3} < p_1 < \infty$ *and*

$$\varkappa(m,p_1,p_2) := \sum_{s=0}^{n} \sum_{\nu \in I_s} k^{\frac{sr}{2}(\frac{1}{3p_2'} - \frac{1}{p_1})} (\prod_{\nu_j > k} \nu_j)^{\frac{r}{2}(\frac{1}{3p_2'} - \frac{1}{p_1})} |m(\nu)|^r < \infty. \tag{19}$$

- $p_1 = \frac{4}{3}$ *and*

$$\varkappa(m,p_1,p_2) := \sum_{s=0}^{n} \sum_{\nu \in I_s} k^{-\frac{sr}{6}(\frac{1}{p_2} + \frac{5}{4})} (\ln(k))^{sr} \prod_{\nu_j > k} [\nu_j^{-\frac{r}{6}(\frac{1}{p_2} + \frac{5}{4})} (\ln(\nu_j))^r] |m(\nu)|^r < \infty. \tag{20}$$

- $1 < p_1 < \frac{4}{3}$ *and*

$$\varkappa(m,p_1,p_2) := \sum_{s=0}^{n} \sum_{\nu \in I_s} k^{\frac{sr}{6}(\frac{1}{p_1} - \frac{1}{p_2} - 2)} \cdot (\prod_{\nu_j > k} \nu_j)^{\frac{r}{6}(\frac{1}{p_1} - \frac{1}{p_2} - 2)} |m(\nu)|^r < \infty. \tag{21}$$

In every case we have $s_r(m,p_1,p_2) \asymp \varkappa(m,p_1,p_2)$.

3 Spectral and Nuclear Traces of Multipliers of the Harmonic Oscillator

3.1 Trace Formulae for Multipliers

If $T : E \to E$ is r-nuclear, with the Banach space E satisfying the approximation property (see [18]), then there exist sequences $(e_n')_{n \in \mathbb{N}_0}$ in E' (the dual space of E) and $(y_n)_{n \in \mathbb{N}_0}$ in E such that

$$Tf = \sum_{n \in \mathbb{N}_0} e_n'(f) y_n, \quad \text{and} \quad \sum_{n \in \mathbb{N}_0} \|e_n'\|_{E'}^r \|y_n\|_F^r < \infty. \tag{22}$$

In this case the nuclear trace of T is given by $\mathrm{Tr}(T) = \sum_{n \in \mathbb{N}_0^n} e_n'(f_n)$. L^p-spaces have the approximation property and as consequence we can compute the nuclear trace of every r-nuclear multiplier $T_m : L^p(\mathbb{R}^n) \to L^p(\mathbb{R}^n)$. In fact, by (10) we have

$$\mathrm{Tr}(T_m) = \sum_{\nu \in \mathbb{N}^n} m(\nu). \tag{23}$$

It was proved by Grothendieck that the nuclear trace of an r-nuclear operator on a Banach space coincides with the spectral trace, provided that $0 < r \leq \frac{2}{3}$. For $\frac{2}{3} \leq r \leq 1$ we recall the following result (see [23]).

Theorem 5. *Let $T : L^p(\mu) \to L^p(\mu)$ be a r-nuclear operator as in (22). If $\frac{1}{r} = 1 + |\frac{1}{p} - \frac{1}{2}|$, then,*

$$Tr(T) := \sum_{n \in \mathbb{N}_0^n} e_n'(f_n) = \sum_n \lambda_n(T) \tag{24}$$

where $\lambda_n(T)$, $n \in \mathbb{N}$ is the sequence of eigenvalues of T with multiplicities taken into account.

As an immediate consequence of the preceding theorem, if $T_m : L^p(\mathbb{R}^n) \to L^p(\mathbb{R}^n)$ is a r-nuclear multiplier and $\frac{1}{r} = 1 + |\frac{1}{p} - \frac{1}{2}|$ then,

$$\mathrm{Tr}(T_m) = \sum_{\nu \in \mathbb{N}^n} m(\nu) = \sum_n \lambda_n(T), \tag{25}$$

where $\lambda_n(T)$, $n \in \mathbb{N}$ is the sequence of eigenvalues of T_m with multiplicities taken into account.

3.2 The Hermite Semigroup

Let us consider the Hermite functions ϕ_ν, $\nu \in \mathbb{N}_0^n$. If P_k, $k \in \mathbb{N}_0$, is the projection on $L^2(\mathbb{R}^n)$ given by

$$P_k f(x) := \sum_{|\nu|=k} \widehat{f}(\phi_\nu)\phi_\nu(x), \tag{26}$$

then, the Hermite semigroup (semigroup associated to the harmonic oscillator) e^{-tH}, $t > 0$ is given by

$$e^{-tH} f(x) = \sum_{k=0}^{\infty} e^{-t(2k+n)} P_k f(x). \tag{27}$$

For every $t > 0$, the operator e^{-tH} has Schwartz kernel given by

$$K_t(x, y) = \sum_{\nu \in \mathbb{N}_0^n} e^{-t(2|\nu|+n)} \phi_\nu(x)\phi_\nu(y). \tag{28}$$

In view of Mehler's formula (see Thangavelu [29]) the above series can be summed up and we obtain $K_t(x, y) = (2\pi)^{-\frac{n}{2}} \sinh(2t)^{-\frac{n}{2}} e^{-(\frac{1}{2}|x|^2+|y|^2)\coth(2t)+x\cdot y\cdot \mathrm{csch}(2t))}$. The symbol of the operator e^{-tH}, $m_t(\nu) = e^{-t(2|\nu|+n)}$, has exponential decaying and we can deduce that e^{-tH} extends to a r-nuclear on every $L^p(\mathbb{R}^n)$, $1 \leq p < \infty$, (this can be verified if we insert the symbol m_t on every condition of the previous subsection) with trace (see (8)) given by

$$\mathrm{Tr}(e^{-tH}) = \sum_{\nu \in \mathbb{N}_0^n} e^{-t(2|\nu|+n)} = (e^t - e^{-t})^{-n}. \tag{29}$$

The trace of e^{-tH} can also be computed by applying (4.2.1), in fact

$$\begin{aligned}\mathrm{Tr}(e^{-tH}) &= \int_{\mathbb{R}^n} K_t(x,x)dx = (2\pi)^{-\frac{n}{2}} \int_{\mathbb{R}^n} \sinh(2t)^{-\frac{n}{2}} e^{-|x|^2(\coth(2t)-\mathrm{csch}(2t))}dx \\ &= (2\pi)^{-\frac{n}{2}} \sinh(2t)^{-\frac{n}{2}} (\pi/(\coth(2t) - \mathrm{csch}(2t)))^{\frac{n}{2}} \\ &= (e^t - e^{-t})^{-n},\end{aligned}$$

where we have use the known formula: $\int_{\mathbb{R}} e^{-ax^2}dx = (\pi/a)^{\frac{1}{2}}$.

References

1. Bagchi, S., Thangavelu, S.: On Hermite pseudo-multipliers. J. Funct. Anal. **268**(1), 140–170 (2015)
2. Cardona, D.: Nuclear pseudo-differential operators in Besov spaces on compact Lie groups; to appear in J. Fourier Anal. Appl. (2017)
3. Delgado, J.: A trace formula for nuclear operators on L^p. In: Schulze, B.W., Wong, M.W. (Eds.), Pseudo-Differential Operators: Complex Analysis and Partial Differential Equations, Operator Theory: Advances and Applications, vol. 205, pp. 181–193. Birkhuser, Basel (2010)
4. Delgado, J., Wong, M.W.: L^p-nuclear pseudo-differential operators on $\mathbb{Z}$ and $\mathbb{S}^1$. Proc. Amer. Math. Soc. **141**(11), 3935–394 (2013)
5. Delgado, J.: The trace of nuclear operators on $L^p(\mu)$ for σ-finite Borel measures on second countable spaces. Integral Equ. Oper. Theor. **68**(1), 61–74 (2010)
6. Delgado, J.: On the r-nuclearity of some integral operators on Lebesgue spaces. Tohoku Math. J. **67**(2), no. 1, 125–135 (2015)
7. Delgado, J., Ruzhansky, M., Wang, B.: Approximation property and nuclearity on mixed-norm L^p, modulation and Wiener amalgam spaces. J. Lond. Math. Soc. **94**, 391–408 (2016)
8. Delgado, J., Ruzhansky, M., Wang, B.: Grothendieck-Lidskii trace formula for mixed-norm L^p and variable Lebesgue spaces. to appear in J. Spectr. Theory
9. Delgado, J., Ruzhansky, M.: L^p-nuclearity, traces, and Grothendieck-Lidskii formula on compact Lie groups., J. Math. Pures Appl. **9**(102)(1), 153–172 (2014)
10. Delgado, J., Ruzhansky, M.: Schatten classes on compact manifolds: kernel conditions. J. Funct. Anal. **267**(3), 772–798 (2014)
11. Delgado, J., Ruzhansky, M.: Kernel and symbol criteria for Schatten classes and r-nuclearity on compact manifolds. C. R. Acad. Sci. Paris. Ser. I. **352**, 779–784 (2014)
12. Delgado, J., Ruzhansky, M.: Fourier multipliers, symbols and nuclearity on compact manifolds. arXiv:1404.6479
13. Delgado, J., Ruzhansky, M.: Schatten-von Neumann classes of integral operators. arXiv:1709.06446
14. Delgado, J., Ruzhansky, M., Tokmagambetov, N.: Schatten classes, nuclearity and nonharmonic analysis on compact manifolds with boundary. arXiv:1505.02261
15. Epperson, J.: Hermite multipliers and pseudo-multipliers. Proc. Amer. Math. Soc. **124**(7), 2061–2068 (1996)
16. Ghaemi, M.B., Jamalpour Birgani, M., Wong, M.W.: Characterizations of nuclear pseudo-differential operators on S1 with applications to adjoints and products. J. Pseudo-Differ. Oper. Appl. **8**(2), 191–201 (2017)

17. Grothendieck, A.: La theorie de Fredholm. Bull. Soc. Math. Fr. **84**, 319–384 (1956)
18. Grothendieck, A.: Produits tensoriels topologiques et espaces nucléaires, Memoirs Amer. Math. Soc. **16**, Providence (1955) (Thesis, Nancy, 1953)
19. Koch, H., Tataru, D.: L^p-eigenfunction bounds for the Hermite operator. Duke Math. J. **128**, 369–392 (2005)
20. Prugovečki, E.: Quantum mechanics in Hilbert space. 2 edn. Pure and Applied Mathematics, 92. Academic Press, Inc, New York-London (1981)
21. Pietsch, A.: Operator ideals. Mathematische Monographien, 16. VEB Deutscher Verlag der Wissenschaften, Berlin (1978)
22. Pietsch, A.: History of Banach Spaces and Linear Operators. Birkhäuser Boston Inc, Boston, MA (2007)
23. Reinov, O.I., Latif, Q.: Grothendieck-Lidskii theorem for subspaces of Lpspaces. Math. Nachr. **286**(2–3), 279–282 (2013)
24. Nicola, F., Rodino, L.: Global Pseudo-differential Calculus on Euclidean Spaces. Pseudo-Differential Operators. Theory and Applications, 4. Birkhuser Verlag, Basel (2010)
25. Simon, B.: Distributions and their Hermite expansions. J. Math. Phys. **12**, 140–148 (1971)
26. Stempak, K.: Multipliers for eigenfunction expansions of some Schrdinger operators. Proc. Amer. Math. Soc. **93**, 477–482 (1985)
27. Stempak, K., Torre, J.L.: On g-functions for Hermite function expansions. Acta Math. Hung. **109**, 99–125 (2005)
28. Stempak, K., Torre, J.L.: BMO results for operators associated to Hermite expansions. Illinois J. Math. **49**, 1111–1132 (2005)
29. Thangavelu, S.: Lectures on Hermite and Laguerre Expansions, Math. Notes, vol. 42, Princeton University Press, Princeton (1993)
30. Thangavelu, S.: Hermite and special Hermite expansions revisited. Duke Math. J. **94**(2), 257–278 (1998)

Hermite Multipliers on Modulation Spaces

Divyang G. Bhimani[1], Rakesh Balhara[2], and Sundaram Thangavelu[2(✉)]

[1] TIFR Centre for Applicable Mathematics, 560065 Bangalore, India
divyang@tifrbng.res.in

[2] Department of Mathematics, Indian Institute of Science, 560 012 Bangalore, India
{rakesh13,veluma}@math.iisc.ernet.in

Abstract. We study multipliers associated to the Hermite operator $H = -\Delta + |x|^2$ on modulation spaces $M^{p,q}(\mathbb{R}^d)$. We prove that the operator $m(H)$ is bounded on $M^{p,q}(\mathbb{R}^d)$ under standard conditions on m, for suitable choice of p and q. As an application, we point out that the solutions to the free wave and Schrödinger equations associated to H with initial data in a modulation space will remain in the same modulation space for all times. We also point out that Riesz transforms associated to H are bounded on some modulation spaces.

Keywords: Hermite multipliers · Modulation spaces · Wave and Schrödinger equations

1 Introduction

The main aim of this article is to study the boundedness properties of Hermite multipliers on modulation spaces. We quickly recall the setup in order to state our results and we refer to Sect. 2 for details. The spectral decomposition of the Hermite operator $H = -\Delta + |x|^2$ on $\mathbb{R}^d$ is given by $H = \sum_{k=0}^{\infty}(2k+d)P_k$ where P_k stands for the orthogonal projectionof $L^2(\mathbb{R}^d)$ onto the eigenspace corresponding to the eigenvalue $(2k+d)$. Given a bounded function m defined on the set of all natural numbers, we define the operator $m(H)$ simply by setting $m(H) = \sum_{k=0}^{\infty} m(2k+d)P_k$. We say that m is an L^p multiplier for the Hermite expansions if $m(H)$ extends to L^p as a bounded operator. Sufficient conditions on m are known so that m is an L^p multiplier, see e.g. [6,7,10,11]. In this article we are interested in multipliers m which define bounded operators on the modulation spaces $M^{p,q}(\mathbb{R}^d)$.

Recall that a tempered distribution f on $\mathbb{R}^d$ belongs to the modulation space $M^{q,p}(\mathbb{R}^d)$ if the Fourier-Wigner transform of f and the Gaussian $\Phi_0(\xi) = \pi^{-d/2}e^{-\frac{1}{2}|\xi|^2}$ defined by

$$\langle \pi(x+iy)f, \Phi_0\rangle = \int_{\mathbb{R}^d} e^{i(x\cdot\xi + \frac{1}{2}x\cdot y)} f(\xi+y)\Phi_0(\xi)d\xi$$

J. Delgado and M. Ruzhansky (Eds.): *Analysis and Partial Differential Equations: Perspectives from Developing Countries*, PROMS 275, pp. 42–64, 2018.
https://doi.org/10.1007/978-3-030-05657-5_5

belongs to the mixed norm space $L^p(\mathbb{R}^d, L^q(\mathbb{R}^d))$ consisting of functions $F(x,y)$ for which the norms

$$\|F\|_{L^{p,q}} = \left(\int_{\mathbb{R}^d} \|F(x,\cdot)\|_q^p dx\right)^{1/p}$$

are finite (See Sect. 2.5 below). These spaces have several interesting properties not shared by the L^p spaces. For example, $M^{p,p}(\mathbb{R}^d)$ are invariant under the Fourier transform and $M^{q,1}(\mathbb{R}^d)$ are algebras under pointwise multiplication.

For the multiplier operators $m(D)$, the problem of establishing sufficient conditions on m that make the operator $m(D)$ bounded on L^p has a long history. As it appears often in various applications, like solving linear dispersive PDE, e.g., wave/Schrödinger equations, for more detail, we refer to [1,11,12,20] and the reference therein. It is well known that the operator (See Definition 4 below) with Fourier multiplier $e^{i|\xi|^\alpha}$ $(\alpha > 2)$ is bounded on $L^p(\mathbb{R}^d)$ if and only if $p = 2$.

The study of Fourier multiplier operators, which are of the form $m(-\Delta)$ in the context of modulation space $M^{2,1}(\mathbb{R}^d)$ was initiated in the works of Wang-Zhao-Guo [19]. In fact, in the consequent year Bényi-Gröchenig-Okoudjou-Rogers [1] have shown that the Fourier multiplier operator with multiplier $e^{i|\xi|^\alpha}$ $(\alpha \in [0,2])$ is bounded on $M^{p,q}(\mathbb{R}^d)$ for all $1 \leq p,q \leq \infty$. The cases $\alpha = 1$ and $\alpha = 2$ are particularly interesting and have been studied intensively in PDE, because they occur in the time evolution of the wave equation $(\alpha = 1)$ and the free Schrödinger operator $(\alpha = 2)$. Thus, the Schödinger and wave propagators are not $L^p (p \neq 2)$-bounded but $M^{p,q}$- bounded for all $1 \leq p,q \leq \infty$. In fact, this leads to fixed-time estimates for Schödinger and wave propagators and some of their applications to well-posedness results on modulation spaces $M^{p,q}(\mathbb{R}^d)$. Modulation spaces have turned out to be very fruitful in numerous applications in various problems in analysis and PDE. And yet there has been a lot of ongoing interest in these spaces from the harmonic analysis and PDE points of view. We refer to the recent survey [12] and the references therein.

Coming back to the Hermite operator, we note that Thangavelu [18] (See also [15, Theorem 4.2.1]) has proved an analogue of the Hörmander-Mikhlin type multiplier theorem for Hermite expansions on $L^p(\mathbb{R}^d)$. Specifically, he showed that under certain conditions on m, the operator $m(H)$ is bounded on $L^p(\mathbb{R}^d)(1 < p < \infty)$. It is well known that the harmonic oscillator $H = -\Delta + |x|^2$ appears in various applications. We refer to the recent article [5] and the reference therein for details.

Taking all these considerations into account, we are motivated to study Hermite multipliers $m(H)$ on modulation spaces $M^{p,q}(\mathbb{R}^d)$. The conditions we impose on the multiplier m is the standard one in terms of local Sobolev spaces. We assume that m is defined on the whole of $\mathbb{R}$. Let $0 \neq \psi \in C_0^\infty(\mathbb{R}^+)$ be a fixed cut-off function with support contained in the interval $[\frac{1}{2}, 1]$, and define the scale invariant localized Sobolev norm of order β of $m \in L^\infty(\mathbb{R}^+)$ by

$$\|m\|_{L^2_{\beta,sloc}} = \sup_{t>0} \|\psi m(t\cdot)\|_{L^2_\beta}.$$

We then have the following theorem.

Theorem 1. *Let $1 < p \leq q \leq 2$ or $2 \leq q \leq p < \infty$. Suppose that $\|m\|_{L^2_{\beta,sloc}} < \infty$ for some $\beta > (2d+1)/2$. Then the operator $m(H)$ is bounded on $M^{q,p}(\mathbb{R}^d)$.*

We remark that this theorem is not sharp. For m to be an L^p multiplier it is sufficient to assume the condition on m with $\beta > d/2$. We believe the same is true in the case of multipliers on modulation spaces though our method of proof requires a stronger assumption on m. However, it is worth noting that $M^{p,p}(\mathbb{R}^d)$ for $p > 2$ is a much wider class than $L^p(\mathbb{R}^d)$ (See Lemma 1(3) below). Thus, Theorem 1 shows that the Hörmander-Mikhlin multiplier type theorem is true for a much wider class than $L^p(\mathbb{R}^d)$.

We deduce Theorem 1 from a corresponding result on the polarised Heisenberg group $\mathbb{H}^d_{pol} = \mathbb{R}^d \times \mathbb{R}^d \times \mathbb{R}$ which is of Euclidean dimension $(2d+1)$. Let $\tilde{\mathcal{L}}$ stand for the sublaplacian on $\mathbb{H}^d_{pol}$ and define $m(\tilde{\mathcal{L}})$ using spectral theorem (See Subsect. 2.2 below for precise definitions). We then have the following transference result.

Theorem 2. *Let $1 < p \leq q \leq 2$ or $2 \leq q \leq p < \infty$. Then $m(H)$ is bounded on $M^{q,p}(\mathbb{R}^d)$ whenever $m(\tilde{\mathcal{L}})$ is bounded on $L^p(\mathbb{H}^d_{pol})$.*

The above theorem allows us to deduce some interesting corollaries for Hermite multipliers. For example, let $R_j = (-\frac{\partial}{\partial \xi_j} + \xi_j)H^{-1/2}, j = 1, 2, ..., d$ be the Riesz transforms associated to H. It is well known that these Riesz transforms are bounded on $L^p(\mathbb{R}^d), 1 < p < \infty$ but their behavior on modulation spaces have not been studied. By considering Riesz transforms on G^d we obtain

Corollary 1. *Let $1 < p \leq q \leq 2$ or $2 \leq q \leq p < \infty$. Then the Riesz transforms R_j are bounded on $M^{q,p}(\mathbb{R}^d)$.*

Another interesting corollary is the following result about solutions of the wave equation associated to H. Consider the following Cauchy problem:

$$\partial_t^2 u(x,t) = -Hu(x,t), \quad u(x,0) = 0, \quad \partial_t u(x,0) = f(x)$$

whose solution is given by $u(x,t) = H^{-1/2} \sin(tH^{1/2}) f(x)$.

Corollary 2. *Let u be the solution of the above Cauchy problem for H. Then for $1 < p \leq q \leq 2$ or $2 \leq q \leq p < \infty$ we have the estimate $\|u(\cdot,t)\|_{M^{q,p}} \leq C_t \|f\|_{M^{q,p}}$ provided $|\frac{1}{p} - \frac{1}{2}| < \frac{1}{2d}$.*

When $p = q$ Theorem 2 as well as Corollary 2 can be improved. This will be achieved using transference as before, but now the transference is from multiplier theorems for multiple Fourier series. Given a function m on $\mathbb{R}$ we define a Fourier multiplier T_m on $L^p(\mathbb{T}^d)$ where $\mathbb{T}^d$ is the $d-$dimensional torus by

$$T_m f(x) = \sum_{\mu \in \mathbb{Z}^d} m(|\mu|) \hat{f}(\mu) e^{i\mu \cdot x}.$$

Here $\hat{f}(\mu)$ are the Fourier coefficients of f and $|\mu| = \sum_{j=1}^d |\mu_j|$. Using the connection (See Proposition 1 below) between Fourier multipliers T_m on $L^p(\mathbb{T}^d)$ and $m(H)$ on $M^{p,p}$ we prove

Theorem 3. *Let $1 < p < \infty$. Suppose that $\|m\|_{L^2_{\beta,sloc}} < \infty$ for some $\beta > d/2$. Then the operator $m(H)$ is bounded on $M^{p,p}(\mathbb{R}^d)$.*

We also have the following improvement of Corollary 2. More generally, we consider multipliers of the form

$$m(2k+d) = \frac{e^{i(2k+d)^{\gamma}}}{(2k+d)^{\beta}}, \quad (\beta > 0, \gamma > 0). \tag{1}$$

Theorem 4. *Let $1 \leq p < \infty$, and $|\frac{1}{p} - \frac{1}{2}| < \frac{\beta}{d\gamma}$, and let m be given by (1). Then $m(H)$ is bounded on $M^{p,p}(\mathbb{R}^d)$. In particular, Corollary 2 is valid on the bigger range $|\frac{1}{p} - \frac{1}{2}| < \frac{1}{d}$ when $p = q$.*

We now turn our attention to a multiplier which occurs in the time evolution of the free Schrödinger equation associated to H. Specifically, we consider the Cauchy problem for the Schrödinger equation associated to H :

$$i\partial_t u(x,t) - Hu(x,t) = 0, \;\; u(x,0) = f(x)$$

whose solution is given by $u(x,t) = e^{itH} f(x)$.

Theorem 5. *The Schrödinger propagator $m(H) = e^{itH}$ is bounded on $M^{p,p}(\mathbb{R}^d)$ for all $1 \leq p < \infty$.*

Recently Cordero-Nicola [2, Sect. 5.1] have studied the operator $m(H) = e^{itH}$ on Wiener amalgam spaces (closely related to modulation spaces). Later Kato-Kobayashi-Ito [8] have given a refinement of Cordero-Nicola's results in the context of Wiener amalgam spaces. We have studied (Theorem 5) the boundeness of $m(H) = e^{itH}$ in the context of modulation spaces. Here we would like to point out that our method of proof is completely different from the method used in the context of Wiener amalgam spaces. We also believe that our method of proof is much simpler than the proofs available in the literature. Our proof relies on properties of Hermite and special Hermite functions and illustrates the importance of these functions in the study of such problems.

Finally, we note that modulation spaces have been used as regularity classes for initial data associated to Cauchy problems for nonlinear dispersive equations (eg., NLS, NLW, etc..) but so far mainly for the nonlinear dispersive equations associated to the Laplacian without potential ($D = \Delta$) [12,19,20]. There is also an ongoing interest to use harmonic analysis tools (specifically, multiplier results) to solve modern nonlinear PDE problems (See e.g., [5]). Thus, we strongly believe that our results will be useful in the future for studying nonlinear dispersive equations associated to H in the realm of modulation spaces.

The paper is organized as follows. In Sect. 2, we introduce notations and preliminaries which will be used in the sequel. Specifically, in Subsect. 2.2, we introduce the Heisenberg and polarised Heisenberg groups, the sublaplacians corresponding to these groups, and spectral multipliers associated to these sub-laplacians. In Subsect. 2.4, we prove the transference result which connects spectral multiplier on polarised Heisenberg groups and reduced polarised Heisenberg groups. In Subsect. 2.5, we introduce modulation spaces and recall some of their basic properties. In Sect. 3, we prove Theorems 1 and 2. In Sect. 4, we prove Theorems 3 and 4. In Sect. 5, we prove Theorem 5.

2 Notation and Preliminaries

2.1 Notations

The notation $A \lesssim B$ means $A \leq cB$ for some constant $c > 0$. The symbol $A_1 \hookrightarrow A_2$ denotes the continuous embedding of the topological linear space A_1 into A_2. If $\alpha = (\alpha_1, ..., \alpha_d) \in \mathbb{N}^d$ is a multi-index, we set $|\alpha| = \sum_{j=1}^d \alpha_j, \alpha! = \prod_{j=1}^d \alpha_j!$. If $z = (z_1, ..., z_d) \in \mathbb{C}^d$, we put $z^\alpha = \prod_{j=1}^d z_j^{\alpha_j}$. We denote the $d-$dimensional torus by $\mathbb{T}^d \equiv [0, 2\pi)^d$, and the $L^p(\mathbb{T}^d)-$norm by

$$\|f\|_{L^p(\mathbb{T}^d)} = \left(\int_{[0,2\pi)^d} |f(t)|^p dt \right)^{1/p}.$$

The class of trigonometric polynomials on $\mathbb{T}^d$ is denoted by $\mathcal{P}(\mathbb{T}^d)$. The mixed $L^p(\mathbb{R}^d, L^q(\mathbb{R}^d))$ norm is denoted by

$$\|f\|_{L^{p,q}} = \left(\int_{\mathbb{R}^d} \left(\int_{\mathbb{R}^d} |f(x,y)|^q dy \right)^{p/q} dx \right)^{1/p} \quad (1 \leq p, q < \infty),$$

the $L^\infty(\mathbb{R}^d)$ norm is $\|f\|_{L^\infty} = \text{ess.sup}_{x \in \mathbb{R}^d} |f(x)|$. The Schwartz class is denoted by $\mathcal{S}(\mathbb{R}^d)$ (with its usual topology), and the space of tempered distributions is denoted by $\mathcal{S}'(\mathbb{R}^d)$. For $x = (x_1, \cdots, x_d), y = (y_1, \cdots, y_d) \in \mathbb{R}^d$, we put $x \cdot y = \sum_{i=1}^d x_i y_i$. Let $\mathcal{F} : \mathcal{S}(\mathbb{R}^d) \to \mathcal{S}(\mathbb{R}^d)$ be the Fourier transform defined by

$$\mathcal{F}f(\xi) = \widehat{f}(\xi) = (2\pi)^{-d/2} \int_{\mathbb{R}^d} f(x) e^{-ix\cdot\xi} dx, \ \xi \in \mathbb{R}^d.$$

Then $\mathcal{F}$ is a bijection and the inverse Fourier transform is given by

$$\mathcal{F}^{-1}f(x) = f^\vee(x) = (2\pi)^{-d/2} \int_{\mathbb{R}^d} f(\xi)\, e^{ix\cdot\xi} d\xi, \quad x \in \mathbb{R}^d.$$

It is well known that the Fourier transform can be uniquely extended to $\mathcal{F} : \mathcal{S}'(\mathbb{R}^d) \to \mathcal{S}'(\mathbb{R}^d)$.

2.2 Heisenberg Group and Fourier Multipliers

We consider the Heisenberg group $\mathbb{H}^d = \mathbb{C}^d \times \mathbb{R}$ with the group law

$$(z,t)(w,s) = (z+w, t+s+\frac{1}{2}\mathrm{Im}(z\cdot\bar{w})).$$

Sometimes we use real coordinates (x,y,t) instead of (z,t) if $z = x+iy, x, y \in \mathbb{R}^d$. In order to define the (group) Fourier transform on the Heisenberg group we briefly recall the following family of irreducible unitary representations. For each non-zero real λ, we have a representation π_λ realised on $L^2(\mathbb{R}^d)$ as follows:

$$\pi_\lambda(z,t)\varphi(\xi) = e^{i\lambda t}e^{i(x\cdot\xi+\frac{1}{2}x\cdot y)}\varphi(\xi+y)$$

where $\varphi \in L^2(\mathbb{R}^d)$. For $f \in L^1(\mathbb{H}^d)$, one defines the operator

$$\hat{f}(\lambda) = \int_{\mathbb{H}^d} f(z,t)\pi_\lambda(z,t)dzdt.$$

The operator valued function $f \to \hat{f}(\lambda)$ is called the group Fourier transform of f on $\mathbb{H}^d$. We refer to [16] for more about the group Fourier transform.

The Fourier transform initially defined on $L^1(\mathbb{H}^d) \cap L^2(\mathbb{H}^d)$ can be extended to the whole of $L^2(\mathbb{H}^d)$ and we have a version of Plancherel theorem. Specifically, when $f \in L^1 \cap L^2(\mathbb{H}^d)$, it can be shown that $\hat{f}(\lambda)$ is a Hilbert-Schmidt operator and the Plancherel theorem holds:

$$\int_{\mathbb{H}^d} |f(z,t)|^2 dzdt = \frac{2^{d-1}}{\pi^{d+1}}\int_{-\infty}^{\infty} \|\hat{f}(\lambda)\|_{HS}^2 |\lambda|^d d\lambda$$

where $\|\cdot\|_{HS}$ is the Hilbert-Schmidt norm given by $\|T\|_{HS}^2 = tr(T^*T)$, for T a bounded operator, T^* being the adjoint operator of T. Under the assumption that $\hat{f}(\lambda)$ is of trace class we have the inversion formula

$$f(z,t) = (2\pi)^{-d-1}\int_{-\infty}^{\infty} tr(\pi_\lambda(z,t)^*\hat{f}(\lambda))|\lambda|^d d\lambda.$$

Let f^λ stand for the inverse Fourier transform of f in the *central variable* t

$$f^\lambda(z) = \int_{-\infty}^{\infty} f(z,t)e^{i\lambda t}dt. \tag{2}$$

By taking the Euclidean Fourier transform of $f^\lambda(z)$ in the variable λ, we obtain

$$f(z,t) = \frac{1}{2\pi}\int_{-\infty}^{\infty} e^{-i\lambda t} f^\lambda(z)\, d\lambda. \tag{3}$$

Recalling the definition of π_λ we see that we can write the group Fourier transform as

$$\hat{f}(\lambda) = \pi_\lambda(f^\lambda)$$

where $\pi_\lambda(f^\lambda) = \int_{\mathbb{C}^d} f^\lambda(z)\pi_\lambda(z,0)dz$. The operator which takes a function g on $\mathbb{C}^d$ into the operator

$$\int_{\mathbb{C}^d} g(z)\pi_\lambda(z,0)dz$$

is called the Weyl transform of g and is denoted by $W_\lambda(g)$. Thus $\widehat{f}(\lambda) = W_\lambda(f^\lambda)$. With these notations we can rewrite the inversion formula as

$$f(z,t) = (2\pi)^{-d-1}\int_{-\infty}^{\infty} e^{-i\lambda t} tr(\pi_\lambda(z,0)^*\pi_\lambda(f^\lambda))|\lambda|^d d\lambda.$$

On the Heisenberg group, we have the vector fields

$$T = \frac{\partial}{\partial t}, \quad X_j = \frac{\partial}{\partial x_j} + \frac{1}{2}y_j\frac{\partial}{\partial t}, \quad Y_j = \frac{\partial}{\partial y_j} - \frac{1}{2}x_j\frac{\partial}{\partial t}, \quad j = 1, ..., d,$$

and they form a basis for the Lie algebra of left invariant vector fields on the Heisenberg group. The second-order operator

$$\mathcal{L} = -\sum_{j=1}^{d}(X_j^2 + Y_j^2)$$

is called the sublaplacian which is self-adjoint and nonnegative and hence admits a spectral decomposition

$$\mathcal{L} = \int_0^\infty \lambda \, dE_\lambda.$$

Given a bounded function m defined on $(0,\infty)$ one can define the operator $m(\mathcal{L})$ formally by setting

$$m(\mathcal{L})f = \int_0^\infty m(\lambda)\, dE_\lambda f.$$

It can be shown that $\widehat{m(\mathcal{L})f}(\lambda) = \hat{f}(\lambda)m(H(\lambda))$ where $H(\lambda) = -\Delta + |\lambda|^2|x|^2$ are the scaled Hermite operators. Hence, the operators $m(\mathcal{L})f$ are examples of Fourier multipliers on the Heisenberg group. More generally, (right) Fourier multipliers on the Heisenberg group are operators defined by $\widehat{T_M f}(\lambda) = \hat{f}(\lambda)M(\lambda)$ where $M(\lambda)$, called the multiplier is a family of bounded linear operators on $L^2(\mathbb{R}^d)$.

The boundedness properties of these operators have been studied by several authors, see e.g. [6,7,11]. We make use of the following result. Let $0 \neq \psi \in C_0^\infty(\mathbb{R}^+)$ be a fixed cut-off function with support contained in the interval $[\frac{1}{2}, 1]$, and define the scale invariant localized Sobolev norm of order β of m by

$$\|m\|_{L^2_{\beta,sloc}} = \sup_{t>0} \|\psi m(t\cdot)\|_{L^2_\beta}.$$

Theorem 6 (Müller-Stein, Hebisch [7,11]). *If $\|m\|_{L^2_{\beta,sloc}} < \infty$ for some $\beta > (2d+1)/2$, then $m(\mathcal{L})$ is bounded on $L^p(\mathbb{H}^d)$ for $1 < p < \infty$, and of weak type $(1,1)$.*

We make use of this theorem in proving our main result. Actually we need an analogue of the above result in the context of polarised Heisenberg group $\mathbb{H}^d_{pol}$ which is just $\mathbb{R}^d \times \mathbb{R}^d \times \mathbb{R}$ with the group law

$$(x, y, t)(x', y', t') = (x + x', y + y', t + t' + x' \cdot y).$$

A basis for the algebra of left invariant vectors fields on this group are given by

$$T = \frac{\partial}{\partial t}, \quad \tilde{X}_j = \frac{\partial}{\partial x_j} + y_j \frac{\partial}{\partial t}, \quad \tilde{Y}_j = \frac{\partial}{\partial y_j}, \quad j = 1, ..., d.$$

The sublaplacian is then defined as the second-order operator

$$\tilde{\mathcal{L}} = -\sum_{j=1}^{d} (\tilde{X}_j^2 + \tilde{Y}_j^2).$$

The group $\mathbb{H}^d_{pol}$ is isomorphic to $\mathbb{H}^d$ and the isomorphism is given by the map $\Phi : \mathbb{H}^d \to \mathbb{H}^d_{pol}$, $\Phi(x, y, t) = (x, y, t + \frac{1}{2} x \cdot y)$. Note that Φ is measure preserving and it is easy to check that

$$\mathcal{L}(f \circ \Phi^{-1}) = \tilde{\mathcal{L}} f \circ \Phi$$

for reasonable functions f on $\mathbb{H}^d_{pol}$. In view of this, an analogue of Theorem 6 is true for $m(\tilde{\mathcal{L}})$. This also follows from the fact that the above theorem is valid in a more general context of H-type groups, see [6, 7, 10].

Using the isomorphism Φ we can define the following family of representations for $\mathbb{H}^d_{pol}$. For each non zero real λ, the representations $\rho_\lambda = \pi_\lambda \circ \Phi^{-1}$ are irreducible and unitary. We can use them to define Fourier transform on the group $\mathbb{H}^d_{pol}$.

2.3 Hermite and Special Hermite Functions

The spectral decomposition of $H = -\Delta + |x|^2$ is given by the Hermite expansion. Let $\Phi_\alpha(x)$, $\alpha \in \mathbb{N}^d$ be the normalized Hermite functions which are products of one dimensional Hermite functions. More precisely, $\Phi_\alpha(x) = \Pi_{j=1}^d h_{\alpha_j}(x_j)$ where

$$h_k(x) = (\sqrt{\pi} 2^k k!)^{-1/2} (-1)^k e^{\frac{1}{2} x^2} \frac{d^k}{dx^k} e^{-x^2}.$$

The Hermite functions Φ_α are eigenfunctions of H with eigenvalues $(2|\alpha| + d)$ where $|\alpha| = \alpha_1 + ... + \alpha_d$. Moreover, they form an orthonormal basis for $L^2(\mathbb{R}^d)$. The spectral decomposition of H is then written as

$$H = \sum_{k=0}^{\infty} (2k + d) P_k, \qquad P_k f(x) = \sum_{|\alpha| = k} \langle f, \Phi_\alpha \rangle \Phi_\alpha$$

where $\langle\cdot,\cdot\rangle$ is the inner product in $L^2(\mathbb{R}^d)$. Given a function m defined and bounded on the set of all natural numbers we can use the spectral theorem to define $m(H)$. The action of $m(H)$ on a function f is given by

$$m(H)f = \sum_{\alpha\in\mathbb{N}^d} m(2|\alpha|+d)\langle f,\Phi_\alpha\rangle\Phi_\alpha = \sum_{k=0}^{\infty} m(2k+d)P_k f.$$

This operator $m(H)$ is bounded on $L^2(\mathbb{R}^d)$. This follows immediately from the Plancherel theorem for the Hermite expansions as m is bounded.

On the other hand, the mere boundedness of m is not sufficient to imply the L^p boundedness of $m(H)$ for $p \neq 2$. So, we need to impose some conditions m to ensure that $m(H)$ is bounded on $L^p(\mathbb{R}^d)$. The boundedness results of $m(H)$ on $L^p(\mathbb{R}^d)$ have been studied by several authors, see e.g. [15,18]. Our aim in this paper is to study the boundedness of $m(H)$ on modulation spaces.

In the sequel, we make use of some properties of special Hermite functions $\Phi_{\alpha,\beta}$ which are defined as follows. Let $\pi(z) = \pi_1(z,t)$ and define

$$\Phi_{\alpha,\beta}(z) = (2\pi)^{-d/2}\langle\pi(z)\Phi_\alpha,\Phi_\beta\rangle. \tag{4}$$

Then it is well known that these so called special Hermite functions form an orthonormal basis for $L^2(\mathbb{C}^d)$. Moreover, they are eigenfunctions of the special Hermite operator L which is defined by the equation

$$\mathcal{L}(f(z)e^{it}) = e^{it}Lf(z).$$

Indeed, we can show ([15, Theorem 1.3.3]) that

$$L\Phi_{\alpha,\beta} = (2|\alpha|+d)\Phi_{\alpha,\beta}.$$

For a function $f \in L^2(\mathbb{C}^d)$ we have the eigenfunction expansion

$$f(z) = \sum_{\alpha\in\mathbb{N}^d}\sum_{\beta\in\mathbb{N}^d}\langle f,\Phi_{\alpha,\beta}\rangle\Phi_{\alpha,\beta}(z)$$

which is called the special Hermite expansion.

Let f and g be two measurable functions on $\mathbb{C}^d$. We recall that the $\lambda-$twisted convolution $(0 \neq \lambda \in \mathbb{R})$ of f and g is the function $f *_\lambda g$ defined by

$$f *_\lambda g(z) = \int_{\mathbb{C}^d} f(z-w)g(w)e^{i\frac{\lambda}{2}\mathrm{Im}(z\cdot\bar{w})}dw$$

for all z such that the integral exits. When $\lambda = 1$, we simply call them twisted convolution and denote them by $f \times g$. It is well known that the twisted convolution of special Hermite functions satisfies [15, Proposition 1.3.2] the following relation

$$\Phi_{\alpha,\beta}\times\Phi_{\mu,v} = (2\pi)^{d/2}\delta_{\beta,\mu}\Phi_{\alpha,v} \tag{5}$$

where $\delta_{\beta,\mu} = 1$ if $\beta = \mu$, otherwise it is 0. Using the identity (5), we can show that the special Hermite expansions can be written ([15, Sect. 2.1]) in the compact form as follows:

$$\begin{aligned} f(z) &= (2\pi)^{-d/2} \sum_{\alpha\in\mathbb{N}^d} f \times \Phi_{\alpha,\alpha}(z) \\ &= (2\pi)^{-d/2} \sum_{k=0}^{\infty} f \times \left(\sum_{|\alpha|=k} \Phi_{\alpha,\alpha}(z) \right) \\ &= (2\pi)^{-d} \sum_{k=0}^{\infty} f \times \phi_k(z) \end{aligned}$$

where φ_k are the Laguerre functions of type $(d-1)$:

$$\varphi_k(z) = (2\pi)^{d/2} \sum_{|\alpha|=k} \Phi_{\alpha,\alpha}(z) = L_k^{d-1}(\frac{1}{2}|z|^2)e^{-\frac{1}{4}|z|^2}.$$

We note that $f \times \Phi_{\alpha,\alpha}$ is an eigenfunction of the operator L with the eigenvalue $(2|\alpha| + d)$. Hence $(2\pi)^{-d} f \times \phi_k$ is the projection of f onto the eigenspace corresponding to the eigenvalue $(2k+d)$. The spectral decomposition of the operator L is given by the special Hermite functions and we can write the same in a compact form as

$$Lf(z) = (2\pi)^{-d} \sum_{k=0}^{\infty} (2k+d) f \times \varphi_k(z).$$

As in the case of the Hermite operators, one can define and study special Hermite multipliers $m(L)$, see [15]. The functions $\Phi_{\alpha,\beta}$ can be expressed in terms of Laguerre functions. In particular, we have ([15, Theorem 1.3.5])

$$\Phi_{\alpha,0}(z) = (2\pi)^{-d/2}(\alpha!)^{-1/2} \left(\frac{i}{\sqrt{2}}\right)^{|\alpha|} \bar{z}^{\alpha} e^{-\frac{1}{4}|z|^2}. \tag{6}$$

We also have to deal with the family of operators $H(\lambda) = -\Delta + \lambda^2|x|^2$ whose eigenfunctions are given by scaled Hermite functions. For $\lambda \in \mathbb{R}^*$ and each $\alpha \in \mathbb{N}^d$, we define the family of scaled Hermite functions

$$\Phi_\alpha^\lambda(x) = |\lambda|^{\frac{d}{4}} \Phi_\alpha(\sqrt{|\lambda|}x), \quad x \in \mathbb{R}^d.$$

They form an orthonormal basis for $L^2(\mathbb{R}^d)$. The spectral decomposition of the scaled Hermite operator $H(\lambda) = -\Delta + |\lambda|^2|x|^2$ is then written as

$$H(\lambda) = \sum_{k=0}^{\infty} (2k+d)|\lambda| P_k(\lambda), \tag{7}$$

for $\lambda \in \mathbb{R}^*$, where $P_k(\lambda)$ are the (finite-dimensional) orthogonal projections defined on $L^2(\mathbb{R}^d)$ by

$$P_k(\lambda)f = \sum_{|\alpha|=k} \langle f, \Phi_\alpha^\lambda \rangle \Phi_\alpha^\lambda,$$

where $f \in L^2(\mathbb{R}^d)$. We can now define scaled special Hermite functions

$$\Phi_{\alpha,\beta}^\lambda(z) = (2\pi)^{-d/2}|\lambda|^{d/2}\langle \pi_\lambda(z)\Phi_\alpha^\lambda, \Phi_\beta^\lambda \rangle. \tag{8}$$

They form a complete orthonormal system in $L^2(\mathbb{C}^d)$. For every $f \in L^2(\mathbb{C}^d)$, we have the special Hermite expansion

$$f = \sum_{\alpha \in \mathbb{N}^d} \sum_{\beta \in \mathbb{N}^d} \langle f, \Phi_{\alpha,\beta}^\lambda \rangle \Phi_{\alpha,\beta}^\lambda. \tag{9}$$

We now define the (scaled) special Hermite operator (twisted Laplacian) by the relation

$$\mathcal{L}(e^{i\lambda t} f(z)) = e^{i\lambda t} L_\lambda f(z). \tag{10}$$

We note that the operators L_λ and $H(\lambda)$ are related via the Weyl transform:

$$\pi_\lambda(L_\lambda f) = \pi_\lambda(f)H(\lambda).$$

In fact, the scaled Hermite functions are eigenfunctions of the operator L_λ :

$$L_\lambda \Phi_{\alpha,\beta}^\lambda(z) = (2|\alpha| + d)|\lambda|\Phi_{\alpha,\beta}^\lambda(z). \tag{11}$$

Note that the eigenvalues of L_λ are of the form $(2k+d)|\lambda|, k = 0, 1, 2, ...,$ and the kth eigenspace corresponding to the eigenvalue $(2k+d)|\lambda|$ is infinite-dimensional being the span of $\{\Phi_{\alpha,\beta}^\lambda : |\alpha| = k, \beta \in \mathbb{N}^d\}$. It is well known that the special Hermite expansion can be written in the compact form as follows:

$$f(z) = (2\pi)^{-d}|\lambda|^d \sum_{k=0}^{\infty} f *_\lambda \varphi_k^\lambda(z) \tag{12}$$

where φ_k^λ is the scaled Laguerre functions of type $(d-1)$

$$\varphi_k^\lambda(z) = L_k^{n-1}\Big(\frac{1}{2}|\lambda||z|^2\Big)e^{-\frac{1}{4}|\lambda||z|^2}. \tag{13}$$

The spectral decomposition of L_λ is then written as

$$L_\lambda f(z) = (2\pi)^{-d}|\lambda|^d \sum_{k=0}^{\infty}(2k + d)|\lambda| f *_\lambda \varphi_k^\lambda(z).$$

In particular, recalling f^λ (see(2)), we have

$$L_\lambda f^\lambda(z) = (2\pi)^{-d}|\lambda|^d \sum_{k=0}^{\infty}(2k + d)|\lambda| f^\lambda *_\lambda \varphi_k^\lambda(z). \tag{14}$$

Taking the Fourier transform in λ variable in equation (14), and using (10) and (3), we obtain the spectral decomposition of the sublaplacian $\mathcal{L}$ on $\mathbb{H}^d$ as follows:

$$\mathcal{L}f(z,t) = (2\pi)^{-d-1} \int_{-\infty}^{\infty} \Big(\sum_{k=0}^{\infty} (2k+d)|\lambda| f^\lambda *_\lambda \varphi_k^\lambda(z)\Big) e^{-i\lambda t} |\lambda|^d d\lambda. \qquad (15)$$

For more on spectral theory of sublaplacian on Heisenberg group, we refer to [17, Chap. 2].

2.4 A Transference Theorem for $m(\mathcal{L})$

Let Γ_0 be the subgroup of $\mathbb{H}^d$ consisting of elements of the form $(0,0,2\pi k)$ where $k \in \mathbb{Z}$. Then Γ_0 is a central subgroup and the quotient $\mathbb{H}^d/\Gamma_0$ whose underlying manifold is $\mathbb{R}^d \times \mathbb{R}^d \times \mathbb{T}$ is called the Heisenberg group with compact center.(See [17, Chap. 4]) Let $\mathcal{L}_0$ be the sublaplacian on $\mathbb{H}^d/\Gamma_0$. Note that $\mathcal{L}_0 f = \mathcal{L}f$, considering f as a function on $\mathbb{H}^d$.

Note that on functions g which are independent of t, $\mathcal{L}_0 g(z) = -\Delta g(z)$ where Δ is the Euclidean Laplacian on $\mathbb{C}^d$. The spectral decomposition of the sublaplacian $\mathcal{L}_0$ on this group is given by

$$\mathcal{L}_0 f(z,t) = (-\Delta) f^0(z) + (2\pi)^{-d-1} \sum_{k\in\mathbb{Z}\setminus\{0\}} \Big(\sum_{j=0}^{\infty} e^{-ikt}((2j+d)|k|) f^k *_k \varphi_j^k(z)\Big) |k|^d$$

where $\varphi_j^k(z) = \varphi_j(\sqrt{|k|}z)$ and f^k is defined by the equation $f * e_j^k(z,t) = e^{-ikt} f^k *_k \varphi_j^k(z)$, $e_j^k(z,t) = e^{ikt}\varphi_j^k(z)$. More generally,

$$m(\mathcal{L}_0) f(z,t) = m(-\Delta) f^0(z) + (2\pi)^{-d-1} \sum_{k\in\mathbb{Z}\setminus\{0\}} \Big(\sum_{j=0}^{\infty} e^{-ikt} m((2j+d)|k|) f^k *_k \varphi_j^k(z)\Big) |k|^d.$$

There is a transference theorem which connects $m(\mathcal{L})$ on $\mathbb{H}^d$ with $m(\mathcal{L}_0)$ on $\mathbb{H}^d/\Gamma_0$ which is the analogue of the classical de Leeuw's theorem (See Theorem 10) on the real line.

Theorem 7. *Let $1 < p < \infty$. Suppose $m(\mathcal{L})$ is a bounded operator on $L^p(\mathbb{H}^d)$. Then the transferred operator $m(\mathcal{L}_0)$ is bounded on $L^p(\mathbb{H}^d/\Gamma_0)$.*

This theorem has been proved and used in [13]. The idea is to realise $m(\mathcal{L})$ ($m(\mathcal{L}_0)$) as an operator valued Fourier multiplier for $\mathbb{R}$ (resp. $\mathbb{T}$.). Indeed, writing the Fourier inversion as

$$f(z,t) = (2\pi)^{-1} \int_{-\infty}^{\infty} e^{-i\lambda t} f^\lambda(z) d\lambda$$

and recalling that L_λ is defined by the equation $\mathcal{L}(f(z)e^{i\lambda t}) = e^{i\lambda t} L_\lambda f(z)$ we have

$$m(\mathcal{L}) f(z,t) = (2\pi)^{-1} \int_{-\infty}^{\infty} e^{-i\lambda t} m(L_\lambda) f^\lambda(z) d\lambda$$

where $m(L_\lambda)$ is the multiplier transform associated to the elliptic operator L_λ which has an explicit spectral decomposition. By identifying $L^p(\mathbb{H}^d)$ with $L^p(\mathbb{R}, L^p(\mathbb{C}^d))$ we see that, in view of the above equation, $m(\mathcal{L})$ can be considered as an operator valued Fourier multiplier for the Euclidean Fourier transform on $\mathbb{R}$. Similarly, $m(\mathcal{L}_0)$ can be thought about as an operator valued multiplier for the Fourier series on $\mathbb{T}$. Now, it is an easy matter to imitate the proof of de Leeuw's theorem to prove the above result.

We need the following analogue of Theorem 7 for the reduced polarised Heisenberg group. Let $\tilde{\mathcal{L}}_0$ be the sublaplacian on $G^d = \mathbb{H}^d_{pol}/\Gamma_0$. We note that $\tilde{\mathcal{L}}_0 f = \tilde{\mathcal{L}} f$ considering f as a function on $\mathbb{H}^d_{pol}$.

Theorem 8. *Let* $1 < p < \infty$*. Suppose* $m(\tilde{\mathcal{L}})$ *is a bounded operator on* $L^p(\mathbb{H}^d_{pol})$*. Then the transferred operator* $m(\tilde{\mathcal{L}}_0)$ *is bounded on* $L^p(\mathbb{H}^d_{pol}/\Gamma_0)$*.*

2.5 Modulation Spaces

In 1983, Feichtinger [3] introduced a class of Banach spaces, the so called modulation spaces, which allow a measurement of space variable and Fourier transform variable of a function or distribution on $\mathbb{R}^d$ simultaneously using the short-time Fourier transform(STFT). The STFT of a function f with respect to a window function $g \in \mathcal{S}(\mathbb{R}^d)$ is defined by

$$V_g f(x, y) = (2\pi)^{-d/2} \int_{\mathbb{R}^d} f(t)\overline{g(t-x)}\, e^{-iy\cdot t}\, dt, \;\; (x, y) \in \mathbb{R}^{2d}$$

whenever the integral exists. For $x, y \in \mathbb{R}^d$ the translation operator T_x and the modulation operator M_w are defined by $T_x f(t) = f(t-x)$ and $M_y f(t) = e^{iy\cdot t} f(t)$. In terms of these operators the STFT may be expressed as

$$V_g f(x, y) = \langle f, M_y T_x g \rangle$$

where $\langle f, g \rangle$ denotes the inner product for L^2 functions, or the action of the tempered distribution f on the Schwartz class function g. Thus $V : (f, g) \to V_g(f)$ extends to a bilinear form on $\mathcal{S}'(\mathbb{R}^d) \times \mathcal{S}(\mathbb{R}^d)$ and $V_g(f)$ defines a uniformly continuous function on $\mathbb{R}^d \times \mathbb{R}^d$ whenever $f \in \mathcal{S}'(\mathbb{R}^d)$ and $g \in \mathcal{S}(\mathbb{R}^d)$.

Definition 1 (modulation spaces). Let $1 \leq p, q \leq \infty$, and $0 \neq g \in \mathcal{S}(\mathbb{R}^d)$. The modulation space $M^{p,q}(\mathbb{R}^d)$ is defined to be the space of all tempered distributions f for which the following norm is finite:

$$\|f\|_{M^{p,q}} = \left(\int_{\mathbb{R}^d} \left(\int_{\mathbb{R}^d} |V_g f(x, y)|^p dx \right)^{q/p} dy \right)^{1/q},$$

for $1 \leq p, q < \infty$. If p or q is infinite, $\|f\|_{M^{p,q}}$ is defined by replacing the corresponding integral by the essential supremum.

Remark 1. The definition of the modulation space given above, is independent of the choice of the particular window function. See [4, Proposition 11.3.2(c), p.233].

Next, we shall see how the Fourier-Wigner transform and the STFT are related. Let π be the Schrödinger representation of the Heisenberg group with the parameter $\lambda = 1$ which is realized on $L^2(\mathbb{R}^d)$ and explicitly given by

$$\pi(x, y, t)\phi(\xi) = e^{it} e^{i(x\cdot\xi + \frac{1}{2}x\cdot y)}\phi(\xi + y)$$

where $x, y \in \mathbb{R}^d, t \in \mathbb{R}, \phi \in L^2(\mathbb{R}^d)$. The Fourier-Wigner transform of two functions $f, g \in L^2(\mathbb{R}^d)$ is defined by

$$W_g f(x, y) = (2\pi)^{-d/2}\langle \pi(x, y, 0)f, g\rangle.$$

For $z = x + iy$, we put $\pi(x, y, 0) = \pi(z, 0) = \pi(z)$. We may rewrite the Fourier-Wigner transform as

$$W_g f(x, y) = \langle f, \pi^*(z)g\rangle,$$

where $\langle f, g\rangle$ denotes the inner product for L^2 functions, or the action of the tempered distribution f on the Schwartz class function g. We recall the representation ρ_1 (See Sect. 2.2) of $\mathbb{H}^d_{pol}$, $\rho_1 = \rho = \pi \circ \Phi^{-1}$, $\rho(x, y, e^{it})$ acting on $L^2(\mathbb{R}^d)$ is given by

$$\rho(x, y, e^{it})\phi(\xi) = e^{it} e^{ix\cdot\xi}\phi(\xi + y), \quad \phi \in L^2(\mathbb{R}^d).$$

We now write the Fourier-Wigner transform in terms of the STFT: Specifically, we put $\rho(x, y)\phi(\xi) = e^{ix\cdot\xi}\phi(\xi + y)$, and have

$$\langle \pi(x, y)f, g\rangle = e^{\frac{i}{2}x\cdot y}\langle \rho(x, y)f, g\rangle = e^{-\frac{i}{2}x\cdot y} V_g f(y, -x). \tag{16}$$

This useful identity (16) reveals that the definition of modulation spaces we have introduced in the introduction and in the present section is essentially the same.

The following basic properties of modulation spaces are well-known and for the proof we refer the reader to [3,4].

Lemma 1. *1. The space $M^{p,q}(\mathbb{R}^d)(1 \le p \le \infty)$ is a Banach space.*
2. $M^{p,p}(\mathbb{R}^d) \hookrightarrow L^p(\mathbb{R}^d)$ for $1 \le p \le 2$, and $L^p(\mathbb{R}^d) \hookrightarrow M^{p,p}(\mathbb{R}^d)$ for $2 \le p \le \infty$.
3. If $q_1 \le q_2$ and $p_1 \le p_2$, then $M^{p_1,q_1}(\mathbb{R}^d) \hookrightarrow M^{p_2,q_2}(\mathbb{R}^d)$.
4. $\mathcal{S}(\mathbb{R}^d)$ is dense in $M^{p,q}(\mathbb{R}^d)(1 \le p, q < \infty)$.

We also refer to Gröchenig's book [4] for the basic definitions and further properties of modulation spaces. Finally, we note that there is also an equivalent definition of modulation spaces using frequency-uniform decomposition techniques (which is quite similar in the spirit of Besov spaces), independently studied by Wang et al. in [19], which has turned out to be very fruitful in PDE, see [20].

3 Hermite Multipliers via Transference Theorems

As mentioned in the introduction we prove our main results via transference techniques. We first investigate the connection between $m(H)$ acting on $M^{q,p}(\mathbb{R}^d)$ and $m(\mathcal{L}_0)$ acting on $L^p(G^d)$ where $G^d = \mathbb{H}^d_{pol}/\Gamma_0$. Recall that $f \in M^{q,p}(\mathbb{R}^d)$ if and only if $\langle \rho(x,y)f, \Phi_0\rangle \in L^p(\mathbb{R}^d, L^q(\mathbb{R}^d))$. Moreover, for any left invariant vector filed X on G^d we have

$$X\langle \rho(x,y,t)f,\Phi_0\rangle = \langle \rho(x,y,t)\rho^*(X)f,\Phi_0\rangle$$

where $\rho^*(X)\varphi = \frac{d}{dt}|_{t=0}\rho(\exp(tX))\varphi$. A simple calculation shows that $\rho^*(X_j) = i\xi_j$ and $\rho^*(Y_j) = \frac{\partial}{\partial \xi_j}$ for $j = 1, 2, ..., d$. Consequently, we get $\mathcal{L}\langle \rho(x,y,t)f,\Phi_0\rangle = \langle \rho(x,y,t)Hf,\Phi_0\rangle$ which leads to the identity

$$m(\tilde{\mathcal{L}}_0)\langle \rho(\cdot)f,\Phi_0\rangle = \langle \rho(x,y,t)m(H)f,\Phi_0\rangle$$

via spectral theorem (See [16, Sect. 2.3] for details). Thus we see that

$$\|m(\tilde{\mathcal{L}}_0)\langle \rho(\cdot)f,\Phi_0\rangle\|_{L^p(\mathbb{R}^d,L^q(\mathbb{R}^d\times\mathbb{T}))} = \|m(H)f\|_{M^{q,p}(\mathbb{R}^d)}. \tag{17}$$

Consequently, the boundedness of $m(H)$ on $M^{q,p}(\mathbb{R}^d)$ is implied by the boundedness of $m(\tilde{\mathcal{L}}_0)$ on $L^p(\mathbb{R}^d, L^q(\mathbb{R}^d \times \mathbb{T}))$. We can now prove Theorem 2.

Proof of Theorem 2. We do this in two steps. Assuming that $m(\tilde{\mathcal{L}})$ is bounded on $L^p(\mathbb{H}^d_{pol})$ it follows from Theorem 8 that $m(\tilde{\mathcal{L}}_0)$ is bounded on $L^p(\mathbb{H}^d_{pol}/\Gamma)$. As the underlying manifold of $\mathbb{H}^d_{pol}/\Gamma$ is $\mathbb{R}^d\times\mathbb{R}^d\times\mathbb{T}$ we have the boundedness of $m(\tilde{\mathcal{L}}_0)$ on the space $L^p(\mathbb{R}^d, L^p(\mathbb{R}^d\times\mathbb{T}))$. If we can show that under the assumptions on p and q stipulated in Theorem 2, the operator $m(\tilde{\mathcal{L}}_0)$ is bounded on the mixed norm space $L^p(\mathbb{R}^d, L^q(\mathbb{R}^d\times\mathbb{T}))$ then we are done simply by applying $m(\tilde{\mathcal{L}}_0)$ to $\langle \pi(z,t)f,\Phi_0\rangle$ (see (17)). To this end, we make use of the following result of Herz and Riviére [9].

Theorem 9 ***(Herz-Rivière*** *[9]).* *Let $G = \Gamma H$ be the semi-direct product of an amenable group H by a locally compact group Γ. Assume that a bounded linear operator $T : L^p(G) \to L^p(G)$ commutes with left-translations. Take q such that $1 \le p \le q \le 2$ or $2 \le q \le p < \infty$. Then for any complex-valued continuous function f of compact support on G, we have*

$$\|Tf\|_{L^p(\Gamma,L^q(H))} \lesssim \|f\|_{L^p(\Gamma,L^q(H))}.$$

Consequently, T has a bounded extension to the mixed norm spaces $L^p(\Gamma, L^q(H))$.

In view of the above theorem all we have to do is to realise G^d as a semidirect product of $\mathbb{R}^d$ with $\mathbb{R}^d\times\mathbb{T}$. As $m(\mathcal{L}_0)$ is invariant under left translations, we can then apply the above theorem to arrive at the required conclusion.

Let us briefly recall the definition of a semidirect product for the convenience of the readers. Let H be a topological group whose operation is written as $+$

and Γ another topological group, written as multiplicatively, such that there is a continuous map $\Gamma \times H \to H$: $(\sigma, x) \mapsto \sigma x$ with $\sigma(x+y) = \sigma x + \sigma y$ and $\tau(\sigma x) = (\tau\sigma)x$. In this situation, we say that Γ acts on H. The semi-direct product ΓH is then the topological space $\Gamma \times H$ with the group operation

$$(\sigma, x) \cdot (\tau, y) = (\sigma\tau, \tau x + y).$$

Let $\Gamma = (\mathbb{R}^d, +)$ be the additive group and $H = (\mathbb{R}^d \times \mathbb{T}, \cdot)$ be the group with following group law:

$$(y, e^{it}) \cdot (y', e^{it'}) = (y + y', e^{i(t+t')}).$$

We define a map $\Gamma \times H \to H$ as $(x, (y, e^{it})) \mapsto (y, e^{i(t+xy)})$. We note that via this map, Γ acts on H. If there is no confusion, we write $(x, (y, e^{it})) = (x, y, e^{it})$ for $(x, (y, e^{it})) \in \Gamma \times H$. Forming the semi-direct product $G = \Gamma H$ we see that the group law is given by

$$(x, y, e^{it})(x', y', e^{it'}) = (x + x', y + y', e^{i(t+t'+x'.y)}).$$

This is precisely the group $G^d = \mathbb{H}^d_{pol}/\Gamma$. Hence, by Theorem 9 we get the boundedness of $m(\mathcal{L}_0)$ on the mixed norm space $L^p(\mathbb{R}^d, L^q(\mathbb{R}^d \times \mathbb{T}))$. This completes the proof of Theorem 2.

Proof of Theorem 1. Using Theorems 6 and 8, we conclude that $m(\tilde{\mathcal{L}}_0)$ is bounded on $L^p(\mathbb{H}^d_{pol})$. We now apply Theorem 2, to complete the proof.

4 Hermite Multipliers via Fourier Multipliers on Torus

In the last section, we have proved the boundedness of Hermite multiplier on modulation spaces via transference results. Specifically, we have proved that the boundedness of multiplier operators on Heisenberg groups ensures the boundeness of corresponding Hermite multiplier operators on modulation spaces.

In this section, we first we prove the transference result which connects Hermite multipliers on modulation spaces and Fourier multipliers on $L^p(\mathbb{T}^d)$. Specifically, our result (Proposition 1) states that boundedness of Fourier multipliers on torus guarantees the boundedness of Hermite multipliers on modulation spaces. In order to find a fruitful application of our result, we need to know the boundedness of Fourier multiplier on $L^p(\mathbb{T}^d)$. To this end, we prove(Proposition 2) boundedness of Fourier multiplier on $L^p(\mathbb{R}^d)$–and then use the celebrated theorem of de Leeuw to come back to the Fourier multiplier on $L^p(\mathbb{T}^d)$. Combining these results, finally in this section we prove Theorem 4. For the sake of convenience of the reader, we recall definitions:

Definition 2 (Hermite multipliers on $M^{p,q}(\mathbb{R}^d)$). Let m be a bounded function on $\mathbb{N}^d$. We say that m is a Hermite multiplier on the space $M^{p,q}(\mathbb{R}^d)$ if the linear operator defined by

$$T_m f = \sum_{\alpha \in \mathbb{N}^d} m(\alpha) \langle f, \Phi_\alpha \rangle \Phi_\alpha, \ (f \in \mathcal{S}(\mathbb{R}^d))$$

extends to a bounded linear operator from $M^{p,q}(\mathbb{R}^d)$ into itself, that is, $\|T_m f\|_{M^{p,q}} \lesssim \|f\|_{M^{p,q}}$.

Definition 3 (Fourier multiplier on $L^p(\mathbb{T}^d)$). Let m be a bounded measurable function defined on $\mathbb{Z}^d$. We say that m is a Fourier multiplier on $L^p(\mathbb{T}^d)$ if the linear operator T_m defined by

$$\widehat{(T_m f)}(\alpha) = m(\alpha)\hat{f}(\alpha), \ (f \in \mathcal{P}(\mathbb{T}^d), \alpha \in \mathbb{Z}^d)$$

where $\hat{f}(\alpha) = \int_{\mathbb{T}^d} f(\theta) e^{-i\theta\cdot\alpha} d\theta$ are the Fourier coefficients of f, extends to a bounded linear operator from $L^p(\mathbb{T}^d)$ into itself, that is, $\|T_m f\|_{L^p(\mathbb{T}^d)} \lesssim \|f\|_{L^p(\mathbb{T}^d)}$.

Now we prove a transference result for Fourier multiplier on torus and Hermite multiplier on modulation spaces–which is of interest in itself. Specifically, we have the following proposition.

Proposition 1. *Let $1 \leq p < \infty$. If $m : \mathbb{Z}^d \to \mathbb{C}$ is a Fourier multiplier on $L^p(\mathbb{T}^d)$, then $m|_{\mathbb{N}^d}$, the restriction of m to $\mathbb{N}^d$, is a Hermite multiplier on $M^{p,p}(\mathbb{R}^d)$.*

Proof. Since $\mathcal{S}(\mathbb{R}^d)$ is dense (Lemma 1(4)) in $M^{p,p}(\mathbb{R}^d)$, it would be sufficient to prove

$$\|T_m f\|_{M^{p,p}} \lesssim \|f\|_{M^{p,p}} \tag{18}$$

for $f \in \mathcal{S}(\mathbb{R}^d)$. In fact, by density argument, it follows that T_m has a bounded extension to $M^{p,p}(\mathbb{R}^d)$, that is, inequality (18) holds true for $f \in M^{p,p}(\mathbb{R}^d)$.

Let $f \in \mathcal{S}(\mathbb{R}^d)$. Then using special Hermite functions (4) and their property (6), we compute the modulation space norm (See Sect. 2.5) of Hermite multiplier operator (See Definition 2), and we obtain

$$\begin{aligned}
\|T_m f\|^p_{M^{p,p}} &= (2\pi)^{-dp/2} \int_{\mathbb{R}^{2d}} |\langle \pi(x,y) T_m f, \Phi_0 \rangle|^p dy\, dx \\
&= (2\pi)^{-dp/2} \int_{\mathbb{R}^{2d}} \left| \sum_{\alpha \in \mathbb{N}^d} m(\alpha) \langle f, \Phi_\alpha \rangle \langle \pi(x,y)\Phi_\alpha, \Phi_0 \rangle \right|^p dy\, dx \\
&= \int_{\mathbb{C}^d} \left| \sum_{\alpha \in \mathbb{N}^d} m(\alpha) \langle f, \Phi_\alpha \rangle \Phi_{\alpha,0}(z) \right|^p dz \\
&= (2\pi)^{-dp/2} \int_{\mathbb{C}^d} \left| \sum_{\alpha \in \mathbb{N}^d} m(\alpha) \langle f, \Phi_\alpha \rangle \frac{i^{|\alpha|} \bar{z}^\alpha}{\sqrt{\alpha!}\, 2^{|\alpha|/2}} e^{-\frac{1}{4}|z|^2} \right|^p dz. \qquad (19)
\end{aligned}$$

By using polar coordinates $z_j = r_j e^{i\theta_j}$, $r_j := |z_j| \in [0,\infty), z_j \in \mathbb{C}$ and $\theta_j \in [0, 2\pi)$, we get

$$z^\alpha = r^\alpha e^{i\alpha\cdot\theta} \ \text{ and } \ dz = r_1 r_2 \cdots r_d d\theta dr \tag{20}$$

where $r=(r_1,\cdots,r_d)$, $\theta=(\theta_1,\cdots,\theta_d)$, $dr=dr_1\cdots dr_d$, $d\theta=d\theta_1\cdots d\theta_d$, $|r|=\sqrt{\sum_{j=1}^d r_j^2}$.
By writing the integral over $\mathbb{C}^d=\mathbb{R}^{2d}$ in polar coordinates in each time-frequency pair and using (20), we have

$$\int_{\mathbb{C}^d}\left|\sum_{\alpha\in\mathbb{N}^d} m(\alpha)\langle f,\Phi_\alpha\rangle\frac{i^{|\alpha|}\bar{z}^\alpha}{\sqrt{\alpha!}\,2^{|\alpha|/2}}e^{-\frac{1}{4}|z|^2}\right|^p dz \tag{21}$$

$$=\prod_{j=1}^d\int_{\mathbb{R}+}\int_{[0,2\pi]}\left|\sum_{\alpha\in\mathbb{N}^d} m(\alpha)\left(\langle f,\Phi_\alpha\rangle\frac{i^{|\alpha|}r^\alpha}{\sqrt{\alpha!}\,2^{|\alpha|/2}}e^{-\frac{1}{4}|r|^2}\right)e^{-i\alpha\cdot\theta}\right|^p r_j\,d\theta_j\,dr_j.$$

We put $a_\alpha=\langle f,\Phi_\alpha\rangle\frac{i^{|\alpha|}r^\alpha}{\sqrt{\alpha!}\,2^{|\alpha|/2}}e^{-\frac{1}{4}|r|^2}$. Since $|\langle f,\Phi_\alpha\rangle|\le\|f\|_{L^2}\|\Phi_\alpha\|_{L^2}\le\|f\|_2$, the series $\sum_{\alpha\in\mathbb{N}^d}|a_\alpha|$ converges. Thus there exists a continuous function $g\in L^p(\mathbb{T}^d)$ with Fourier coefficients $\hat{g}(\alpha)=a_\alpha$ for $\alpha\in\mathbb{N}^d$ and $\hat{g}(\alpha)=0$ for $\alpha\in\mathbb{Z}^d\setminus\mathbb{N}^d$. In fact, the Fourier series of this g is absolutely convergent, and therefore we may write

$$g(e^{i\theta})=\sum_{\alpha\in\mathbb{Z}^d}a_\alpha e^{-i\alpha\cdot\theta}. \tag{22}$$

Since m is a Fourier multiplier on $L^p(\mathbb{T}^d)$, (22) gives

$$\int_{[0,2\pi]^d}\left|\sum_{\alpha\in\mathbb{N}^d}m(\alpha)a_\alpha e^{i\alpha\cdot\theta}\right|^p d\theta\lesssim\int_{[0,2\pi]^d}\left|\sum_{\alpha\in\mathbb{N}^d}a_\alpha e^{-i\alpha\cdot\theta}\right|^p d\theta. \tag{23}$$

Now taking (19), (21), and (23) into account, we obtain

$$\begin{aligned}\|T_mf\|_{M^{p,p}}&\lesssim(2\pi)^{-d/2}\left(\prod_{j=1}^d\int_{\mathbb{R}+}\int_{[0,2\pi]}\left|\sum_{\alpha\in\mathbb{N}^d}\left(\langle f,\Phi_\alpha\rangle\frac{i^{|\alpha|}r^\alpha}{\sqrt{\alpha!}\,2^{|\alpha|/2}}e^{-\frac{1}{4}|r|^2}\right)e^{-i\alpha\cdot\theta}\right|^p r_j\,d\theta_j\,dr_j\right)^{\frac{1}{p}}\\ &=(2\pi)^{-d/2}\left(\int_{\mathbb{C}^d}\left|\sum_{\alpha\in\mathbb{N}^d}\langle f,\Phi_\alpha\rangle\frac{i^{|\alpha|}\bar{z}^\alpha}{\sqrt{\alpha!}\,2^{|\alpha|/2}}e^{-\frac{1}{4}|z|^2}\right|^p dz\right)^{\frac{1}{p}}\\ &=\left(\int_{\mathbb{C}^d}\left|\sum_{\alpha\in\mathbb{N}^d}\langle f,\Phi_\alpha\rangle\Phi_{\alpha,0}(z)\right|^p dz\right)^{\frac{1}{p}}\\ &=(2\pi)^{-d/2}\left(\int_{\mathbb{R}^{2d}}\left|\sum_{\alpha\in\mathbb{N}^d}\langle f,\Phi_\alpha\rangle\langle\pi(x,y)\Phi_\alpha,\Phi_0\rangle\right|^p dy\,dx\right)^{\frac{1}{p}}\\ &=\|f\|_{M^{p,p}}.\end{aligned} \tag{24}$$

Thus, we conclude that $\|T_mf\|_{M^{p,p}}\lesssim\|f\|_{M^{p,p}}$ for $f\in\mathcal{S}(\mathbb{R}^d)$. This completes the proof of Proposition 1.

Next we recall the celebrated theorem of de Leeuw, which gives the relation between Fourier multipliers on Euclidean spaces and tori. To do this, we start with

Definition 4 (Fourier multiplier on $L^p(\mathbb{R}^d)$). Let m be a bounded measurable function defined on $\mathbb{R}^d$. We say that m is a Fourier multiplier on $L^p(\mathbb{R}^d)$ if the linear operator T_m defined by

$$\widehat{(T_mf)}(\xi)=m(\xi)\hat{f}(\xi),\ (f\in\mathcal{S}(\mathbb{R}^d),\ \xi\in\mathbb{R}^d)$$

where $\hat{f}$ is the Fourier transform, extends to a bounded linear operator from $L^p(\mathbb{R}^d)$ into itself, that is, $\|T_mf\|_{L^p}\lesssim\|f\|_{L^p}$.

Theorem 10 ***(de Leeuw).*** *If $m : \mathbb{R}^d \to \mathbb{C}$ is continuous and Fourier multiplier on $L^p(\mathbb{R}^d)(1 \leq p < \infty)$, then $m|_{\mathbb{Z}^d}$, the restriction of m to $\mathbb{Z}^d$, is a Fourier multiplier on $L^p(\mathbb{T}^d)$.*

In order prove Theorem 4, we also need the boundedness of a following Fourier multiplier

$$m(\xi) = \frac{e^{i(2|\xi|_1+d)^\gamma}}{(2|\xi|_1 + d)^\beta} \quad (\beta > 0, \gamma > 0, \xi \in \mathbb{R}^d) \tag{25}$$

where $|\xi|_1 := \sum_{j=1}^d |\xi_j|$, on $L^p(\mathbb{R}^d)$. This we shall prove in the next proposition. We note that in [14, Theorem 1], it is proved that the function $m(\xi)$ which is 0 near the origin and $|\xi|^{-\beta}e^{i|\xi|^\alpha}$ $(|\xi| = \sqrt{\sum_{j=1}^d \xi_j^2})$ outside a compact set is a Fourier multiplier on $L^p(\mathbb{R}^d)$ for a suitable choice of α, β, p, d. Our proof of Proposition 2 is motivated by this result.

Proposition 2. *Let m be given by (25). Then m is a Fourier multiplier on $L^p(\mathbb{R}^d)$ for $|\frac{1}{p} - \frac{1}{2}| < \frac{\beta}{d\gamma}$.*

To prove this proposition we need the following technical lemma. We will prove this lemma at the end.

Lemma 2. *Let $\sigma > 0, \gamma > 0$ and*

$$k_\sigma(x) = \frac{1}{(2\pi)^{d/2}} \int_{\mathbb{R}^d} e^{i(2|\xi|_1+d)^\gamma} e^{-\sigma(2|\xi|_1+d)^\gamma} e^{ix\cdot\xi} \, d\xi \ (x \in \mathbb{R}^d). \tag{26}$$

Then

$$\|k_\sigma\|_{L^1} \lesssim \sigma^{-d/2} e^{-\frac{1}{2}\sigma d^\gamma}.$$

Proof of Proposition 2 Performing a simple change of variables in the gamma function, we write

$$(2|\xi|_1 + d)^{-\beta} = \frac{1}{\Gamma(\beta/\gamma)} \int_0^\infty \sigma^{\frac{\beta}{\gamma}-1} \exp(-\sigma(2|\xi|_1 + d)^\gamma) \, d\sigma. \tag{27}$$

In view of Definition 4, (25) and (27), we write

$$\begin{aligned} T_m f(x) &= (2\pi)^{-d/2} \int_{\mathbb{R}^d} m(\xi) \hat{f}(\xi) e^{ix\cdot\xi} d\xi \\ &= \frac{1}{\Gamma(\beta/\gamma)} \int_0^\infty \sigma^{(\beta/\gamma)-1} (k_\sigma * f)(x) \, d\sigma \end{aligned} \tag{28}$$

where k_σ is as defined in Lemma 2. Using (28), we have

$$\|T_m f\|_{L^p} \leqslant \frac{1}{\Gamma(\beta/\gamma)} \int_0^\infty \sigma^{(\beta/\gamma)-1} \|k_\sigma * f\|_{L^p} \, d\sigma. \tag{29}$$

The boundedness of the multiplier operator $T_m f$ on L^p will follow if we could show that the operator $f \to k_\sigma * f$ is bounded on L^p. We shall achieve this by using Riesz-Thorin interpolation theorem and the standard duality argument. To this end, we use Lemma 2 and Young's inequality, to obtain

$$\|k_\sigma * f\|_{L^1} \lesssim \sigma^{-d/2} e^{-\frac{1}{2}\sigma d^\gamma} \|f\|_{L^1}. \tag{30}$$

Since $|\hat{k}_\sigma(\xi)| \leqslant e^{-\frac{1}{2}\sigma d^\gamma}$, Plancherel theorem gives

$$\|k_\sigma * f\|_{L^2} \leqslant e^{-\frac{1}{2}\sigma d^\gamma} \|f\|_{L^2}. \tag{31}$$

Taking (30), and (31) into account, Riesz-Thorin interpolation theorem gives

$$\|k_\sigma * f\|_{L^p} \leqslant C_1 \sigma^{-\lambda} e^{-\frac{1}{2}\sigma d^\gamma} \|f\|_{L^p} \tag{32}$$

where $\lambda = d\left(\frac{1}{p} - \frac{1}{2}\right)$, for $1 \leq p \leq 2$. Finally using (32) and (29), we see that T_m is bounded from $L^p(\mathbb{R}^d) \to L^p(\mathbb{R}^d)$ if

$$\int_0^\infty \sigma^{\beta/\gamma - 1 - \lambda} e^{-\frac{1}{2}\sigma d^\gamma}\, d\sigma < \infty,$$

which happens if and only if $\frac{1}{p} - \frac{1}{2} < \frac{\beta}{d\gamma}$. This proves the theorem for the case when $1 < p < 2$ and the case $p \geqslant 2$ follows from the duality.

Now we shall prove our Lemma 2.

Proof of Lemma 2. Since k_σ (see (26)) is the inverse Fourier transform of the function $\exp\left((i-\sigma)(2|\xi|_1 + d)^\gamma\right)$, we have

$$\widehat{k_\sigma}(\xi) = e^{(i-\sigma)(2|\xi|_1 + d)^\gamma}. \tag{33}$$

Let $\alpha = (\alpha_1, \cdots, \alpha_d) \in \mathbb{N}^d$ be a multi-index of length l, that is, $\sum_{j=1}^d \alpha_j = l$, and put $D_\xi^\alpha = \left(\frac{\partial}{\partial \xi_1}\right)^{\alpha_1} \cdots \left(\frac{\partial}{\partial \xi_d}\right)^{\alpha_d}$, $\xi = (\xi_1, ..., \xi_d) \in \mathbb{R}^d$. Then in view of (33), we have

$$|D_\xi^\alpha \widehat{k_\sigma}(\xi)| \lesssim (2|\xi|_1 + d)^{-l(1-\gamma)} e^{-\frac{1}{2}\sigma(2|\xi|_1 + d)^\gamma} e^{-\frac{1}{2}\sigma d^\gamma}. \tag{34}$$

To obtain inequality (34), we have used these ideas: After taking partial derivatives of higher order for $\widehat{k_\sigma}$, we have estimated all the powers of $(2|\xi|_1 + d)$ by the highest appearing power–which is $(2|\xi|_1 + d)^{-l(1-\gamma)}$, this we could do because $(2|\xi|_1 + d) \geqslant 1$. We have also dominated $e^{-\frac{1}{2}\sigma(2|\xi|_1+d)^\gamma}$ by $e^{-\frac{1}{2}\sigma d^\gamma}$, which is obvious.

Next, by Plancherel's theorem and (34), we get

$$\||x|^l k_\sigma\|_{L^2} \lesssim \left(\int_{\mathbb{R}^d} (2|\xi|_1 + d)^{-2l(1-\gamma)} e^{-\sigma(2|\xi|_1 + d)^\gamma}\, d\xi\right)^{\frac{1}{2}} e^{-\frac{1}{2}\sigma d^\gamma}.$$

Performing a change of variable, we get

$$\||x|^l k_\sigma\|_{L^2} \lesssim \sigma^{-\frac{d}{2\gamma} + \frac{l(1-\gamma)}{\gamma}} e^{-\frac{1}{2}\sigma d^\gamma} \left(\int_{\mathbb{R}^d} g(2|\xi|_1 + \sigma^{1/\gamma} d)\, d\xi\right)^{1/2},$$

where $g(t) = t^{-2l(1-\gamma)}e^{-t^\gamma}$ $(t > 0)$. Noting that $\int g(2|\xi|_1 + \sigma^{1/\gamma}d)\, d\xi < \infty$ for all γ and l, we conclude

$$\||x|^l k_\sigma\|_{L^2} \lesssim \sigma^{-\frac{d}{2\gamma}+\frac{l(1-\gamma)}{\gamma}} e^{-\frac{1}{2}\sigma d^\gamma}.$$

Next we use the fact that for $h \in L^2(\mathbb{R}^d) \cap L^1(\mathbb{R}^d)$, $l > d/2$ and $R > 0$, we have

$$\|h\|_{L^1} \lesssim R^{d/2}(\|h\|_{L^2} + R^{-l}\||x|^l h\|_{L^2})$$

for all $R > 0$. Taking $h = k_\sigma$ and $R = \sigma^{\frac{1-\gamma}{\gamma}}$, we obtain

$$\|k_\sigma\|_{L^1} \leqslant C\sigma^{-d/2}e^{-\frac{1}{2}\sigma d^\gamma}.$$

This completes the proof.

We now use Proposition 2 and Theorem 10 to obtain the following corollary.

Corollary 3. *Let* $\left|\frac{1}{p} - \frac{1}{2}\right| < \frac{\beta}{d\gamma}$, $\beta > 0, \gamma > 0$, $|\alpha|_1 = \sum_{j=1}^d |\alpha_j|$. *Then the sequence* $\{(2|\alpha|_1 + d)^{-\beta}\exp(i(2|\alpha|_1 + d)^\gamma)\}_{\alpha\in\mathbb{Z}^d}$ *defines a multiplier on* $L^p(\mathbb{T}^d)$.

Proof of Theorem 4 Using Corollary 3 and Proposition 2, we may deduce that $m(H)$ is bounded on $M^{p,p}(\mathbb{R}^d)$. This completes the proof.

5 Hermite Multiplier for Schrödinger Propagator

In this section, we prove the boundedness of Schrödinger propagator $m(H) = e^{itH}$ using the properties of Hermite and special Hermite functions. Our approach of proof illustrates how these functions nicely fit into modulation spaces–and prove useful estimate.

Proof of Theorem 5. Let $f \in \mathcal{S}(\mathbb{R}^d)$. Then we have the Hermite expansion of f as follows:

$$f = \sum_{\alpha\in\mathbb{N}^d} \langle f, \Phi_\alpha\rangle \Phi_\alpha. \tag{35}$$

Now using (35) and (4), we obtain

$$\begin{aligned}\langle \pi(z)f, \Phi_0\rangle &= \sum_{\alpha\in\mathbb{N}^d} \langle f, \Phi_\alpha\rangle\langle \pi(z)\Phi_\alpha, \Phi_0\rangle \\ &= \sum_{\alpha\in\mathbb{N}^d} \langle f, \Phi_\alpha\rangle \Phi_{\alpha,0}(z).\end{aligned} \tag{36}$$

Since $\{\Phi_\alpha\}$ forms an orthonormal basis for $L^2(\mathbb{R}^d)$, (36) gives

$$\begin{aligned}\langle \pi(z)m(H)f, \Phi_0\rangle &= \sum_{\alpha\in\mathbb{N}^d} \langle m(H)f, \Phi_\alpha\rangle \Phi_{\alpha,0}(z) \\ &= \sum_{\alpha\in\mathbb{N}^d} m(2|\alpha| + d)\langle f, \Phi_\alpha\rangle \Phi_{\alpha,0}(z).\end{aligned}$$

Therefore, for $m(H) = e^{itH}$, we have

$$\langle \pi(z)e^{itH} f, \Phi_0 \rangle = e^{itd} \sum_{\alpha \in \mathbb{N}^d} e^{2it|\alpha|} \langle f, \Phi_\alpha \rangle \Phi_{\alpha,0}(z)$$
$$= e^{itd} (2\pi)^{-d/2} \sum_{\alpha \in \mathbb{N}^d} e^{2it|\alpha|} \langle f, \Phi_\alpha \rangle (\alpha!)^{-1/2} \left(\frac{i}{\sqrt{2}}\right)^{|\alpha|} \bar{z}^\alpha e^{-\frac{1}{4}|z|^2}. \quad (37)$$

In view of (16) and (37), we have

$$\|e^{itH} f\|^p_{M^{p,p}} = \frac{1}{(2\pi)^{d/2}} \int_{\mathbb{C}^d} \left| \sum_{\alpha \in \mathbb{N}^d} e^{2it|\alpha|} \langle f, \Phi_\alpha \rangle (\alpha!)^{-1/2} \left(\frac{i}{\sqrt{2}}\right)^{|\alpha|} \bar{z}^\alpha e^{-\frac{1}{4}|z|^2} \right|^p dz. \quad (38)$$

Using polar coordinates as above (see (20)), we have

$$\int_{\mathbb{C}^d} \left| \sum_{\alpha \in \mathbb{N}^d} e^{2it|\alpha|} \langle f, \Phi_\alpha \rangle (\alpha!)^{-1/2} \left(\frac{i}{\sqrt{2}}\right)^{|\alpha|} \bar{z}^\alpha e^{-\frac{1}{4}|z|^2} \right|^p dz \quad (39)$$
$$= \prod_{j=1}^d \int_{\mathbb{R}^+} \int_{[0,2\pi]} \left| \sum_{\alpha \in \mathbb{N}^d} \langle f, \Phi_\alpha \rangle (\alpha!)^{-1/2} \left(\frac{i}{\sqrt{2}}\right)^{|\alpha|} r^\alpha e^{i \sum_{j=1}^d (2t-\theta_j)\alpha_j} e^{-\frac{1}{4}|r|^2} \right|^p r_j dr_j d\theta_j.$$

By a simple change of variable $(\theta_j - 2t) \to \theta_j$, we obtain

$$\prod_{j=1}^d \int_{\mathbb{R}^+} \int_{[0,2\pi]} \left| \sum_{\alpha \in \mathbb{N}^d} \langle f, \Phi_\alpha \rangle (\alpha!)^{-1/2} \left(\frac{i}{\sqrt{2}}\right)^{|\alpha|} r^\alpha e^{i \sum_{j=1}^d (2t-\theta_j)\alpha_j} e^{-\frac{1}{4}|r|^2} \right|^p r_j dr_j d\theta_j \quad (40)$$
$$= \prod_{j=1}^d \int_{\mathbb{R}^+} \int_{[0,2\pi]} \left| \sum_{\alpha \in \mathbb{N}^d} \langle f, \Phi_\alpha \rangle (\alpha!)^{-1/2} \left(\frac{i}{\sqrt{2}}\right)^{|\alpha|} r^\alpha e^{-i\theta \cdot \alpha} e^{-\frac{1}{4}|r|^2} \right|^p r_j dr_j d\theta_j.$$

Combining (38), (39), (40), and Lemma 1(4), we have $\|e^{itH} f\|_{M^{p,p}} = \|f\|_{M^{p,p}}$ for $f \in M^{p,p}(\mathbb{R}^d)$. This completes the proof of Theorem 5.

Acknowledgment. The work leading to this article began while DGB was a project assistant with Professor Thangavelu in IISc. DGB is very grateful to Professor Thangavelu for providing the funding and arranging research facilities during his stay at IISc. DGB is also thankful to DST-INSPIRE and TIFR CAM for the current support. RB wishes to thank UGC-CSIR for financial support. ST is supported by J C Bose National Fellowship from D.S.T., Govt. of India.

References

1. Bényi, A., Gröchenig, K., Okoudjou, K.A., Rogers, L.G.: Unimodular fourier multipliers for modulation spaces. J. Funct. Anal. **246**(2), 366–384 (2007)
2. Cordero, E., Nicola, F.: Metaplectic representation on Wiener amalgam spaces and applications to the Schrödinger equation. J. Funct. Anal. **254**(2), 506–534 (2008)
3. Feichtinger, H.G.: Modulation spaces on locally compact abelian groups, Technical Report, University of Vienna (1983)
4. Gröchenig, K.: Foundations of Time-Frequency Analysis. Birkhäuser Boston Inc., Boston, MA (2001)
5. Jao, C.: The energy-critical quantum harmonic oscillator. Commun. Partial Differ. Equ. **41**(1), 79–133 (2016)

6. Hebisch, W.: Multiplier theorem on generalized Heisenberg groups. Colloq. Math. **65**(2), 231–239 (1993)
7. Hebisch, W., Zienkiewicz, J.: Multiplier theorem on generalized Heisenberg groups. II. Colloq. Math. **69**(1), 29–36 (1995)
8. Kato, K., Kobayashi, M., Ito, S.: Remarks on Wiener Amalgam Space Type Estimates for Schrödinger Equation, Harmonic Analysis and Nonlinear Partial Differential Equations, pp. 41–48. RIMS Kokyuroku Bessatsu, B33, Res. Inst. Math. Sci. (RIMS), Kyoto (2012)
9. Herz, C., Rivière, N.: Estimates for translation invariant operators on spaces with mixed norms. Studia Math. **44**, 511–515 (1972)
10. Müller, D., Ricci, F., Stein, E.: Marcinkiewicz multipliers and multi-parameter structure on Heisenberg (-type) groups. II. Math. Z. **221**(2), 267–291 (1996)
11. Müller, D., Stein, E.M.: On spectral multipliers for Heisenberg and related groups. J. Math. Pures Appl. **73**(4), 413–440 (1994)
12. Ruzhansky, M., Sugimoto, M., Wang, B.: Modulation spaces and nonlinear evolution equations. In: Evolution Equations of Hyperbolic and Schrödinger Type. Progress in Mathematics, vol. 301, pp. 267–283. Birkhäuser/Springer Basel AG, Basel (2012)
13. Sanjay, P.K., Thangavelu, S.: Revisiting Riesz transforms on Heisenberg groups. Rev. Mat. Iberoam. **28**(4), 1091–1108 (2012)
14. Schonbek, T.P.: L^p−multipliers: a new proof of an old theorem. Proc. Am. Math. Soc. **102**(2), 361–364 (1988)
15. Thangavelu, S.: Lectures on Hermite and Laguerre Expansions. Mathematical Notes, vol. 42. Princeton University Press, Princeton (1993)
16. Thangavelu, S.: An Introduction to the Uncertainty Principle. Hardy's theorem on Lie groups. Progress in Mathematics, vol. 217, Birkhäuser., Boston, MA (2004)
17. Thangavelu, S.: Harmonic Analysis on the Heisenberg Group. Progress in Mathematics, vol. 159. Birkhäuser, Boston, MA (1998)
18. Thangavelu, S.: Multipliers for Hermite expansions. Rev. Mat. Iberoam. **3**(1), 1–24 (1987)
19. Wang, B.X., Zhao, L., Guo, B.: Isometric decomposition operators, function spaces $E^{\lambda}_{p,q}$ and applications to nonlinear evolution equations. J. Funct. Anal. **233**(1), 1–39 (2006)
20. Wang, B.X., Zhaohui, H., Chengchun, H., Zihua, G.: Harmonic Analysis Method for Nonlinear Evolution Equations I. World Scientific Publishing Co., Pte. Lt (2011)

Potential Theory Results for a Class of PDOs Admitting a Global Fundamental Solution

Andrea Bonfiglioli(✉)

Dipartimento di Matematica, Università degli Studi di Bologna,
Piazza di Porta San Donato, 5, 40126 Bologna, Italy
andrea.bonfiglioli6@unibo.it

Abstract. We outline several results of Potential Theory for a class of linear partial differential operators $\mathcal{L}$ of the second order in divergence form. Under essentially the sole assumption of hypoellipticity, we present a non-invariant homogeneous Harnack inequality for $\mathcal{L}$; under different geometrical assumptions on $\mathcal{L}$ (mainly, under global doubling/Poincaré assumptions), it is described how to obtain an invariant, non-homogeneous Harnack inequality. When $\mathcal{L}$ is equipped with a global fundamental solution Γ, further Potential Theory results are available (such as the Strong Maximum Principle). We present some assumptions on $\mathcal{L}$ ensuring that such a Γ exists.

Keywords: Fundamental solutions · Hypoelliptic operators
Harnack inequalities · Potential theory

1 Introduction

The keystones of classic Potential Theory were established in the century elapsing between the 20's of the nineteenth century and the 20's of the twentieth century. During this period, the following topics were studied, regarding the classic Laplace operator Δ:

- 1823: the Dirichlet problem (Poisson);
- 1828: the Green's function (Green);
- 1839: the Maximum Principle (Earnshaw);
- 1840: mean value properties (Gauss);
- 1870: further results on the Dirichlet problem (Schwarz);
- 1887: convergence properties (Harnack);
- 1890: barrier functions (Poincaré);
- 1923 ca: weak solvability (Perron, Wiener, Brelot);
- 1928: representation theory and the potential function (Riesz);
- 1929: axiomatic Potential Theory (Kellogg).

J. Delgado and M. Ruzhansky (Eds.): *Analysis and Partial Differential Equations: Perspectives from Developing Countries*, PROMS 275, pp. 65–83, 2018.
https://doi.org/10.1007/978-3-030-05657-5_6

All these topics are deeply interrelated, and each of them could hardly exist without the others. However, if we were to select a restricted number of them according to their most capital role in Potential Theory, we would point out the undisputable importance of the *Green function*, the *Maximum Principle*, and the *convergence properties*. In this paper we outline some recent (or very recent) results concerning the latter topics, for classes of linear partial differential operators (PDOs, in the sequel) of the second order in divergence form.

As an outline of the paper, the exposition is thusly presented:

1. in Sect. 2, we introduce a rather general class of *hypoelliptic* PDOs $\mathcal{L}$, and we show how to obtain a non-invariant homogenous Harnack inequality for $\mathcal{L}$; by restricting the class considered, we then describe how to prove an invariant and non-homogenous inequality (Sect. 4);
2. when $\mathcal{L}$ possesses a well-behaved global Green function Γ (also called *fundamental solution*), more Potential Theoretic results are available, including Maximum Principles (Sect. 3);
3. due to the importance of Γ, under stronger assumptions on $\mathcal{L}$, we provide sufficient conditions for its existence (Sect. 5).

2 A Non-invariant Harnack Inequality

On Euclidean space $\mathbb{R}^N$, we consider a PDO $\mathcal{L}$ of the second order in the following divergence form ("weighted" by a density V)

$$\mathcal{L} := \frac{1}{V(x)} \sum_{i,j=1}^{N} \frac{\partial}{\partial x_i}\Big(V(x)\, a_{i,j}(x)\, \frac{\partial}{\partial x_j}\Big), \tag{1}$$

where the functions V and $a_{i,j}$ fulfil the following assumptions:

- V and any $a_{i,j}$ are C^∞ from $\mathbb{R}^N$ to $\mathbb{R}$;
- $V(x) > 0$, $A(x) := (a_{i,j}(x))_{i,j}$ is symmetric and positive semi-definite at any $x \in \mathbb{R}^N$;
- **assumption (NTD)**: $\mathcal{L}$ is non-totally-degenerate at any point, i.e., $A(x) \neq 0$ for any $x \in \mathbb{R}^N$;
- **assumption (HY)**: $\mathcal{L}$ is C^∞-hypoelliptic on every open subset of $\mathbb{R}^N$.

There are examples of operators satisfying our assumptions in: Lie group theory (e.g., sub-Laplacian operators [26]); CR Geometry and Riemannian Geometry; several complex variables; Brownian motion; Stochastic PDEs.

Remark 1. We observe that there are examples of operators (with infinitely-degenerate coefficients) fulfilling all of our assumptions but outside the well-studied class of Hörmander operators; for example:

$$\frac{\partial^2}{\partial x_1^2}+\Big(\exp\Big(-\frac{1}{x_1^2}\Big)\frac{\partial}{\partial x_2}\Big)^2 \quad \text{in } \mathbb{R}^2,$$

$$\frac{\partial^2}{\partial x_1^2}+\Big(\exp\Big(-\frac{1}{\sqrt{|x_1|}}\Big)\frac{\partial}{\partial x_2}\Big)^2+\frac{\partial^2}{\partial x_3^2} \quad \text{in } \mathbb{R}^3,$$

$$\frac{\partial^2}{\partial x_1^2}+\Big(\exp(-1/|x_1|)\frac{\partial}{\partial x_2}\Big)^2+\Big(\exp(-1/|x_1|)\frac{\partial}{\partial x_3}\Big)^2 \quad \text{in } \mathbb{R}^3,$$

$$\frac{\partial^2}{\partial x_2^2}+\Big(x_2\frac{\partial}{\partial x_1}\Big)^2+\frac{\partial^2}{\partial x_4^2}+\Big(\exp(-1/\sqrt[3]{|x_1|})\frac{\partial}{\partial x_3}\Big)^2 \quad \text{in } \mathbb{R}^4,$$

respectively studied by Fediĭ [41]; Kusuoka, Stroock [63]; Christ [34]; Morimoto [65], who proved that the above PDOs are indeed C^∞-hypoelliptic.

To specify the role of V, if $\mathrm{d}x$ is Lebesgue measure in $\mathbb{R}^N$, in the sequel we set

$$\mathrm{d}\mu(x) := V(x)\,\mathrm{d}x. \tag{2}$$

Clearly, $\mathcal{L}$ is formally self-adjoint in $L^2(\mathbb{R}^N,\mathrm{d}\mu)$ on the smooth compactly-supported functions:

$$\int \varphi\,\mathcal{L}\psi\,\mathrm{d}\mu = \int \psi\,\mathcal{L}\varphi\,\mathrm{d}\mu, \qquad \forall\,\varphi,\psi\in C_0^\infty(\mathbb{R}^N).$$

The opportunity to permit V to be non-identically 1 comes from our interest in second-order operators arising in Lie group theory; indeed, one can find in [6, Example 1.1] some examples of operators of the form (1), where $V\neq 1$ is the density of the Haar measure μ of a Lie group G whose underlying manifold is $\mathbb{R}^N$, and $X_1,\dots,X_m$ is a family of generators of the Lie algebra of G. In this case, it appears very natural to consider sub-Laplace operators of the form

$$\mathcal{L} = -\sum_{j=1}^m X_j^{*\mu}X_j,$$

where $X_j^{*\mu}$ denotes the adjoint of the vector field X_j in $L^2(G,\mathrm{d}\mu)$. Such an $\mathcal{L}$ can always be written in the form (1); it satisfies (HY) in force of Hörmander's hypoellipticity theorem (since $X_1,\dots,X_m$ are Lie-generators of Lie(G)).

For another explicit example, see the next one:

Example 1. Consider the Lie group $G=(\mathbb{R}^3,*)$ where

$$(x_1,x_2,x_3)*(y_1,y_2,y_3)=(x_1+y_1,x_2+y_2,x_3+y_3e^{x_1});$$

it is natural to consider the Haar measure μ on G defined by $\mu=V(x)\,\mathrm{d}x$, with $V(x)=e^{-x_1}$; if we also consider the Lie generators X_1,X_2 of Lie(G) given by $X_1=\partial_{x_1}$ and $X_2=\partial_{x_2}+e^{x_1}\,\partial_{x_3}$, then the operator $\mathcal{L}=-X_1^{*\mu}X_1-X_2^{*\mu}X_2$ has the form

$$\mathcal{L}=(\partial_{x_1})^2+(\partial_{x_2})^2+e^{2x_1}(\partial_{x_3})^2+2e^{x_1}\partial_{x_2,x_3}-\partial_{x_1}.$$

$\mathcal{L}$ enjoys the form (1) with the above $V(x)$ and with

$$A(x) = \begin{pmatrix} 1 & 0 & 0 \\ 0 & 1 & e^{x_1} \\ 0 & e^{x_1} & e^{2x_1} \end{pmatrix},$$

but it cannot be written as a pure divergence form operator $\operatorname{div}(B(x)\,\nabla)$, whence the opportunity to consider operators of the more general form (1). $\mathcal{L}$ also enjoys assumptions (NTD) and (HY) (the latter is due to Hörmander's hypoellipticity theorem, as already remarked).

Apart from the elliptic case, in the existing literature, as regards the Maximum Principles and the Harnack inequality for (divergence-form) degenerate-elliptic PDOs under a subelliptic assumption (or assuming that the degeneracy be controlled by some Muckenhoupt weights), there are several references starting from the 1980's: Chanillo and Wheeden [33]; Fabes, Jerison and Kenig [37]; Fabes, Kenig and Serapioni [38]; Fefferman and Phong [40]; Franchi and Lanconelli [45]; Gutiérrez [52] Jerison and Sánchez-Calle [58].

It is also well known that a natural framework for the Harnack inequality is that of doubling metric spaces (where a doubling condition and a Poincaré inequality are available): Aimar, Forzani and Toledano [3]; Barlow and Bass [5]; Di Fazio, Gutiérrez and Lanconelli [36]; Grigor'yan and Saloff-Coste [51]; Gutiérrez and Lanconelli [53]; Hebisch and Saloff-Coste [54]; Indratno, Maldonado and Silwal [57]; Kinnunen, Marola, Miranda and Paronetto [59]; Mohammed [64]; Saloff-Coste [70].

Without making any subellipticity or Muckenhoupt assumption, nor requiring any underlying doubling/Poincaré metric space setting, in [6] we proved the following Strong Maximum Principle, a key step for the Harnack inequality:

Theorem 1. (Strong Maximum Principle for $\mathcal{L}$) *Suppose that $\mathcal{L}$ is an operator of the form (1), with C^∞ coefficients $V > 0$ and $(a_{i,j})_{i,j} \geq 0$, and that it satisfies* (NTD) *and* (HY). *Let $O \subseteq \mathbb{R}^N$ be a connected open set. Then, the following facts hold.*

1. *Any function $u \in C^2(O,\mathbb{R})$ satisfying $\mathcal{L}u \geq 0$ on O and attaining a maximum in O is constant throughout O.*
2. *If $c \in C^\infty(\mathbb{R}^N,\mathbb{R})$ is nonnegative on $\mathbb{R}^N$, and if we set*

$$\mathcal{L}_c := \mathcal{L} - c,$$

then any function $u \in C^2(O,\mathbb{R})$ satisfying $\mathcal{L}_c u \geq 0$ on O and attaining a nonnegative maximum in O is constant throughout O.

We proved Theorem 1 following the approach by Bony in his celebrated 1969 paper [27] concerning with Hörmander sums of squares. However, in view of Remark 1, since our assumptions are general enough to comprise *non-Hörmander* PDOs we cannot simply recover Bony's argument, which is based on the well-known connectivity theorem of Carathéodory, Chow, Hermann and Rashevskiĭ.

Luckily for us, a theorem by Amano [4] on a connectivity/controllability property for hypoelliptic PDOs allows us to recover Bony's ideas in their full strength (thus circumventing the issue caused by non-Hörmander PDOs). Our argument for Theorem 1 is essentially as follows. We set

$$A_i := \sum_{j=1}^{N} a_{i,j} \frac{\partial}{\partial x_j} \quad (i = 1, \dots, N), \quad A_0 := \frac{1}{V} \sum_{i=1}^{N} \frac{\partial V}{\partial x_i} A_i; \tag{3}$$

then we argue in this way (for the details and the nomenclature, see [6]):

1. $A_1, \dots, A_N$ are tangent to the set $F(u) := \{x \in \Omega : u(x) = \max_\Omega u\}$, provided $\mathcal{L}u \geq 0$ (Hopf-type lemma);
2. A_0 is also tangent to $F(u)$ (it is a smooth combination of $A_1, \dots, A_N$);
3. the tangentiality of $A_0, A_1 \dots, A_N$ ensures the propagation of $F(u)$ along the integral curves of any linear combination of $A_0, A_1, \dots, A_N$ (Bony);
4. the C^∞-hypoellipticity of $\mathcal{L}$ ensures the controllability (hence, a suitable connectivity property) of the set $\operatorname{span}\{A_0, A_1, \dots, A_N\}$ (this is the point where we invoke Amano's crucial result).

These facts plainly prove that $F(u)$ is forced to coincide with Ω, as soon as $F(u)$ is non-void. As for (4), we review Amano's geometric result on hypoellipticity: once one has rewritten $\mathcal{L}$ in the following way

$$\mathcal{L} = \sum_{i=1}^{N} \frac{\partial}{\partial x_i}(A_i) + A_0,$$

then assumptions (NTD) and (HY) implies a connectivity property of any pair of points of the underlying domain Ω, along the linear combinations of $A_0, A_1, \dots, A_N$. See [4] for all the details. In order to prove our non-invariant Harnack inequality, we need a further assumption:

- **assumption (HY)$_\varepsilon$**: for every $\varepsilon > 0$, $\mathcal{L} - \varepsilon$ is C^∞-hypoelliptic on every open set in $\mathbb{R}^N$.

The problem of establishing, in general, whether or not our previous assumption (HY) implies (HY)$_\varepsilon$ is non-trivial; it appears that having some quantitative information on the loss of derivatives may help in facing this question (personal communication by Parmeggiani). Concerning this issue it is known that, for example, in the complex coefficient case the presence of a zero-order term (even a small ε) may drastically alter hypoellipticity (see Stein, [71]; see also the very recent paper [66] by Parmeggiani for related topics).

The role of (HY)$_\varepsilon$ is crucial in establishing a comparison argument (analogous to that in [27, Proposition 7.1, p. 298]) giving the lower bound

$$u(x_0) \geq \varepsilon \int_\Omega u(y)\, G_\varepsilon(x_0, y)\, V(y)\, \mathrm{d}y \qquad \forall\, x_0 \in \Omega, \tag{4}$$

for every nonnegative $\mathcal{L}$-harmonic function u on the open set Ω which possesses a Green kernel $G_\varepsilon(x, y)$ relative to the perturbed operator $\mathcal{L} - \varepsilon$.

With the aid of (4), with the use of the Green functions related to $\mathcal{L}$ and $\mathcal{L}_\varepsilon$, and by means of Theorem 1, in [6] we proved the following Harnack inequality:

Theorem 2. (Harnack Inequality). *Suppose that $\mathcal{L}$ is an operator of the form (1), with C^∞ coefficients $V > 0$ and $(a_{i,j}) \geq 0$, and suppose it satisfies hypotheses* (NTD), (HY) *and* (HY)$_\varepsilon$.

Then, for every connected open set $O \subseteq \mathbb{R}^N$ and every compact subset K of O, there exists a constant $M = M(\mathcal{L}, O, K) \geq 1$ such that

$$\sup_K u \leq M \inf_K u, \tag{5}$$

for every nonnegative $\mathcal{L}$-harmonic function u in O. If $\mathcal{L}$ is subelliptic or if it has C^ω coefficients, then assumption (HY)$_\varepsilon$ *can be dropped.*

We remark that the operators in Remark 1 are *not* subelliptic (nor C^ω), yet they satisfy hypotheses (NTD), (HY) and (HY)$_\varepsilon$. The lack of subellipticity is a consequence of the characterization of the subelliptic PDOs due to Fefferman and Phong [39,40] (see also [62, Prop.1.3] or [58, Th.2.1 and Prop.2.1]); the validity of (HY)$_\varepsilon$ derives from a result by Kohn, [62]. Thus our Theorem 2 is not a consequence of Bony's results in [27], nor of the results on Harnack inequalities for subelliptic PDOs cited after Example 1.

Technically, we obtained Theorem 2 from the following result (due to Mokobodzki and Brelot, [30, Chap. I]):

Lemma 1. *Let L be a second order linear PDO in $\mathbb{R}^N$ with smooth coefficients. Suppose the following conditions are satisfied.*
(Regularity)*: There exists a basis $\mathcal{B}$ for the Euclidean topology of $\mathbb{R}^N$ (consisting of bounded open sets) such that, for every $\Omega \in \mathcal{B} \setminus \{\varnothing\}$ and for every $\varphi \in C(\partial\Omega, \mathbb{R})$, there exists a unique L-harmonic function $H_\varphi^\Omega \in C^2(\Omega) \cap C(\overline{\Omega})$ solving the Dirichlet problem*

$$\begin{cases} Lu = 0 & \text{in}\, \Omega \\ u = \varphi & \text{on}\, \partial\Omega, \end{cases}$$

and satisfying $H_\varphi^\Omega \geq 0$ whenever $\varphi \geq 0$.
(Weak Harnack Inequality)*: For every connected open set $O \subseteq \mathbb{R}^N$, every compact subset K of O and every $y_0 \in O$, there exists a constant $C(y_0) = C(L, O, K, y_0) > 0$ such that*

$$\sup_K u \leq C(y_0)\, u(y_0),$$

for every nonnegative L-harmonic function u in O.

Then, the thesis of Theorem 2 holds for L.

When one takes $L = \mathcal{L}$, the validity of condition '(Regularity)' can be obtained as a consequence of the hypoellipticity of $\mathcal{L}$ (as in [27]); besides, the validity of condition '(Weak Harnack Inequality)' is a consequence of (4) plus some basic Functional Analysis.

Introducing the topics of the following section, we underline that the Harnack inequality (5) implies the following *Brelot convergence property*: every increasing sequence of $\mathcal{L}$-harmonic functions (on a connected open set) either converges to $\equiv \infty$ or to an $\mathcal{L}$-harmonic function. This is one of the most crucial axiom (and often one of the most difficult to prove) of the Axiomatic Potential Theory for $\mathcal{L}$ (see [26, Chap. 6]).

3 Further Potential Theory Results

A most convenient assumption for potential-theoretic problems for operators $\mathcal{L}$ as in (1) is that $\mathcal{L}$ possesses a smooth, global and positive *fundamental solution*: see the series of papers [1,7,24,25]. For the use of the fundamental solution in obtaining the Harnack inequality for Hörmander sums of squares, see: Citti, Garofalo and Lanconelli [35]; Garofalo and Lanconelli [48,49]; Pascucci and Polidoro [67,68]; Bramanti, Brandolini, Lanconelli and Uguzzoni [29].

First of all we clarify what we mean by a fundamental solution: for the (formally) self-adjoint operator $\mathcal{L}$ as in (1), we assume the existence of a *(global) fundamental solution*, that is a function

$$\Gamma : \{(x,y) \in \mathbb{R}^N \times \mathbb{R}^N : x \neq y\} \longrightarrow \mathbb{R} \qquad \text{such that:}$$

1. for every fixed $x \in \mathbb{R}^N$ (the so-called 'pole'), $\Gamma_x := \Gamma(x, \cdot)$ belongs to $L^1_{\text{loc}}(\mathbb{R}^N)$ and it solves $-\mathcal{L}\Gamma_x = \text{Dirac}_x$, i.e.,
$$\int_{\mathbb{R}^N} \Gamma(x,y)\, \mathcal{L}\varphi(y)\, \mathrm{d}\mu(y) = -\varphi(x), \quad \forall\, \varphi \in C_0^\infty(\mathbb{R}^N).$$
2. for every fixed $x \in \mathbb{R}^N$ one has $\lim_{|y|\to\infty} \Gamma(x,y) = 0$.

For geometrical purposes it may occasionally be convenient to further require that:

$$\lim_{y \to x} \Gamma(x,y) = \infty.$$

This "blow-up" pole-condition is not part of our definition of a global fundamental solution, but will be required only if need be. A fundamental solution for $\mathcal{L}$, if it exists, is unique (a.e.): this follows from the Hörmander hypoellipticity theorem and from the Strong Maximum Principle. Following [35], we define the Γ-ball of centre x and radius r:

$$\Omega_r(x) := \big\{ y \in \mathbb{R}^N \,:\, \Gamma_x(y) > 1/r \big\}.$$

Under the blow-up pole-condition, $\{\Omega_r(x)\}_r$ forms a basis of neighborhoods (as $r \downarrow 0$), invading $\mathbb{R}^N$ (as $r \uparrow \infty$).

The Divergence Theorem and the V-divergence form of $\mathcal{L}$ imply the following *surface mean-integral formula*: whenever u is C^2 on $\overline{\Omega(x,r)}$ it holds that

$$u(x) = m_r(u)(x) - \int_{\Omega_r(x)} \Big(\Gamma_x(y) - \frac{1}{r} \Big)\, \mathcal{L}u(y)\, \mathrm{d}\mu(y),$$

where

$$m_r(u)(x) := \int_{\partial\Omega_r(x)} u\,\mathcal{K}_x\, V \mathrm{d}H^{N-1},$$

with the kernel $\mathcal{K}_x$ being defined on $\mathbb{R}^N \setminus \{x\}$ by

$$\mathcal{K}_x := \frac{\langle A\nabla\Gamma_x, \nabla\Gamma_x\rangle}{|\nabla\Gamma_x|}.$$

Unlike the classic Laplacian case, the kernel $\mathcal{K}_x(y)$ may vanish: this is indeed the case with the first Heisenberg group, where the associated $y \mapsto \mathcal{K}_x(y)$ vanishes precisely on the line $\{y_1 = x_1,\ y_2 = x_2\}$.

In general, since $\mathcal{K}_x(y) = 0$ if and only if $\langle A(y)\nabla\Gamma_x(y), \nabla\Gamma_x(y)\rangle = 0$, and the latter holds true if and only if $A(y)\nabla\Gamma_x(y) = 0$, we see that $\mathcal{K}_x(y) = 0$ precisely when

$$A_1\Gamma_x(y) = \cdots = A_N\Gamma_x(y) = 0,$$

with the notation in (3). Thus, by Amano's controllability result [4], we infer that

$$\mathcal{K}_x(y) > 0 \text{ on an open dense set in } \mathbb{R}^N.$$

Arguing as in [1], this last fact can lead to an alternative proof of the Strong Maximum Principle for an $\mathcal{L}$ endowed with a global fundamental solution. Indeed, if $\mathcal{L}u \geq 0$ and if $x_0 \in \Omega$ is a maximum point of u, we have

$$u(x_0) \leq m_r(u)(x_0) = \int_{\partial\Omega_r(x_0)} u\,\mathcal{K}_{x_0}\,\mathrm{d}H^{N-1} \leq u(x_0).$$

This immediately gives

$$0 = \int_{\partial\Omega_r(x_0)} (u(x_0) - u)\,\mathcal{K}_{x_0}\,\mathrm{d}H^{N-1};$$

if we integrate with respect to $r \in [0, R]$ and we use co-area formula, we get

$$0 = \int_{\Omega_R(x_0)} (u(x_0) - u)\,\langle A\nabla\Gamma_{x_0}, \nabla\Gamma_{x_0}\rangle \mathrm{d}\mu.$$

Since $u(x_0) - u \geq 0$ and $\langle A\nabla\Gamma_{x_0}, \nabla\Gamma_{x_0}\rangle > 0$ in an open dense set, we see that u is constantly equal to $u(x_0)$ in $\Omega_R(x_0)$, so that u is constant throughout Ω (which is connected).

The applications of mean-value formulas to Potential Theory are countless; among them we mention the applications to:

- Maximum Principles, [1];
- Gauss&Koebe-type Theorems (characterization of the $\mathcal{L}$-harmonicity), [7];
- Montel-type Theorems (concerning the topology of the $\mathcal{L}$-harmonic functions), [7];
- many possible different characterizations of the $\mathcal{L}$-subharmonicity, [24,25].

When sub-Laplacians on Carnot groups are involved, we also have applications to:

- Harnack inequalities and Liouville-type theorems, [21];
- Maximum Principles on unbounded domains, [22];
- Bôcher-type Theorems (on the removability of singularities), [23].

The vast applicability of mean-value formulas (deriving from the sole existence of Γ) makes it (very!) desirable to enlarge the class of PDOs for which the existence of a global fundamental solution (with the above well-behaved properties) is known. See Sect. 5.

4 An Invariant Harnack Inequality

The aim of this section is to describe how we could obtain an *invariant and non-homogeneous* Harnack inequality for PDOs $\mathcal{L}$ of the form (1) but under different assumptions than those in Sect. 2. For instance, we drop our previous hypoellipticity assumptions (HY) and $(\mathrm{HY})_\varepsilon$, in favor of more geometrical hypotheses. We shall skip any detail here, referring the interested reader to [8].

As we already discussed, a convenient framework to prove an invariant Harnack inequality for (1) is that of doubling metric spaces; hence, we suppose that the quadratic form of $\mathcal{L}$ can be controlled by a family of locally-Lipschitz vector fields $X = \{X_1, \ldots, X_m\}$ on $\mathbb{R}^N$, in the sense the X-elliptic operators introduced by Kogoj and Lanconelli in [60]: this means that there exist two constants $\lambda, \Lambda > 0$ such that

$$\lambda \sum_{j=1}^{m} \langle X_j(x), \xi\rangle^2 \leq \sum_{i,j=1}^{N} a_{i,j}(x)\, \xi_i \xi_j \leq \Lambda \sum_{j=1}^{m} \langle X_j(x), \xi\rangle^2, \qquad \forall\, x, \xi \in \mathbb{R}^N.$$

Moreover, we assume that, with the associated Carnot-Carathéodory metric d, N-dimensional Euclidean space is endowed by d with the structure of a doubling space (globally) and a Poincaré inequality holds true on any d-ball in space (this can be seen as another 'global' assumption). Here the doubling measure is again the measure μ introduced in (2), naturally associated with $\mathcal{L}$. These assumptions read as follows:

- **assumption (D)**: there exists $Q > 0$ such that

$$\mu(B(x, 2\,r)) \leq 2^Q\, \mu(B(x, r)), \qquad \text{for every}\, x \in \mathbb{R}^N \text{and every}\, r > 0;$$

- **assumption (P)**: there exists a constant $C_P > 0$ such that, for every $x \in \mathbb{R}^N$, $r > 0$ and every u which is C^1 in a neighborhood of $B(x, 2r)$ one has

$$\fint_{B(x,r)} \Big| u(y) - u_{B(x,r)} \Big|\, \mathrm{d}\mu(y) \leq C_P\, r \fint_{B(x,2r)} \Big| Xu(y) \Big|\, \mathrm{d}\mu(y).$$

Here we used the following notations: $|Xu| := \sqrt{\sum_{j=1}^m |X_j u|^2}$ and

$$\fint_{B(x,r)} \{\cdots\}\,\mathrm{d}\mu = \frac{1}{\mu(B(x,r))}\int_{B(x,r)} \{\cdots\}\,\mathrm{d}\mu, \qquad u_{B(x,r)} := \fint_{B(x,r)} u(y)\,\mathrm{d}\mu(y).$$

Under these assumptions (plus some further mild technical hypotheses on the d-topology[1]), one can derive Poincaré-Sobolev-type inequalities on the d-balls, following the ideas by Hajłasz and Koskela in [55].

Since we are interested in relating the global doubling condition and the global Poincaré inequality to the theory of subelliptic operators, it is natural to work in the setting of the Sobolev spaces associated with the family of vector fields X with respect to the doubling measure μ. Accordingly, we take for granted the associated notion of weak solution to $-\mathcal{L}u = g$ (when g is in L^p).

Thanks to a density result of Meyers-Serrin type (see [44,47]), one can consider the Sobolev space $W^{1,2}(\Omega, X, \mu)$ as the closure of the space of the smooth functions in the norm of $W^{1,2}$, similarly to the classic case. This result, together with some Sobolev-type inequality on the d-balls (see also [20,46,50,55]), is an important tool in order to obtain an adaptation of the well-established *Moser iterative technique*; as is well known, this powerful technique permits to obtain the following invariant and non-homogeneous Harnack inequality for $\mathcal{L}$ (with the consequent inner and boundary Hölder estimates):

Theorem 3. (Non-homogeneous, invariant Harnack inequality) *Let the assumptions (D) and (P) be satisfied for $\mathcal{L}$ and for the doubling metric space $(\mathbb{R}^N, d, \mu)$. Let $\Omega \subseteq \mathbb{R}^N$ be an open set, and let $g \in L^p(\Omega)$, with $p > Q/2$.*

Then there exists a structural constant $C > 0$ (only depending on the doubling/Poincaré constants Q, C_P, on the X-ellipticity constants λ, Λ and on p) such that, for every d-ball $B(x,R)$ satisfying $\overline{B(x,4R)} \subset \Omega$, one has

$$\sup_{B(x,R)} u \le C\left(\inf_{B(x,R)} u + R^2 \|g\|^*_{L^p(B(x,4R))}\right), \tag{6}$$

for any nonnegative W^1_{loc}-weak solution u of $-\mathcal{L}u = g$ in Ω.

Here we used the notation

$$\|g\|^*_{L^p(A)} := \left(\fint_A |g|^p\,\mathrm{d}\mu\right)^{1/p}.$$

In the particular case when $g \equiv 0$, from Theorem 3 one obtains the homogeneous, invariant Harnack inequalities obtained by Kogoj and Lanconelli in [60,61]. Again in the homogeneous case $g \equiv 0$, an invariant Harnack inequality under *local* doubling/Poincaré has been proved by Gutiérrez and Lanconelli in [53], for balls of *small* radii. In the same paper [53], the authors obtain a non-homogeneous invariant Harnack inequality, under the presence of some dilation-invariance property on the vector fields X involved.

[1] We assume that the metric space $(\mathbb{R}^N, d)$ is complete and its topology coincides with the Euclidean one.

The summand $R^2 \|g\|^*_{L^p(B(x,4R))}$ is bounded from above by

$$\frac{R^2}{\mu(B(x,R))^{1/p}}\, \|g\|_{L^p(\Omega)};$$

when R is small and x lies in a compact set $K \subset \Omega$, there exists a constant $C(Q,K) > 0$ such that (due to the doubling inequality (D)) the latter does not exceed

$$C(Q,K)\, R^{2-Q/p} \|g\|_{L^p(\Omega)}.$$

Thus, our inequality (6) contains the analogous non-homogeneous, invariant Harnack inequality by Uguzzoni in [72], where it is considered the particular case when x is confined in some compact set $K \subset \Omega$ and R is very small. Roughly put, these more restrictive requirements on the d-balls involved are the drawback of the local doubling/Poincaré assumptions made in [72]; in the same spirit, our *global* inequality (6) (in the sense that it holds for any d-ball) is the product of our more onerous assumptions that doubling and Poincaré inequalities hold *globally* for any d-ball.

In a future investigation, we intend to apply Theorem 3 in order to prove the existence of a global fundamental solution for our X-elliptic operators $\mathcal{L}$: this will require a deep insight of the Green function related to $\mathcal{L}$, plus the maximum principles by Gutiérrez and Lanconelli in [53].

5 The Existence of the Fundamental Solution for a Class of PDOs

In order to explain the main idea behind our technique for obtaining a fundamental solution for a (selected) class od PDOs, we describe an example. Let $\Delta_n = \sum_{j=1}^n (\partial_{x_j})^2$ be the classic Laplace operator on $\mathbb{R}^n$; then we may say that Δ_{n+p} is a 'lifting' of Δ_n (for every $p \geq 1$) in the sense that

$$\Delta_{n+p}(f \circ \pi) = (\Delta_n f) \circ \pi \quad \text{for every } f \in C^\infty(\mathbb{R}^n, \mathbb{R}),$$

where $\pi : \mathbb{R}^{n+p} \to \mathbb{R}^n$ is the projection of $\mathbb{R}^{n+p}$ onto $\mathbb{R}^n$. If $n > 2$, the fundamental solution of Δ_n (a constant multiple of $\|x\|^{2-n}$) can be recovered by that of Δ_{n+p} by a process that we may call 'saturation', in the sense that we integrate over the remaining p coordinates: this comes from the integral identity[2]

$$c\left(\sqrt{x_1^2 + \cdots + x_n^2}\right)^{2-n} = \int_{\mathbb{R}^p} \left(\sqrt{x_1^2 + \cdots + x_n^2 + t_1^2 + \cdots + t_p^2}\right)^{2-n-p} \mathrm{d}t_1 \ldots \mathrm{d}t_p,$$

holding true for some constant c, whenever $n > 2$. Incidentally, the above integral is convergent precisely when $n > 2$. We aim to obtain an analogous process of lifting/saturation for other classes of PDOs.

[2] The identity follows from the change of variable $t = \|x\|\, \tau$ (with $\tau \in \mathbb{R}^p$).

To this end, we consider a set $\mathcal{X} = \{X_1, \dots, X_m\}$ of linearly independent[3] smooth vector fields on $\mathbb{R}^n$ satisfying the following assumptions:

(H1) $X_1, \dots, X_m$ are homogeneous of degree 1 with respect to the non-isotropic dilations

$$\delta_\lambda(x) = (\lambda^{\sigma_1} x_1, \dots, \lambda^{\sigma_n} x_n),$$

where $1 = \sigma_1 \leq \ldots \leq \sigma_n$ and $q := \sum_{j=1}^n \sigma_j > 2$ (the so-called homogeneous dimension related to δ_λ); this means that (for any $i = 1, \dots, m$)

$$X_i(f \circ \delta_\lambda) = \lambda\,(X_i f) \circ \delta_\lambda \qquad \forall\,\lambda > 0, \quad \forall\, f \in C^\infty(\mathbb{R}^n, \mathbb{R}).$$

(H2) $X_1, \dots, X_m$ satisfy the Hörmander rank condition:

$$\dim\big\{X(x) : X \in \mathrm{Lie}\{X_1, \dots, X_m\}\big\} = n, \quad \text{for every}\, x \in \mathbb{R}^n.$$

We then introduce the operator $\mathcal{L}$ naturally associated with $\mathcal{X}$, that is,

$$\mathcal{L} = \sum_{j=1}^m X_j^2. \tag{7}$$

Example 2. Let us consider, on Euclidean space $\mathbb{R}^2$, the Grushin vector fields:

$$X_1 = \partial_{x_1}, \quad X_2 = x_1\, \partial_{x_2}.$$

Clearly, X_1, X_2 are independent (as differential operators[4], even if not pointwise independent) and smooth; they are homogeneous of degree 1 with respect to $\delta_\lambda(x) = (\lambda x_1, \lambda^2 x_2)$; they satisfy the Hörmander rank condition. A lifting of $\mathcal{G} := X_1^2 + X_2^2$ is given by the PDO on $\mathbb{R}^3$ defined by

$$\widetilde{\mathcal{G}} := (\partial_{x_1})^2 + (\partial_{x_3} + x_1\, \partial_{x_2})^2.$$

The idea of obtaining a fundamental solution for the Grushin operator $\mathcal{G}$ via a saturation argument applied to the (explicit!) fundamental solution of $\widetilde{\mathcal{G}}$ has already been exploited in the literature (e.g., in Bauer, Furutani, Iwasaki [9]; see also Calin, Chang, Furutani, Iwasaki [31]; for more general PDOs on 2-step Carnot groups, see Beals, Gaveau, Greiner, Kannai [14]). Other references related to the (difficult) problem of obtaining explicit/integrally-represented fundamental solutions are the following ones: [2,10–13,15,18,19,28,32,56]. See [16] for a wider list.

[3] Throughout, we consider the set of the smooth vector fields on $\mathbb{R}^N$ as a real vector space and not as a C^∞-module; therefore, the vector fields $X_1, \dots, X_m$ are linearly dependent iff there exist $c_1, \dots, c_m \in \mathbb{R}$, not all vanishing, such that $c_1 X_1 + \cdots + c_m X_m = 0$ as a first order linear PDO (i.e., all of its coefficient functions with respect to the usual coordinate partial derivatives are identically equal to 0).

[4] This is not in contrast with the fact their coefficient vectors $(1, 0)$ and $(0, x_1)$ are linearly dependent at $x_1 = 0$; see the previous footnote.

We intend to show that a process of lifting/saturation can always be performed for our PDO $\mathcal{L}$ in (7), when (H1) and (H2) are satisfied. Indeed, $\mathcal{L}$ admits a lifting which is a sub-Laplacian $\mathcal{L}_{\mathbb{G}}$ on some Carnot group $\mathbb{G}$; most importantly, since $\mathcal{L}_{\mathbb{G}}$ admits a global fundamental solution $\Gamma_{\mathbb{G}}$ (see Folland [42]), we shall show that a global fundamental solution Γ for $\mathcal{L}$ can always be obtained by a saturation argument from $\Gamma_{\mathbb{G}}$. This gives the existence of Γ, together with an integral representation for it.

We describe more closely our procedure. The lifting method introduced by Rothschild and Stein [69] (which is, roughly put, a local fact) does not seem to be applicable. What is suited to our case is the *global* lifting for homogeneous operators proved by Folland in [43], which we now describe. We set $\mathfrak{a} := \mathrm{Lie}\{X_1, \dots, X_m\}$. Then $\mathfrak{a}$ is nilpotent and stratified:

$$\mathfrak{a} = \mathfrak{a}_1 \oplus \mathfrak{a}_2 \oplus \cdots \oplus \mathfrak{a}_r,$$

with $[\mathfrak{a}_1, \mathfrak{a}_{i-1}] = \mathfrak{a}_i$ if $2 \le i \le r = \sigma_n$, and $[\mathfrak{a}_1, \mathfrak{a}_r] = \{0\}$, where we have also set $\mathfrak{a}_1 := \mathrm{span}\{X_1, \dots, X_m\}$. We equip $\mathfrak{a}$ with the *Campbell-Hausdorff* operation

$$X \diamond Y = X + Y + \frac{1}{2}[X, Y] + \frac{1}{12}[X, [X, Y]] - \frac{1}{12}[Y, [X, Y]] + \cdots,$$

and with the family of dilations

$$\Delta_\lambda(X) = \sum_{k=1}^{r} \lambda^k V_k, \quad \text{where} X = V_1 + \cdots + V_r \text{ and } V_k \in \mathfrak{a}_k.$$

Folland's theorem ensures that the triple $\mathbb{A} := (\mathfrak{a}, \diamond, \Delta_\lambda)$ is a homogeneous Carnot group, in the sense of [26]. Moreover, setting[5]

$$\pi : \mathbb{A} \longrightarrow \mathbb{R}^n, \quad \pi(X) := \exp(X)(0),$$

the following properties hold true:

- π is surjective;
- there exists a system $\{J_1, \dots, J_m\}$ of Lie-generators of $\mathrm{Lie}(\mathbb{A})$ such that

$$\mathrm{d}_a \pi(J_i) = (X_i)_{\pi(a)}, \quad a \in \mathbb{A} \quad \text{and} \quad i = 1, \dots, m.$$

In [16] we proved that something more can be done: we perform a suitable change of variable $T : \mathbb{R}^N \equiv \mathbb{A} \to \mathbb{R}^N$ turning π into the canonical projection $\pi : \mathbb{R}^n \times \mathbb{R}^p \to \mathbb{R}^n$,

$$\pi(x_1, \dots, x_n, \xi_1, \dots, \xi_p) = (x_1, \dots, x_n),$$

[5] Here $\exp(tX)(x)$ denotes the solution $t \mapsto \gamma(t)$ of the Cauchy problem

$$\dot{\gamma}(t) = X(\gamma(t)), \quad \gamma(0) = x.$$

where $p = N - n = \dim(\mathrm{Lie}\{X_1, \dots, X_m\}) - n$; this gives an explicit knowledge of the fibers of π (all equal to $\mathbb{R}^p$).

Putting together Folland's lifting and our change of variable, we infer that there exists a homogeneous Carnot group $\mathbb{G} = \big(\mathbb{R}^N, \star, D^\star_\lambda\big)$ and there exists a system $\{Z_1, \dots, Z_m\}$ of generators of Lie($\mathbb{G}$) such that

$$Z_i = X_i + \sum_{j=1}^{p} r_{i,j}(x, \xi) \frac{\partial}{\partial \xi_j}, \quad \text{for every } i = 1, \dots, m.$$

We also have

$$D^\star_\lambda(x, \xi) = \big(\delta_\lambda(x), \delta^*_\lambda(\xi)\big) = \big(\delta_\lambda(x), \lambda^{s_1}\xi_1, \dots, \lambda^{s_p}\xi_p\big).$$

Due to the mentioned result by Folland, the sub-Laplacian $\mathcal{L}_\mathbb{G} := \sum_{j=1}^m Z_j^2$ has a unique global fundamental solution Γ of the form

$$\Gamma_\mathbb{G}\big((x,\xi),(y,\eta)\big) = d^{2-q-\widetilde{q}}\left((x,\xi)^{-1} \star (y,\eta)\right), \quad (x,\xi) \neq (y,\eta),$$

where d is a suitable homogeneous norm on $\mathbb{G}$ and $\widetilde{q} = \sum_{j=1}^p s_j$ is the homogeneous dimension of $\mathbb{R}^p$ with respect to δ^*_λ.

Now we can perform the announced saturation argument: it can be proved that the following is a fundamental solution for $\mathcal{L}$:

$$\Gamma(x,y) := \int_{\mathbb{R}^p} d^{2-q-\widetilde{q}}\left((x,0)^{-1} \star (y,\eta)\right) \mathrm{d}\eta \qquad (x \neq y).$$

To prove this, the main issue is to show that, for every fixed $x, y \in \mathbb{R}^n$ with $x \neq y$, one has

$$(\mathrm{S}): \quad \mathbb{R}^p \ni \eta \mapsto d^{2-q-\widetilde{q}}\left((x,0)^{-1} \star (y,\eta)\right) \text{ is in } L^1(\mathbb{R}^p).$$

Some key facts for the proof of property (S) are the following:

- for every $x, y \in \mathbb{R}^n$ with $x \neq y$, we have

$$\Gamma_\mathbb{G} = d^{2-q-\widetilde{q}}\left((x,0)^{-1} \star (y,\eta)\right) \in L^1_{\mathrm{loc}}(\mathbb{R}^p);$$

- there exists a real $\mathbf{c} > 0$ such that

$$d^{2-q-\widehat{q}}(z,\zeta) \le \mathbf{c} \left(\sum_{i=1}^{p} |\zeta_i|^{1/s_i}\right)^{2-q-\widetilde{q}} =: \varphi(\zeta), \quad \text{for all } (z,\zeta) \in \mathbb{R}^N;$$

- via a suitable change of variable, $\Gamma_\mathbb{G}\big((x,0)^{-1}\star(y,\eta)\big)$ is η-integrable at infinity if the same is true of φ;
- φ is integrable at infinity (by a homogeneous argument on annuli).

Once (S) is established, it is not too difficult to recognize that Γ is a fundamental solution for $\mathcal{L}$, further satisfying the following properties:

1. setting $\mathcal{O} := \{(x, y) \in \mathbb{R}^n \times \mathbb{R}^n : x \neq y\}$, then $\Gamma \in C^\infty(\mathcal{O}, \mathbb{R})$;
2. $\Gamma(x, y) > 0$ on $\mathcal{O}$;
3. $\Gamma(x, y) = \Gamma(y, x)$ for every $x \neq y$;
4. $\Gamma \in L^1_{\text{loc}}(\mathbb{R}^n \times \mathbb{R}^n)$;
5. $\Gamma\big(\delta_\lambda(x), \delta_\lambda(y)\big) = \lambda^{2-q}\, \Gamma(x, y)$.

In the forthcoming paper [17], we shall consider *the parabolic case* as well. More precisely, we consider once again a set $\mathcal{X} = \{X_1, \ldots, X_m\}$ of linearly independent smooth vector fields on $\mathbb{R}^n$ satisfying the above assumptions (H1) and (H2). We define

$$\mathcal{H} := \mathcal{L} - \partial_t = \sum_{j=1}^{m} X_j^2 - \partial_t, \quad \text{on } \mathbb{R}^{1+n} = \mathbb{R}_t \times \mathbb{R}^n_x.$$

We prove the existence of a "well-behaved" global fundamental solution (usually referred to as a heat kernel) for the parabolic PDO $\mathcal{H}$. A lifting/saturation approach can be performed. Indeed, let $Z_1, \ldots, Z_m$ be as above; then the operator

$$\mathcal{H}_{\mathbb{G}} := \mathcal{L}_{\mathbb{G}} - \partial_t = \sum_{j=1}^{m} Z_j^2 - \partial_t, \quad \text{on } \mathbb{R}^{1+N} = \mathbb{R}_t \times \mathbb{R}^N_{(x,\xi)}$$

is a lifting of $\mathcal{H}$ on $\mathbb{R} \times \mathbb{G} \equiv \mathbb{R}^{1+N}$. As regards the operator $\mathcal{H}_{\mathbb{G}}$, it is well-know that there exists a smooth map $\gamma_{\mathbb{G}} : \mathbb{R}^{1+N} \setminus \{0\} \to \mathbb{R}$ such that the function

$$\Gamma_{\mathbb{G}}(t, x, \xi; s, y, \eta) := \gamma_{\mathbb{G}}\big((s - t, (x, \xi)^{-1} \star (y, \eta)\big), \quad (t, x, \xi) \neq (s, y, \eta)$$

is a global fundamental solution for $\mathcal{H}_{\mathbb{G}}$ (here, $\star$ is the composition law of $\mathbb{G}$). Since $\mathcal{H}_{\mathbb{G}}$ is a lifting for $\mathcal{H}$, it is natural to define

$$\Gamma(t, x; s, y) := \int_{\mathbb{R}^p} \gamma_{\mathbb{G}}\big(s - t, (x, 0)^{-1} \star (y, \eta)\big)\, d\eta.$$

The main issue here is to prove that, for every fixed $(t, x) \neq (s, y) \in \mathbb{R}^{1+n}$, one has

$$\text{(S)':} \quad \mathbb{R}^p \ni \eta \mapsto \gamma_{\mathbb{G}}\big(s - t, (x, 0)^{-1} \star (y, \eta)\big) \text{ is in } L^1(\mathbb{R}^p).$$

Key tools for proving property (S)' are:

- the existence of a constant $\mathbf{c} > 0$ such that, for every $(x, \xi) \in \mathbb{R}^N$ and every $t > 0$, the following uniform Gaussian estimates hold true:

$$\frac{1}{\mathbf{c}\, t^{Q/2}} \exp\left(-\frac{\mathbf{c}\, d^2(x, \xi)}{t}\right) \leq \gamma_{\mathbb{G}}(t, x, \xi) \leq \frac{\mathbf{c}}{t^{Q/2}} \exp\left(-\frac{d^2(x, \xi)}{\mathbf{c}\, t}\right);$$

  here, d is the same homogeneous norm introduced in the stationary case;
- the estimates for d exploited in the stationary case.

Once (S)' is established, one can show that the function Γ is well-defined and is a global fundamental solution for $\mathcal{H}$, further satisfying the following properties:

1. $\Gamma \geq 0$ and $\Gamma(t, x; s, y) = 0$ if and only if $s \leq t$;
2. Γ depends on t and s only through the difference $s - t$; moreover, Γ is symmetric with respect to the space variables;
3. $\Gamma(\lambda^2 t, \delta_\lambda(x); \lambda^2 s, \delta_\lambda(y)) = \lambda^{-q}\, \Gamma(t, x; s, y)$;
4. Γ is smooth out of the diagonal of $\mathbb{R}^{1+n} \times \mathbb{R}^{1+n}$;
5. $\Gamma \in L^1_{\mathrm{loc}}(\mathbb{R}^{1+n} \times \mathbb{R}^{1+n})$;
6. for every fixed $(t, x) \in \mathbb{R}^{1+n}$ we have

$$\int_{\mathbb{R}^n} \Gamma(t, x; s, y)\, \mathrm{d}y = 1, \quad \text{for every } s > t;$$

Acknowledgments. The results of this paper were presented by the author at the Conference "Noncommutative Analysis and Partial Differential Equations", 11–15 April, 2016, Imperial College, London; the author wishes to express his gratitude to the Organizing Committee of the Conference for the hospitality.

References

1. Abbondanza, B., Bonfiglioli, A.: On the Dirichlet problem and the inverse mean value theorem for a class of divergence form operators. J. Lond. Math. Soc. 1–26 (2012). https://doi.org/10.1112/jlms/jds050.
2. Agrachev, A., Boscain, U., Gauthier, J.-P., Rossi, F.: The intrinsic hypoelliptic Laplacian and its heat kernel on unimodular Lie groups. J. Funct. Anal. **256**, 2621–2655 (2009)
3. Aimar, H., Forzani, L., Toledano, R.: Hölder regularity of solutions of PDE's: a geometrical view. Commun. Partial Differ. Equ. **26**, 1145–1173 (2001)
4. Amano, K.: A necessary condition for hypoellipticity of degenerate elliptic-parabolic operators. Tokyo J. Math. **2**, 111–120 (1979)
5. Barlow, M.T., Bass, R.F.: Stability of parabolic Harnack inequalities. Trans. Am. Math. Soc. **356**, 1501–1533 (2004)
6. Battaglia, E., Biagi, S., Bonfiglioli, A.: The strong maximum principle and the Harnack inequality for a class of hypoelliptic non-Hörmander operators. Ann. Inst. Fourier (Grenoble) **66**(2), 589–631 (2016)
7. Battaglia, E., Bonfiglioli, A.: Normal families of functions for subelliptic operators and the theorems of Montel and Koebe. J. Math. Anal. Appl. **409**, 1–12 (2014)
8. Battaglia, E., Bonfiglioli, A.: An invariant Harnack inequality for a class of subelliptic operators under global doubling and Poincaré assumptions, and applications. J. Math. Anal. Appl. **460**, 302–320 (2018)
9. Bauer, W., Furutani, K., Iwasaki, C.: Fundamental solution of a higher step Grushin type operator. Adv. Math. **271**, 188–234 (2015)
10. Beals, R.: Exact fundamental solutions, "Equations aux D'eriv'ees Partielles" (Saint-Jean-de-Monts, 1998). University of Nantes, Nantes. Exp. No. I, pp. 9 (1998)
11. Beals, R., Gaveau, B., Greiner, P.: On a geometric formula for the fundamental solution of subelliptic Laplacians. Math. Nachr. **181**, 81–163 (1996)
12. Beals, R., Gaveau, B., Greiner, P.: The green function of model step two hypoelliptic operators and the analysis of certain tangential Cauchy Riemann complexes. Adv. Math. **121**(2), 288–345 (1996)

13. Beals, R., Gaveau, B., Greiner, P., Kannai, Y.: Exact fundamental solutions for a class of degenerate elliptic operators. Commun. Partial Differ. Equ. **24**(3–4), 719–742 (1999)
14. Beals, R., Gaveau, B., Greiner, P., Kannai, Y.: Transversally elliptic operators. Bull. Sci. Math. **128**(7), 531–576 (2004)
15. Beals, R., Greiner, P., Gaveau, B.: Fundamental solutions for some highly degenerate elliptic operators. J. Funct. Anal. **165**(2), 407–429 (1999)
16. Biagi, S., Bonfiglioli, A.: The existence of a global fundamental solution for homogeneous Hörmander operators via a global lifting method. Proc. Lond. Math. Soc. **114**, 855–889 (2017)
17. Biagi, S., Bonfiglioli, A.: Global heat kernels for homogeneous Hörmander operators and applications, in preparation (2017)
18. Bieske, T.: Fundamental solutions to the p-Laplace equation in a class of Grushin vector fields. Electron. J. Differ. Equ. **84**, 1–10 (2011)
19. Bieske, T., Gong, J.: The p-Laplace equation on a class of Grushin-type spaces. Proc. Am. Math. Soc. **134**, 3585–3594 (2006)
20. Biroli, M., Mosco, U.: Sobolev inequalities on homogeneous spaces. Potential theory and degenerate partial differential operators (Parma). Potential Anal. **4**, 311–324 (1995)
21. Bonfiglioli, A., Lanconelli, E.: Liouville-type theorems for real sub-Laplacians. Manuscr. Math. **105**, 111–124 (2001)
22. Bonfiglioli, A., Lanconelli, E.: Subharmonic functions on Carnot groups. Math. Ann. **325**, 97–122 (2003)
23. Bonfiglioli, A., Lanconelli, E.: Gauge functions, Eikonal equations and Bôcher's theorem on stratified Lie groups. Calc. Var. Partial Differ. Equ. **30**, 277–291 (2007)
24. Bonfiglioli, A., Lanconelli, E.: Subharmonic functions in sub-Riemannian settings. J. Eur. Math. Soc. **15**, 387–441 (2013)
25. Bonfiglioli, A., Lanconelli, E., Tommasoli, A.: Convexity of average operators for subsolutions to subelliptic equations. Anal. PDE **7**, 345–373 (2014)
26. Bonfiglioli, A., Lanconelli, E., Uguzzoni, F.: Stratified Lie groups and potential theory for their sub-Laplacians. Springer Monographs in Mathematics. Springer, Berlin (2007)
27. Bony, J.-M.: Principe du maximum, inégalité de Harnack et unicité du problème de Cauchy pour les opérateurs elliptiques dégénérés. Ann. Inst. Fourier (Grenoble) **19**, 277–304 (1969)
28. Boscain, U., Gauthier, J.-P., Rossi, F.: The hypoelliptic heat kernel over three-step nilpotent Lie groups. Sovrem. Mat. Fundam. Napravl. **42**, 48–61 (2011) (Transl. in: J. Math. Sci. (N.Y.) **199**(6), 614–628, 2014)
29. Bramanti, M., Brandolini, L., Lanconelli, E., Uguzzoni, F.: Non-divergence equations structured on Hörmander vector fields: heat kernels and Harnack inequalities. Mem. Am. Math. Soc. **204**(961) (2010)
30. Brelot, M.: Axiomatique des fonctions harmoniques, Séminaire de Mathématiques Supérieures, 14 (té, 1965). Les Presses de l'Université de Montréal, Montréal (1969)
31. Calin, O., Chang, D.-C., Furutani, K., Iwasaki, C.: Heat kernels for elliptic and sub-elliptic operators. Methods and techniques. Applied and Numerical Harmonic Analysis. Birkhäuser/Springer, New York (2011)
32. Capogna, L., Danielli, D., Garofalo, N.: Capacitary estimates and subelliptic equations. Am. J. Math. **118**, 1153–1196 (1996)
33. Chanillo, S., Wheeden, R.L.: Harnack's inequality and mean-value inequalities for solutions of degenerate elliptic equations. Commun. Partial Differ. Equ. **11**, 1111–1134 (1986)

34. Christ, M.: Hypoellipticity in the infinitely degenerate regime. In: J.D. McNeal (ed.) Complex Analysis and Geometry, Ohio State University Mathematical Research Institute Publication, vol. 9, pp. 59–84. Walter de Gruyter, Berlin (2001)
35. Citti, G., Garofalo, N., Lanconelli, E.: Harnack's inequality for sum of squares of vector fields plus a potential. Am. J. Math. **115**, 699–734 (1993)
36. Di Fazio, G., Gutiérrez, C.E., Lanconelli, E.: Covering theorems, inequalities on metric spaces and applications to PDE's. Math. Ann. **341**, 255–291 (2008)
37. Fabes, E.B., Jerison, D., Kenig, C.E.: The Wiener test for degenerate elliptic equations. Ann. Inst. Fourier (Grenoble) **32**, 151–182 (1982)
38. Fabes, E.B., Kenig, C.E., Serapioni, R.P.: The local regularity of solutions of degenerate elliptic equations. Commun. Partial Differ. Equ. **7**, 77–116 (1982)
39. Fefferman, C., Phong, D.H.: Subelliptic eigenvalue problems, Wadsworth Math. Ser. (Chicago, III., 1981), Wadsworth, Belmont, CA, pp. 590–606 (1983)
40. Fefferman, C., Phong, D.H.: The uncertainty principle and sharp Garding inequalities. Commun. Pure Appl. Math. **34**, 285–331 (1981)
41. Fediĭ, V.S.: On a criterion for hypoellipticity. Math. USSR Sb. **14**, 15–45 (1971)
42. Folland, G.B.: Subelliptic estimates and function spaces on nilpotent lie groups. Ark. MAt. **13**, 161–207 (1975)
43. Folland, G.B.: On the Rothschild-Stein lifting theorem. Commun. Partial Differ. Equ. **2**, 161–207 (1977)
44. Franchi, B., Hajłasz, P., Koskela, P.: Definitions of Sobolev classes on metric spaces. Ann. Inst. Fourier (Grenoble) **49**, 1903–1924 (1999)
45. Franchi, B., Lanconelli, E.: Une condition géométrique pour l'inégalité de Harnack. J. Math. Pures Appl. **64**, 237–256 (1985)
46. Franchi, B., Lu, G., Wheeden, R.L.: A relationship between Poincaré-type inequalities and representation formulas in spaces of homogeneous type. Int. Math. Res. Not. **1**, 1–14 (1996)
47. Franchi, B., Serapioni, R., Serra Cassano, F.: Approximation and imbedding theorems for weighted Sobolev spaces associated with Lipschitz continuous vector fields. Boll. U.M.I B (7) **11**, 83–117 (1997)
48. Garofalo, N., Lanconelli, E.: Asymptotic behavior of fundamental solutions and potential theory of parabolic operators with variable coefficients. Math. Ann. **283**, 211–239 (1989)
49. Garofalo, N., Lanconelli, E.: Level sets of the fundamental solution and Harnack inequality for degenerate equations of Kolmogorov type. Trans. Am. Math. Soc. **321**, 775–792 (1990)
50. Garofalo, N., Nhieu, D.-M.: Isoperimetric and Sobolev inequalities for Carnot-Carathéodory spaces and the existence of minimal surfaces. Commun. Pure Appl. Math. **49**, 1081–1144 (1996)
51. Grigor'yan, A., Saloff-Coste, L.: Stability results for Harnack inequalities. Ann. Inst. Fourier (Grenoble) **55**, 825–890 (2005)
52. Gutiérrez, C.E.: Harnack's inequality for degenerate Schrödinger operators. Trans. Am. Math. Soc. **312**, 403–419 (1989)
53. Gutiérrez, C.E., Lanconelli, E.: Maximum principle, nonhomogeneous Harnack inequality, and Liouville theorems for X-elliptic operators. Commun. Partial Differ. Equ. **28**, 1833–1862 (2003)
54. Hebisch, W., Saloff-Coste, L.: On the relation between elliptic and parabolic Harnack inequalities. Ann. Inst. Fourier (Grenoble) **51**, 1437–1481 (2001)
55. Hajłasz, P., Koskela, P.: Sobolev met Poincaré. Mem. Am. Math. Soc. **145** (2000)
56. Heinonen, J., Holopainen, I.: Quasiregular maps on Carnot groups. J. Geom. Anal. **7**, 109–148 (1997)

57. Indratno, S., Maldonado, D., Silwal, S.: On the axiomatic approach to Harnack's inequality in doubling quasi-metric spaces. J. Differ. Equ. **254**, 3369–3394 (2013)
58. Jerison, D.S., Sánchez-Calle, A.: Subelliptic, second order differential operators. In: Complex Analysis III, Proceeding Spec. Year, College Park 1985–1986. Lecture Notes Mathematics, vol. 1277, pp. 46–77 (1987)
59. Kinnunen, J., Marola, N., Miranda, M., Paronetto, F.: Harnack's inequality for parabolic De Giorgi classes in metric spaces. Adv. Differ. Equ. **17**, 801–832 (2012)
60. Kogoj, A.E., Lanconelli, E.: X-elliptic operators and X-elliptic distances, Contributions in honor of the memory of Ennio De Giorgi. Ricerche Mat. **49**, 223–243 (2000)
61. Kogoj, A.E., Lanconelli, E.: Liouville theorem for X-elliptic operators. Nonlinear Anal. **70**, 2974–2985 (2009)
62. Kohn, J.J.: Hypoellipticity of some degenerate subelliptic operators. J. Funct. Anal. **159**, 203–216 (1998)
63. Kusuoka, S., Stroock, D.: Applications of the Malliavin calculus II. J. Fac. Sci. Univ. Tokyo Sect. IA Math. **32**, 1–76 (1985)
64. Mohammed, A.: Harnack's inequality for solutions of some degenerate elliptic equations. Rev. Mat. Iberoamericana **18**, 325–354 (2002)
65. Morimoto, Y.: A criterion for hypoellipticity of second order differential operators. Osaka J. Math. **24**, 651–675 (1987)
66. Parmeggiani, A.: A remark on the stability of C^∞-hypoellipticity under lower-order perturbations. J. Pseudo-Differ. Oper. Appl. **6**, 227–235 (2015)
67. Pascucci, A., Polidoro, S.: A Gaussian upper bound for the fundamental solutions of a class of ultraparabolic equations. J. Math. Anal. Appl. **282**, 396–409 (2003)
68. Pascucci, A., Polidoro, S.: Harnack inequalities and Gaussian estimates for a class of hypoelliptic operators. Trans. Am. Math. Soc. **358**, 4873–4893 (2006)
69. Rothschild, L.P., Stein, E.M.: Hypoelliptic differential operators and nilpotent groups. Acta Math. **137**, 247–320 (1976)
70. Saloff-Coste, L.: Parabolic Harnack inequality for divergence-form second-order differential operators. Potential Anal. **4**, 429–467 (1995)
71. Stein, E.M.: An example on the Heisenberg group related to the Lewy operator. Invent. Math. **69**, 209–216 (1982)
72. Uguzzoni, F.: Estimates of the Green function for X-elliptic operators. Math. Ann. **361**, 169–190 (2015)

Semiclassical Analysis of Dispersion Phenomena

Victor Chabu[1], Clotilde Fermanian-Kammerer[2(✉)], and Fabricio Macià[3]

[1] Universidade de São Paulo, IF-USP, DFMA, CP 66.318, São Paulo, SP 05314-970, Brazil
vbchabu@if.usp.br

[2] LAMA, UMR CNRS 8050, Université Paris Est, 61, avenue du Général de Gaulle, 94010 Créteil Cedex, France
Clotilde.Fermanian@u-pec.fr

[3] Universidad Politécnica de Madrid, DCAIN, ETSI Navales, Avda. de la Memoria 4, 28040 Madrid, Spain
fabricio.macia@upm.es

Abstract. Our aim in this work is to give some quantitative insight on the dispersive effects exhibited by solutions of a semiclassical Schrödinger-type equation in $\mathbf{R}^d$. We describe quantitatively the localisation of the energy in a long-time semiclassical limit within this non compact geometry and exhibit conditions under which the energy remains localized on compact sets. We also explain how our results can be applied in a straightforward way to describe obstructions to the validity of smoothing type estimates.

Keywords: Semiclassical analysis · Schrödinger equations · Wigner measures · Degenerate critical points

1 Introduction

1.1 Description of the Problem

Our aim in this work is to revisit some of the results obtained in [5] in order to give some quantitative insight on the dispersive effects exhibited by solutions of the semiclassical Schrödinger-type equation:

$$\begin{cases} i\varepsilon\partial_t v^\varepsilon(t,x) = \lambda(\varepsilon D_x)v^\varepsilon(t,x) + \varepsilon^2 V(x) v^\varepsilon(t,x), & (t,x) \in \mathbf{R}\times\mathbf{R}^d, \\ v^\varepsilon|_{t=0} = u_0^\varepsilon, \end{cases} \tag{1}$$

Above, $\lambda, V \in \mathcal{C}^\infty(\mathbf{R}^d;\mathbf{R})$; the function λ is the symbol of the semiclassical Fourier multiplier defined by:

$$\forall v \in L^2(\mathbf{R}^d), \quad \lambda(\varepsilon D_x)v(x) := \int_{\mathbf{R}^d} \lambda(\varepsilon\xi)\widehat{v}(\xi)e^{i\xi\cdot x}\frac{d\xi}{(2\pi)^d},$$

J. Delgado and M. Ruzhansky (Eds.): *Analysis and Partial Differential Equations: Perspectives from Developing Countries*, PROMS 275, pp. 84–108, 2018.
https://doi.org/10.1007/978-3-030-05657-5_7

where, in general, the integral has to be understood in distributional sense. The following convention for the Fourier transform is used:

$$\widehat{v}(\xi) := \int_{\mathbf{R}^d} v(x) e^{-i\xi \cdot x} d\xi.$$

Our goal is to understand the behavior as $\varepsilon \to 0^+$ of solutions to (1) corresponding to sequences of initial data (u_0^ε) whose characteristic length-scale of oscillations is of order at least ε (see (7) for a precise definition) at very long times, of the order of $1/\varepsilon$.

To this aim, we scale in time the solutions to (1) and define:

$$u^\varepsilon(t,\cdot) := v^\varepsilon\left(\frac{t}{\varepsilon},\cdot\right).$$

Note that these functions solve the following problem.

$$\begin{cases} i\varepsilon^2 \partial_t u^\varepsilon(t,x) = \lambda(\varepsilon D_x) u^\varepsilon(t,x) + \varepsilon^2 V(x) u^\varepsilon(t,x), & (t,x) \in \mathbf{R} \times \mathbf{R}^d, \\ u^\varepsilon|_{t=0} = u_0^\varepsilon, \end{cases} \tag{2}$$

If the symbol λ happens to be homogeneous of degree two, (2) reduces to the non-semiclassical equation:

$$i\partial_t u^\varepsilon(t,x) = \lambda(D_x) u^\varepsilon(t,x) + V(x) u^\varepsilon(t,x).$$

In what follows we shall consider sequences of initial data (u_0^ε) that are bounded in $L^2(\mathbf{R}^d)$. Denote by (u^ε) the corresponding sequence of solutions to (2) and construct the position densities:

$$n^\varepsilon(t,x) := |u^\varepsilon(t,x)|^2.$$

For every $t \in \mathbf{R}$, the sequence $(n^\varepsilon(t,\cdot))$ is bounded in $L^1(\mathbf{R}^d)$, since

$$||n^\varepsilon(t,\cdot)||_{L^1(\mathbf{R}^d)} = ||u^\varepsilon(t,\cdot)||_{L^2(\mathbf{R}^d)} = ||u_0^\varepsilon||_{L^2(\mathbf{R}^d)};$$

it is not difficult to show, using the fact that u^ε solve (2), that there exists a subsequence $\varepsilon_n \to 0^+$ and a t-measurable family of finite positive Radon measures $\nu_t(dx)$ on $\mathbf{R}^d$ such that the space-time averages of the position densities (n^{ε_n}) converge:

$$\lim_{n\to\infty} \int_a^b \int_{\mathbf{R}^d} \phi(x) |u^{\varepsilon_n}(t,x)|^2 dx dt = \int_a^b \int_{\mathbf{R}^d} \phi(x) \nu_t(dx) dt, \tag{3}$$

for every $a < b$ and every $\phi \in \mathcal{C}_0(\mathbf{R}^d)$. The limiting measure ν_t is sometimes called a *defect measure* of the sequence (u^ε). It will follow from our results that defect measures give indeed a quantitative description of the lack of dispersion for solutions to (1).

The long-time semiclassical limit has been studied with some detail in the context of Schrödinger equations in compact geometries, see for instance

[1,2,19,20]. In the compact setting, the dispersive nature of the equation manifests through more subtle mechanisms, and is intimately related to the global dynamics of the underlying classical system.

When the potential V in (2) is identically equal to zero, simple calculations can be implemented for specific initial data. Construct for example, for $\xi_0 \in \mathbf{R}^d$, and $\theta \in \mathcal{S}(\mathbf{R}^d)$ with $\|\theta\|_{L^2(\mathbf{R}^d)} = 1$:

$$u^\varepsilon_{\xi_0}(x) = \theta(x)\mathrm{e}^{\frac{i}{\varepsilon}\xi_0\cdot x}. \tag{4}$$

These sequences of initial data are highly oscillating as soon as $\xi_0 \neq 0$ and therefore converge weakly to zero in $L^2(\mathbf{R}^d)$; however $|u^\varepsilon_{\xi_0}|^2 = |\theta|^2$ is independent of ε and ξ_0.

A direct application of the stationary/non-stationary phase principle gives for any $\phi \in \mathcal{C}_c(\mathbf{R}^d)$ and any $a < b$ that the following limits hold.

- If ξ_0 is not a critical point of λ, then

$$\lim_{\varepsilon\to 0^+} \int_a^b \int_{\mathbf{R}^d} \phi(x)|\mathrm{e}^{\frac{it}{\varepsilon^2}\lambda(\varepsilon D_x)} u^\varepsilon_{\xi_0}(x)|^2 dx dt = 0.$$

- If ξ_0 is a critical point of λ, then

$$\lim_{\varepsilon\to 0^+} \int_a^b \int_{\mathbf{R}^d} \phi(x)|\mathrm{e}^{\frac{it}{\varepsilon^2}\lambda(\varepsilon D_x)} u^\varepsilon_{\xi_0}(x)|^2 dx dt = \int_a^b \int_{\mathbf{R}^d} \phi(x) \left|\mathrm{e}^{-\frac{it}{2}\nabla^2\lambda(\xi_0)D_x\cdot D_x}\theta(x)\right|^2 dx dt,$$

  where $\nabla^2\lambda(\xi_0)$ denotes the Hessian of λ at the point ξ_0.

When ξ_0 is not a critical point of the symbol λ, the fact that no energy remains on any compact set in the high frequency limit is precisely a manifestation of dispersive behavior of the semiclassical problem (1). However, when ξ_0 happens to be a critical point of λ, such a dispersive behavior fails, and a fraction of the energy remains localized on compact sets of $\mathbf{R}^d$. Note that $\xi_0 = 0$ plays a special role in this setting, since it corresponds to initial data that are not oscillating. Therefore, even if $\xi_0 = 0$ is a critical point of λ,[1] the fact that the local energy does not escape from every compact set as $\varepsilon \to 0^+$ in this case should not be interpreted as a lack of dispersion of (1).

The situation can be more intricate for initial data which are superposition of oscillating functions of the form above:

$$u^\varepsilon_0(x) = \theta_1(x)\mathrm{e}^{\frac{i}{\varepsilon}\xi_1\cdot x} + \theta_2(x)\mathrm{e}^{\frac{i}{\varepsilon}\xi_2\cdot x} \tag{5}$$

with $\theta_1, \theta_2 \in \mathcal{S}(\mathbf{R}^d)$, both non-zero, and $\xi_1, \xi_2 \in \mathbf{R}^d$ such that ξ_2 is a critical point of λ while ξ_1 is not. One easily checks that:

$$\lim_{\varepsilon\to 0^+} \int_a^b \int_{\mathbf{R}^d} \phi(x)|u^\varepsilon(t,x)|^2 dx dt = \int_a^b \int_{\mathbf{R}^d} \phi(x) \left|\mathrm{e}^{-\frac{it}{2}\nabla^2\lambda(\xi_2)D_x\cdot D_x}\theta_2(x)\right|^2 dx dt, \tag{6}$$

[1] Think for instance of $\lambda(\xi) = \|\xi\|^2$, for which (2) corresponds to the standard, non-semiclassical, Schrödinger equation, one of the most studied dispersive equations.

which shows that only a fraction of the (asymptotic) total mass $\|\theta_1\|^2_{L^2(\mathbf{R}^d)} + \|\theta_2\|^2_{L^2(\mathbf{R}^d)}$ of the sequence of solutions is dispersed in this case.

Our aim here, is to provide a general description of these high frequency effects. In particular, we will generalize the analysis done in the previous examples to arbitrary sequence of initial data, and investigate the effects produced by the presence of a bounded non-zero potential V. We will also show in Corollary 1 how our results can be applied in a straightforward way to describe obstructions to the validity of smoothing type estimates for equations of the form (2) in the presence of critical points of the symbol λ.

1.2 Non Dispersive Effects Associated to Isolated Critical Points

We are first going to show that, in the presence of isolated critical points of λ, some of the high frequency effects exhibited by the sequence of initial data persist after applying the time evolution (2). As we said before, we give a complete description of the asymptotic behavior of the densities $|u^\varepsilon(t,x)|^2$ associated to a sequence of solutions to (2) issued from a sequence of initial data $(u_0^\varepsilon)_{\varepsilon>0}$ bounded in $L^2(\mathbf{R}^d)$. When the critical points of the symbol λ are non-degenerate, we present an explicit procedure to compute all weak-$\star$ accumulation points of the sequence of time-dependent positive measures $\left(|u^\varepsilon(t,\cdot)|^2\right)_{\varepsilon>0}$ in terms of quantities that only depend on the sequence of initial data.

In order to prevent that all the mass of the sequence $\left(|u^\varepsilon(t,\cdot)|^2\right)_{\varepsilon>0}$ trivially escapes to infinity, we must make sure that the characteristic length scale of the oscillations of the sequence of initial data is at least of order ε. The following, now standard, assumption is sufficient for our purposes:

A0 The family $(u_0^\varepsilon)_{\varepsilon>0}$ is uniformly bounded in $L^2(\mathbf{R}^d)$ and ε-oscillating, in the sense that its energy is concentrated on frequencies smaller or equal than $1/\varepsilon$:

$$\limsup_{\varepsilon\to 0} \int_{\|\xi\|>R/\varepsilon} |\widehat{u_0^\varepsilon}(\xi)|^2 d\xi \underset{R\to+\infty}{\longrightarrow} 0, \tag{7}$$

In order to keep the presentation relatively simple, we also impose smoothness and growth conditions on λ and V. More precisely:

A1 $V \in \mathcal{C}^\infty(\mathbf{R}^d)$ is bounded together with all its derivatives and that $\lambda \in \mathcal{C}^\infty(\mathbf{R}^d)$ is a symbol of order $N > 0$ (as in [7], Definition 7.5):

$$\forall \alpha \in \mathbb{N}^d, \quad \sup_{\xi\in\mathbf{R}^d} |\partial_\xi^\alpha \lambda(\xi)| \, (1 + \|\xi\|)^{-N} < \infty.$$

Our last hypothesis deals with the set of critical points of the symbol:

$$\Lambda := \{\xi \in \mathbf{R}^d : \nabla\lambda(\xi) = 0\}.$$

In our first result, we assume the following.

A2 Λ is a countable set of $\mathbf{R}^d$.

Theorem 1. *Assume that the sequence of initial data* $(u_0^\varepsilon)_{\varepsilon>0}$ *verifies* **A0** *and that* λ *and* V *satisfy* **A1** *and* ***A2****; denote by* $(u^\varepsilon)_{\varepsilon>0}$ *the corresponding family of solutions to (2). Suppose* $(\varepsilon_n)_{n\in\mathbb{N}}$ *is a subsequence along which* $(|u^{\varepsilon_n}|^2)_{n\in\mathbb{N}}$ *converges, in the sense of* (3)*, to some defect measure* $\nu_t(dx)dt$*. Then, for almost every* $t\in\mathbf{R}$ *the following holds:*

$$\nu_t(dx) \geq \sum_{\xi\in\Lambda} |u_\xi(t,x)|^2 dx, \tag{8}$$

where u_ξ *is a solution to the following Schrödinger equation:*

$$\begin{cases} i\partial_t u_\xi(t,x) = \frac{1}{2}\nabla^2\lambda(\xi)D_x\cdot D_x u_\xi(t,x) + V(x)u_\xi(t,x), \\ u_\xi|_{t=0} = u_\xi^0, \end{cases} \tag{9}$$

and u_ξ^0 *is the limit, for the weak topology on* $L^2(\mathbf{R}^d)$*, of the sequence* $(\mathrm{e}^{-\frac{i}{\varepsilon_n}\xi\cdot x}u_0^{\varepsilon_n})_{n\in\mathbb{N}}$.
If in addition, all critical points of λ *are non-degenerate, then inequality (8) becomes an equality.*

We will show below (see Proposition 1) that, when at least one of the critical points of λ is degenerate, there exist sequences of initial data for which inequality (8) is strict. However, even when the non-degeneracy condition is violated, there are simple conditions on the sequence of initial data that ensure that (8) is an equality. In order to state those, let us consider a cut-off function $\chi\in\mathcal{C}_0^\infty(\mathbf{R}^d)$ such that

$$0\leq\chi\leq 1,\ \ \chi(\eta)=1 \ \text{ for } \ \|\eta\|\leq 1 \ \text{ and } \ \chi(\eta)=0 \ \text{ for } \ \|\eta\|\geq 2. \tag{10}$$

Theorem 2. *Assume that the same hypotheses as in Theorem 1 hold, and that the following additional condition on the sequence* $(u_0^\varepsilon)_{\varepsilon>0}$ *of initial data is satisfied: for all* $\xi\in\Lambda$*,*

$$\limsup_{\delta\to 0^+}\limsup_{R\to+\infty}\limsup_{\varepsilon\to 0^+}\left\|(1-\chi)\left(\frac{\varepsilon D_x-\xi}{\varepsilon R}\right)\chi\left(\frac{\varepsilon D_x-\xi}{\delta}\right)u_0^\varepsilon\right\|_{L^2(\mathbf{R}^d)} = 0.$$

Then, the inequality in formula (8) becomes an equality.

Note that for the sequence initial data (5) introduced previously, one has $u_\xi^0=0$ for $\xi\notin\{\xi_1,\xi_2\}$ and $u_{\xi_j}^0=\theta_j$, $j=1,2$. Identity (6) is a consequence of Theorem 1 in that setting.

Note also that u_ξ may be identically equal to zero even if the family $(u_0^\varepsilon)_{\varepsilon>0}$ oscillates in the direction ξ. To see this, simply modulate the waves in example (4) by an amplitude that concentrates around some point $x_0\in\mathbf{R}^d$:

$$u_0^\varepsilon(x) = \frac{1}{\varepsilon^{d/4}}\theta\left(\frac{x-x_0}{\sqrt{\varepsilon}}\right)\mathrm{e}^{\frac{i}{\varepsilon}\xi_0\cdot x} \tag{11}$$

This corresponds to a coherent state centered at the point (x_0, ξ_0) in phase-space. In this case $u_\xi = 0$ for every $\xi \in \mathbf{R}^d$. Thus, Theorem 1 allows us to conclude that the corresponding solutions $(u^\varepsilon)_{\varepsilon>0}$ converge to zero in $L^2_{\text{loc}}(\mathbf{R} \times \mathbf{R}^d)$ as $\varepsilon \to 0^+$.

To conclude this section, let us investigate what kind of behavior can be expected when λ has degenerate critical points. Suppose $\xi_0 \in \Lambda$ and that $\omega_0 \in \mathbf{R}^d$ exists such that $\omega_0 \in \ker \nabla^2\lambda(\xi_0)$ and $\|\omega_0\| = 1$. Let us slightly modify the initial data (11) for $x_0 = 0$ by introducing a phase shift:

$$u_0^\varepsilon(x) = \frac{1}{\varepsilon^{\alpha d/2}} \theta\left(\frac{x}{\varepsilon^\alpha}\right) \mathrm{e}^{\frac{i}{\varepsilon} x \cdot (\xi_0 + \varepsilon^\beta \omega_0)}, \tag{12}$$

where $\theta \in \mathcal{S}(\mathbf{R}^d)$, $\alpha \in [0,1)$ and $\beta \in (0,1)$ satisfies $\alpha + \beta < 1$. A simple computation shows that these data do not satisfy the assumptions of Theorem 2. Again, one has $u_\xi^0 = 0$ for any $\xi \in \mathbf{R}^d$; therefore, if ξ_0 were a non-degenerate critical point, Theorem 1 would imply $\nu_t(dx) = 0$, this means that the energy of the modified coherent state (12) would be dispersed to infinity. However, when $\xi_0 \in \Lambda$ is degenerate this is no longer the case.

Proposition 1. *Assume $\omega_0 \in \ker \nabla^2\lambda(\xi_0)$, $|\omega_0| = 1$, $\beta > \frac{2}{3}$ and $V = 0$. Let $(u^\varepsilon)_{\varepsilon_0}$ denote the sequence of solutions to (2) issued from the initial data (12). Then, for every $a < b$ and every $\phi \in \mathcal{C}_0(\mathbf{R}^d)$ the following holds.*

(i) If $\alpha = 0$, then

$$\lim_{\varepsilon\to 0} \int_a^b \int_{\mathbf{R}^d} \phi(x) \left|u^\varepsilon(t,x)\right|^2 dx dt = \int_a^b \int_{\mathbf{R}^d} \phi(x) \left|\mathrm{e}^{-\frac{it}{2}\nabla^2\lambda(\xi_0) D_x \cdot D_x} \theta(x)\right|^2 dx dt.$$

(ii) If $\alpha \neq 0$, then

$$\lim_{\varepsilon\to 0} \int_a^b \int_{\mathbf{R}^d} \phi(x) |u^\varepsilon(t,x)|^2 dx dt = (b-a)\, \phi(0)\, \|\theta\|^2_{L^2(\mathbf{R}^d)}.$$

This example also shows that defect measures can be singular when critical points of the symbol are degenerate. In the example above we have:

$$\nu_t(dx) = \|\theta\|^2_{L^2(\mathbf{R}^d)} \delta_0(dx).$$

Of course, this can never occur if Λ consists only of non-degenerate critical points, as Theorem 1 shows. The proofs of the results in this section are given in Sect. 3.

1.3 Non Dispersive Effects Associated to a Manifold of Critical Points

A natural generalization of the results of the previous section consists in assuming that the set of critical points is a smooth submanifold of $\mathbf{R}^d$. This situation has been examined in detail in [4, 5]. Here, in order to keep the presentation reasonably self-contained, we describe the results in the geometrically simpler case in which Λ is an affine variety of codimension $0 < p \le d$. After performing a linear change of coordinates in momentum space, we can assume that Λ takes the following form.

A2' The set Λ of critical points of λ is of the form:

$$\Lambda = \{(\xi = (\xi', \xi'') \in \mathbf{R}^r \times \mathbf{R}^p \,:\, \xi'' = \xi_0''\},$$

for some $\xi_0'' \in \mathbf{R}^p$. Above we have $0 < p \le d$ and $r := d - p$.

Before stating the main result in this case, we must introduce some notations. We decompose the physical space as $x = (x', x'') \in \mathbf{R}^r \times \mathbf{R}^p$. Given a function $\phi \in L^\infty(\mathbf{R}^d)$, we write $m_\phi(x')$, where $(x') \in \mathbf{R}^r$, to denote the operator acting on $L^2(\mathbf{R}^p)$ by multiplication by $\phi(x', \cdot)$:

$$m_\phi(x') f(y) = \phi(x', y) f(y), \quad \text{for } f \in L^2(\mathbf{R}^p). \tag{13}$$

Note that assumption **A2'** implies that for any $\xi \in \Lambda$ the non-trivial part of the Hessian of λ at ξ defines a differential operator $\nabla^2_{\xi''}\lambda(\xi) D_y \cdot D_y$ acting on function defined on $\mathbf{R}^p$.

In our next result, the sum over critical points appearing in the statement of Theorem 1 is replaced by an integral with respect to a certain measure over $\mathbf{R}^r \times \Lambda$, and the Schrödinger equation (9) becomes a Heisenberg equation for a time-dependent family M of trace-class operators acting on $L^2(\mathbf{R}^p)$. More precisely, the operators M depend on $t \in \mathbf{R}$ and on $(x', \xi') \in \mathbf{R}^r \times \mathbf{R}^r$; for every choice of these parameters, $M_t(x', \xi')$ is an element of $\mathcal{L}^1_+\left(L^2(\mathbf{R}^p)\right)$, *i.e.*, it is a positive, Hermitian, trace-class operator acting on $L^2(\mathbf{R}^p)$.

Theorem 3. *Assume that the sequence of initial data $(u_0^\varepsilon)_{\varepsilon>0}$ verifies* **A0** *and that λ and V satisfy* **A1** *and* **A2'***; denote by $(u^\varepsilon)_{\varepsilon>0}$ the corresponding family of solutions to (2). Suppose $(\varepsilon_n)_{n\in\mathbb{N}}$ is a subsequence along which $(|u^{\varepsilon_n}|^2)_{n\in\mathbb{N}}$ converges, in the sense of (3), to some defect measure $\nu_t(dx)dt$. Then there exist a positive Radon measure ν^0 defined on $\mathbf{R}^r \times \mathbf{R}^r$ and a measurable family of self-adjoint, positive, trace-class operators*

$$M_0 : \mathbf{R}^r \times \mathbf{R}^r \ni (x', \xi') \longmapsto M_0(x', \xi') \in \mathcal{L}^1_+(L^2(\mathbf{R}^r)), \quad \operatorname{Tr}_{L^2(\mathbf{R}^p)} M_0(x', \xi') = 1,$$

such that, for almost every $t \in \mathbf{R}$ and every $\phi \in \mathcal{C}_0(\mathbf{R}^d)$ the following holds:

$$\int_{\mathbf{R}^d} \phi(x)\nu_t(dx) \ge \int_{\mathbf{R}^r\times\mathbf{R}^r} \operatorname{Tr}_{L^2(\mathbf{R}^p)}\left[m_\phi(x', \xi') M_t(x', \xi')\right] \nu^0(dx', d\xi'), \tag{14}$$

and $M \in \mathcal{C}(\mathbf{R}, \mathcal{L}^1_+(L^2(\mathbf{R}^p))$ solves the following Heisenberg equation:

$$\begin{cases} i\partial_t M_t(x', \xi') = \left[\frac{1}{2}\nabla^2_{\xi''}\lambda(\xi', \xi_0'') D_y \cdot D_y + m_V(x'), M_t(x', \xi')\right], \\ M|_{t=0} = M_0. \end{cases} \tag{15}$$

Moreover, the measure ν^0 and the family of operators M_0 are computed in terms of the sequence initial data $(u_0^\varepsilon)_{\varepsilon>0}$. In particular, they do not depend on λ or V.

The nature of the objects involved in this result is described in Sect. 2.2. As before, a certain non-degeneracy condition on the points of Λ implies that the inequality (14) is in fact an identity.

Theorem 4. *Suppose all the hypotheses of Theorem 3 are satisfied. If in addition to those, for every $\xi \in \Lambda$ the rank of the Hessian $\nabla^2\lambda(\xi)$ is equal to p then* (14) *is an identity.*

When $\Lambda = \{\xi_0\}$ consists of a single critical point, the statements of Theorems 1 and 3 turn out to be completely equivalent. In this case, $r = 0$, which forces $\nu^0(dx', d\xi') = \|u^0_{\xi_0}\|^2_{L^2(\mathbf{R}^d)}\delta_0(dx')\delta_0(d\xi')$. In addition, $p = d$, and the operator M_t (which will not depend on (x', ξ')) will be the orthogonal projection onto $u_{\xi_0}(t, \cdot)$ in $L^2(\mathbf{R}^d)$. Since u_{ξ_0} solves the Schrödinger equation (9), these orthogonal projections satisfy the Heisenberg equation (15). As it will be clear from the proofs, Theorem 3 generalises in a straightforward way to the case that Λ is a disjoint union of affine varieties of $\mathbf{R}^d$.

Remark 1. As soon as the dimension of Λ is strictly positive, the measure ν^0 may be singular with respect to the variable x'. This fact implies that the limiting measure of the sequence $\left(|u^\varepsilon|^2\right)_{\varepsilon>0}$ may be singular in the variable x. Indeed, assume for example $\Lambda = \{\xi'' = 0\}$, $p \neq 0$, and

$$u_0^\varepsilon(x) = \varepsilon^{\frac{\alpha(p-d)}{2}}\theta(x'')\varphi\left(\frac{x' - z_0}{\varepsilon^\alpha}\right)e^{i\frac{x'\cdot\zeta_0}{\varepsilon}},$$

where $\alpha \in (0,1)$, $z_0, \zeta_0 \in \mathbf{R}^r$, $\varphi \in C_0^\infty(\mathbf{R}^r)$, $\theta \in C_0^\infty(\mathbf{R}^p)$ and $\|\theta\|_{L^2(\mathbf{R}^p)} = 1$. Then the measure ν^0 and the operator M_0 of Theorem 3 will be:

$$\nu^0(dx', d\xi') = \|\varphi\|^2_{L^2(\mathbf{R}^r)}\delta_{z_0}(dx')\delta_{\zeta_0}(d\xi') \qquad \text{and} \qquad M_0(x', \xi') = |\theta\rangle\langle\theta|, \quad (16)$$

see Corollary 3 in Sect. 4.

1.4 Link with Smoothing-Type Estimates

Since the pioneering works [3,6,16,17,24,25] it is well-known that dispersive-type equations develop some kind of smoothing effect. Usually, this is described by means of smoothing-type estimates. Theorems 1 and 3 can be used, in a rather straightforward way, to describe obstructions to the validity of smoothing-type estimates in the presence of non-zero critical points of the symbol λ. Note that this type of behavior was already described in [15]; smoothing-type estimates outside the critical points of λ were recently presented in [23]). We present a simple application of Theorem 1 to this setting.

Corollary 1. *Suppose* **A1**, **A2**, *hold and that λ has a non-zero critical point ξ_0. Then, given any $\delta, s > 0$ and any ball $B \subset \mathbf{R}^d$ it is not possible to find a constant $C > 0$ such that the estimate*

$$\int_0^\delta \||D_x|^s u^\varepsilon(t,\cdot)\|^2_{L^2(B)}dt \leq C\|u_0^\varepsilon\|^2_{L^2(\mathbf{R}^d)}, \quad (17)$$

holds uniformly for every solution u^ε of (2) with initial datum $u_0^\varepsilon \in C_0^\infty(\mathbf{R}^d)$.

Proof. We argue by contradiction. Suppose the estimate (17) holds for some $\delta, s, C > 0$ and some ball B. Let $\theta \in \mathcal{C}_0^\infty(\mathbf{R}^d)$ with $\|\theta\|_{L^2(\mathbf{R}^d)} = 1$ and consider the sequence of initial data:

$$u_0^\varepsilon(x) := \theta(x) e^{i\frac{\xi_0}{\varepsilon}\cdot x}.$$

Clearly $\|u_0^\varepsilon\|_{L^2(\mathbf{R}^d)} = 1$ and (u_0^ε) converges weakly to zero since $\xi_0 \neq 0$. Estimate (17) then implies that (u^ε) is bounded in $L^2((0,\delta); H^s(B))$ and Rellich's theorem gives that a subsequence of (u^ε) converges strongly in $L^2((0,\delta)\times B)$. This limit must be zero, since (u^ε) weakly converges to zero in that space.
Now, Theorem 1 implies that:

$$0 \geq |u_{\xi_0}(t,\cdot)|^2 dx;$$

in particular, $u_{\xi_0}(t,\cdot) = 0$ for every $t \in \mathbf{R}$. But this is a contradiction, since, as u_{ξ_0} is the solution of the Schrödinger equation (9) with initial datum $u_{\xi_0}^0 = \theta$, one necessarily has $\|u_{\xi_0}(t,\cdot)\|_{L^2(\mathbf{R}^d)} = 1$.

Of course, Theorem 3 gives an analogous consequence when the set of critical points is not isolated.

2 The Microlocal Approach to the Problem

2.1 Wigner Measures

Wigner distributions provide a useful way for computing weak-$\star$ accumulation points of a sequence of densities $\left(|u^\varepsilon|^2\right)_{\varepsilon>0}$ constructed from a L^2-bounded sequence $(u^\varepsilon)_{\varepsilon>0}$ of solutions to a semiclassical (pseudo) differential equation. They provide a joint position and momentum description of the L^2-mass distribution of functions. The (momentum scaled) Wigner distribution of a function $f \in L^2(\mathbf{R}^d)$ is defined as:

$$W_f^\varepsilon(x,\xi) = \int_{\mathbf{R}^d} f\left(x - \frac{\varepsilon v}{2}\right)\overline{f\left(x + \frac{\varepsilon v}{2}\right)} e^{i\xi\cdot v}\frac{dv}{(2\pi)^d}.$$

It enjoys several interesting properties:

- $W_f^\varepsilon \in L^2(\mathbf{R}^d \times \mathbf{R}^d)$.
- Projecting W_f^ε on x or ξ gives the position or momentum densities of f, respectively:

$$\int_{\mathbf{R}^d} W_f^\varepsilon(x,\xi)d\xi = |f(x)|^2, \quad \int_{\mathbf{R}^d} W_f^\varepsilon(x,\xi)dx = \frac{1}{(2\pi\varepsilon)^d}\left|\widehat{f}\left(\frac{\xi}{\varepsilon}\right)\right|^2.$$

Note that in spite of this, W_f^ε is not positive in general.

- For every $a \in \mathcal{C}_0^\infty(\mathbf{R}^d \times \mathbf{R}^d)$ one has:

$$\int_{\mathbf{R}^d \times \mathbf{R}^d} a(x,\xi) W_f^\varepsilon(x,\xi) dx\, d\xi = (\mathrm{op}_\varepsilon(a) f, f)_{L^2(\mathbf{R}^d)}, \tag{18}$$

where $\mathrm{op}_\varepsilon(a)$ is the semiclassical pseudodifferential operator of symbol a obtained through the Weyl quantization rule:

$$\mathrm{op}_\varepsilon(a) f(x) = \int_{\mathbf{R}^d \times \mathbf{R}^d} a\left(\frac{x+y}{2}, \varepsilon\xi\right) e^{i\xi\cdot(x-y)} f(y) dy \frac{d\xi}{(2\pi)^d}.$$

See, for instance, [11] for proofs of these results.

If $(f^\varepsilon)_{\varepsilon>0}$ is a bounded sequence in $L^2(\mathbf{R}^d)$, then $(W_{f^\varepsilon}^\varepsilon)_{\varepsilon>0}$ is a bounded sequence of tempered distributions in $\mathcal{S}'(\mathbf{R}^d \times \mathbf{R}^d)$. In addition, every accumulation point of $(W_{f^\varepsilon}^\varepsilon)_{\varepsilon>0}$ in $\mathcal{S}'(\mathbf{R}^d \times \mathbf{R}^d)$ is a positive distribution and, therefore, by Schwartz's theorem, an element of $\mathcal{M}_+(\mathbf{R}^d \times \mathbf{R}^d)$, the set of positive Radon measures on $\mathbf{R}^d \times \mathbf{R}^d$. These measures are called *semiclassical* or *Wigner measures*. See Refs. [12–14,18] for different proofs of the results we have presented in here.

Now, if $\mu \in \mathcal{M}_+(\mathbf{R}^d \times \mathbf{R}^d)$ is an accumulation point of $(W_{f^\varepsilon}^\varepsilon)_{\varepsilon>0}$ along some sequence $(\varepsilon_k)_{k\in\mathbb{N}}$ and $\left(|f^{\varepsilon_k}|^2\right)_{k\in\mathbb{N}}$ converges weakly-$\star$ towards a measure ν on $\mathbf{R}^d$, then one has:

$$\int_{\mathbf{R}^d} \mu(\cdot, d\xi) \leq \nu.$$

Equality holds if and only if $(f^{\varepsilon_k})_{k\in\mathbb{N}}$ is ε-oscillating in the sense of **A0** (see [12–14]). Note also that this implies that μ is always a finite measure and its total mass is bounded by $\sup_\varepsilon \|f^\varepsilon\|^2_{L^2(\mathbf{R}^d)}$.

This fact justifies the idea of replacing the analysis of energy densities by that of Wigner distributions, which allows one to use a larger set of test functions and to take into account in a more precise way the effects of oscillation of the studied functions, by considering the Fourier variable.

When the sequence under consideration consists of solutions to the dispersive equation (2), the convergence of the corresponding Wigner distributions towards a Wigner measure still holds provided one averages in time. More precisely, let $(u^\varepsilon)_{\varepsilon>0}$ be a sequence of solutions to (2) issued from a sequence of initial data $(u_0^\varepsilon)_{\varepsilon>0}$ satisfying **A0**. Then there exist a subsequence (ε_k) tending to zero as $k \to \infty$ and a t-measurable family $\mu_t \in \mathcal{M}_+(\mathbf{R}^d \times \mathbf{R}^d)$ of finite measures, with total mass essentially uniformly bounded in $t \in \mathbf{R}$, such that, for every $\Xi \in L^1(\mathbf{R})$ and $a \in \mathcal{S}(\mathbf{R}^d \times \mathbf{R}^d)$:

$$\lim_{k\to\infty} \int_{\mathbf{R}\times\mathbf{R}^d\times\mathbf{R}^d} \Xi(t) a(x,\xi) W^{\varepsilon_k}_{u^{\varepsilon_k}(t)}(x,\xi) dx\, d\xi\, dt = \int_{\mathbf{R}\times\mathbf{R}^d\times\mathbf{R}^d} \Xi(t) a(x,\xi) \mu_t(dx, d\xi) dt.$$

Moreover, for every $\Xi \in L^1(\mathbf{R})$ and $\phi \in \mathcal{C}_0(\mathbf{R}^d)$:

$$\lim_{k\to\infty} \int_{\mathbf{R}} \int_{\mathbf{R}^d} \Xi(t)\phi(x) |u^{\varepsilon_k}(t,x)|^2 dx\, dt = \int_{\mathbf{R}} \int_{\mathbf{R}^d\times\mathbf{R}^d} \Xi(t)\phi(x)\mu_t(dx, d\xi) dt.$$

It turns out that the fact that $(u^{\varepsilon_k})_{k\in\mathbb{N}}$ is a sequence of solutions to (2) imposes certain restrictions on the measures μ_t that can be attained as a limit. In the region of the phase space $\mathbf{R}^d_x \times \mathbf{R}^d_\xi$ where equation (2) is dispersive (*i.e.*, away from the non-zero critical points of λ), the energy of the sequence $(u^{\varepsilon_k})_{k\in\mathbb{N}}$ is dispersed at infinite speed towards infinity. More precisely, Wigner measures μ_t satisfy:

$$\operatorname{supp}\mu_t \subset \mathbf{R}^d \times \Lambda. \tag{19}$$

Proofs of these results can be found in [5].

In what follows, we investigate the precise structure of Wigner measures μ_t. In order to get a better description of μ_t on $\mathbf{R}^d \times \Lambda$ we shall perform a second microlocalisation of the solutions above $\mathbf{R}^d \times \Lambda$.

2.2 Two-Microlocal Wigner Measures

Two-microlocal Wigner measures are objects designed to describe in a precise way oscillation and concentration effects exhibited by sequences of functions on a submanifold $X \subset \mathbf{R}^d \times \mathbf{R}^d$ of phase space. Roughly speaking, the idea consists in working in an enlarged phase space by adding an additional variable that will give a more precise description of the behavior of the Wigner functions close to the set X. These measures were introduced in [8,9,21,22] and further developed in [10] in a slightly different framework.

Here, we are particularly interested in the situation where $X = \Lambda$, the set of critical points of the symbol λ. In any case, the theory can be developed without assuming that we are dealing with solutions to an evolution equation, and the submanifold X is not required to have some dynamical meaning. It is convenient to present first the results in this more general framework.

We are first going to assume that X is an affine manifold of $\mathbf{R}^d$ with codimension p given by the equations:

$$\xi_{r+1} = \xi_0^1, \ \dots, \xi_d = \xi_0^p, \quad \text{for} \quad \xi_0'' = (\xi_0^1, \dots, \xi_0^p) \in \mathbf{R}^p, \quad r := d - p,$$

and, given $\xi \in \mathbf{R}^d$, we will set $\xi = (\xi', \xi'')$ with $\xi'' = (\xi_{r+1}, \dots, \xi_d)$.

Remark 2. Note that any submanifold of codimension p in $\mathbf{R}^d$ can be locally identified to a linear space $\{\xi'' = 0\}$ by using a suitable coordinate system, which may be used to extend the analysis of this section to this more general setting. However, in doing so, it turns out that the dependence on the choice of local coordinates becomes an issue and requires special care. We refer the reader to [5] for precise results in that setting.

Now we will extend the phase space $\mathbf{R}^d_x \times \mathbf{R}^d_\xi$ with a new variable $\eta \in \overline{\mathbf{R}^p}$, where $\overline{\mathbf{R}^p}$ is the compactification of $\mathbf{R}^p$ obtained by adding a sphere $\mathbf{S}^{p-1}$ at infinity. The test functions associated to this extended phase space are functions $a \in \mathcal{C}^\infty(\mathbf{R}^d_x \times \mathbf{R}^d_\xi \times \mathbf{R}^p_\eta)$ which satisfy the two following properties:

1. there exists a compact $K \subset \mathbf{R}^{2d}$ such that, for all $\eta \in \mathbf{R}^p$, the map $(x, \xi) \longmapsto a(x, \xi, \eta)$ is a smooth function compactly supported in K;

2. there exists a function a_∞ defined on $\mathbf{R}^d \times \mathbf{R}^d \times \mathbf{S}^{p-1}$ and $R_0 > 0$ such that,

$$\text{if } \|\eta\| > R_0, \text{ then } a(x,\xi,\eta) = a_\infty\left(x,\xi,\frac{\eta}{\|\eta\|}\right).$$

We denote by $\mathcal{S}^0(p)$ the set of such functions; to every $a \in \mathcal{S}^0(p)$ we associate a pseudodifferential operator $\mathrm{op}_\varepsilon^\sharp(a)$ as follows:

$$\mathrm{op}_\varepsilon^\sharp(a) = \mathrm{op}_\varepsilon(a_\varepsilon^\sharp), \quad \text{where} \quad a_\varepsilon^\sharp(x,\xi) = a\left(x,\xi,\frac{\xi''-\xi''_0}{\varepsilon}\right). \tag{20}$$

In the above formula, the additional variable $\eta = \frac{\xi''-\xi''_0}{\varepsilon}$ is introduced to capture in greater detail the concentration properties of a sequence of functions onto the set $\{\xi'' = \xi''_0\}$. Moreover, notice that

$$\mathrm{op}_\varepsilon^\sharp(a) = \mathrm{e}^{-i\frac{x''\cdot\xi''_0}{\varepsilon}}\mathrm{op}_1(a(x,\varepsilon\xi',\xi''_0+\varepsilon\xi'',\xi''))\mathrm{e}^{i\frac{x''\cdot\xi''_0}{\varepsilon}}, \tag{21}$$

which implies in particular that the family $(\mathrm{op}_\varepsilon^\sharp(a))_{\varepsilon>0}$ is uniformly bounded in $\mathcal{L}\left(L^2(\mathbf{R}^d)\right)$.

Now, let $(u^\varepsilon)_{\varepsilon>0}$ be a sequence in $\mathcal{C}^0\left(\mathbf{R}, L^2(\mathbf{R}^d)\right)$ (so each u^ε is a continuous function of time into $L^2(\mathbf{R}^d)$) satisfying the uniform bounds:

$$\exists C_0 > 0, \ \forall t \in \mathbf{R}, \ \|u^\varepsilon(t,\cdot)\|_{L^2(\mathbf{R}^d)} \le C_0.$$

Note that this is the case if $(u^\varepsilon)_{\varepsilon>0}$ is a family of solutions to 2 evolved from a sequence of initial data bounded in $L^2(\mathbf{R}^d)$. We will use these functions to define a linear functional $I^\varepsilon_{u^\varepsilon}$ acting on $\mathcal{S}^0(p) \times L^1(\mathbf{R})$ as:

$$I^\varepsilon_{u^\varepsilon}(a,\Xi) = \int_{\mathbf{R}} \Xi(t)\left(\mathrm{op}_\varepsilon^\sharp(a)u^\varepsilon(t,\cdot), u^\varepsilon(t,\cdot)\right)_{L^2(\mathbf{R}^d)} dt.$$

These functionals are actually lifts to the extended phase space $\mathbf{R}^d_x \times \mathbf{R}^d_\xi \times \overline{\mathbf{R}^p}_\eta$ of the Wigner distributions $W^\varepsilon_{u^\varepsilon(t,\cdot)}$. To see this, note that any function $a \in \mathcal{C}_0^\infty(\mathbf{R}^d_x \times \mathbf{R}^d_\xi)$ can be identified to an elements of $\mathcal{S}^0(p)$ that is constant in the variable η; clearly, under this identification one has:

$$\mathrm{op}_\varepsilon^\sharp(a) = \mathrm{op}_\varepsilon(a), \tag{22}$$

which implies, for these kind of a independent of η:

$$I^\varepsilon_{u^\varepsilon}(a,\Xi) = \int_{\mathbf{R}}\int_{\mathbf{R}^d\times\mathbf{R}^d} \Xi(t)a(x,\xi)W^\varepsilon_{u^\varepsilon(t,\cdot)}(x,\xi)\,dx\,d\xi, \quad \forall a \in \mathcal{C}_0^\infty(\mathbf{R}^d\times\mathbf{R}^d).$$

Therefore, letting μ_t denote the Wigner measures of $(u^\varepsilon(t,\cdot))_{\varepsilon>0}$ as described in Sect. 2.1, we have by dominated convergence and the definition of μ_t:

$$I^\varepsilon_{u^\varepsilon}(a,\Xi) \underset{\varepsilon\to 0}{\longrightarrow} \int_{\mathbf{R}} \Xi(t)\int_{\mathbf{R}^d\times\mathbf{R}^d} a(x,\xi)\,\mu_t(dx,d\xi)dt, \quad \forall a \in \mathcal{C}_0^\infty(\mathbf{R}^d\times\mathbf{R}^d). \tag{23}$$

Nevertheless, when the convergence of $(I^\varepsilon_{u^\varepsilon})_{\varepsilon>0}$ is tested against general functions on the extended phase, the resulting accumulation points have some additional structure:

Proposition 2. *Suppose that $(u^\varepsilon)_{\varepsilon>0}$ and μ_t are as above. Then, up to the extraction of a sequence $(\varepsilon_k)_{k\in\mathbb{N}}$, there exist a L^∞-map $\gamma : t \longmapsto \gamma_t$ taking values in the set of positive Radon measures on $\mathbf{R}^d \times \mathbf{R}^r \times \mathbf{S}^{p-1}$ and a L^∞-map $\mathtt{M} : t \longmapsto \mathtt{M}_t$ into the set of operator-valued positive measures on $\mathbf{R} \times \mathbf{R}^{2r}$ that are trace class operators on $L^2(\mathbf{R}^p_y)$ such that, for every $a \in \mathcal{S}^0(p)$ and $\Xi \in L^1(\mathbf{R})$,*

$$
\begin{aligned}
I^\varepsilon_{u^\varepsilon}(a,\Xi) \underset{\varepsilon\to 0}{\longrightarrow} & \int_{\mathbf{R}} \Xi(t) \int_{\{\xi''\neq\xi''_0\}} a_\infty\left(x,\xi,\frac{\xi''-\xi''_0}{\|\xi''-\xi''_0\|}\right)\mu_t(dx,d\xi)dt \\
&+ \int_{\mathbf{R}} \Xi(t) \int_{\mathbf{R}^d\times\mathbf{R}^r\times\mathbf{S}^{p-1}} a_\infty(x,\xi',\xi''_0,\omega)\gamma_t(dx,d\xi',d\omega)dt \\
&+ \int_{\mathbf{R}} \Xi(t) \int_{\mathbf{R}^{2r}} \mathrm{Tr}_{L^2(\mathbf{R}^p)}\left[a^W(x',y,\xi',\xi''_0,D_y)\mathtt{M}_t(dx',d\xi')\right] dt,
\end{aligned}
\tag{24}
$$

where, for every $(x',\xi') \in \mathbf{R}^{2r}$, $a^W(x',y,\xi',\xi''_0,D_y)$ denotes the pseudodifferential operator acting on $L^2(\mathbf{R}^p)$ obtained by the Weyl quantization of the symbol $(y,\eta) \longmapsto a(x',y,\xi',\xi''_0,\eta)$.

The proof of this result is essentially identical to that of Theorem 1 in [9] (except for the fact that here everything depends on t); see also [1,2,8] for very closely related results in a slightly different context.

In order to enlighten the nature of the different objects involved in formula (24), we emphasize the following characterization: let $\chi \in \mathcal{C}^\infty_0(\mathbf{R}^p)$ be a cut-off function such that $0 \le \chi \le 1$, $\chi(\eta) = 1$ for $\|\eta\| \le 1$ and $\chi(\eta) = 0$ for $\|\eta\| \ge 2$; then, one has: (see again the proof of Theorem 1 in [9], or [1,2,8]):

(i) the measure γ_t in Proposition 2 is obtained through the limiting procedure

$$
\int_{\mathbf{R}} \Xi(t) \int_{\mathbf{R}^d\times\mathbf{R}^r\times\mathbf{S}^{p-1}} a_\infty(x,\xi',\xi''_0,\omega)\gamma_t(dx,d\xi',d\omega)dt = \lim_{\delta\to 0}\lim_{R\to\infty}\lim_{\varepsilon\to 0} I^\varepsilon_{u^\varepsilon}(a^{R,\delta},\Xi),
$$

where

$$
a^{R,\delta}(x,\xi,\eta) = a(x,\xi,\eta)\chi\left(\frac{\xi''-\xi''_0}{\delta}\right)\left(1-\chi\left(\frac{\eta}{R}\right)\right); \tag{25}
$$

(ii) the measure $\mathtt{M}_t$ in Proposition 2 is obtained as the iterated limits

$$
\int_{\mathbf{R}} \Xi(t) \int_{\mathbf{R}^{2r}} \mathrm{Tr}_{L^2(\mathbf{R}^p)}[a^W(x',y,\xi',\xi''_0,D_y)\mathtt{M}_t(dx',d\xi')]dt = \lim_{\delta\to 0}\lim_{R\to\infty}\lim_{\varepsilon\to 0} I^\varepsilon_{u^\varepsilon}(a_{R,\delta},\Xi),
$$

where

$$
a_{R,\delta}(x,\xi,\eta) = a(x,\xi,\eta)\chi\left(\frac{\xi''-\xi''_0}{\delta}\right)\chi\left(\frac{\eta}{R}\right). \tag{26}
$$

The presence of the cut-off $\chi(\eta/R)$ explains the different roles played by γ_t and $\mathtt{M}_t$. The measure $\mathtt{M}_t$ captures the fraction of the L^2-mass of the sequence (u^ε) that concentrates onto $\{\xi'' = \xi''_0\}$ at rate precisely ε. The measure γ_t, on the other hand, describes how the L^2-mass of the sequences concentrates on $\{\xi'' = \xi''_0\}$ at a slower rate.

Besides, it is convenient to use a decomposition of $\mathtt{M}_t$ based on the Radon-Nikodym Theorem. Define the map $\nu : t \longmapsto \nu_t$ by

$$\nu_t(dx', d\xi') = \mathrm{Tr}_{L^2(\mathbf{R}^p)}\mathtt{M}_t(dx', d\xi').$$

This L^∞-map is valued in the set of positive measures on $\mathbf{R}^{2r}$ and there exists a measurable map $M : (t, x', \xi') \longmapsto M_t(x', \xi')$ valued in the set of self-adjoint, positive, trace-class operators on $L^2(\mathbf{R}^p)$ such that

$$\mathtt{M}_t(dx', d\xi') = M_t(x', \xi')\nu_t(dx', d\xi').$$

Note that, by construction, we have $\mathrm{Tr}_{L^2(\mathbf{R}^p)}M_t(x', \xi') = 1$. We are then left with three objects, γ_t, ν_t and M_t.

Note that the results we have presented so far hold without assuming that u^ε solves an evolution equation, nor that $\{\xi'' = \xi_0''\}$ is the set of critical points of the function λ. The fact that the sequence $(u^{\varepsilon_k}(t, \cdot))_{k\in\mathbb{N}}$ generating μ_t consists of solutions to Eq. (2) and that Λ is the set of critical points of its symbol implies additional regularity and propagation properties on the measures γ_t and $\mathtt{M}_t$ that we will use in the next section. Let us anticipate that the latter propagates following a Heisenberg equation, whereas γ_t, enjoys an additional geometric invariance. Finally, it is not hard to prove (though we will not do that here, see [5]) that ν_t does not depend on t, and in fact $\nu_t = \nu^0$, which is the measure appearing in the statement of Theorem 3, only depends on the sequence of initial data.

Let us mention that the use of two-microlocal semiclassical measures for dispersive equations was initiated in [20], in the context of the Schrödinger equation on the torus. These results were largely extended and improved in subsequent works [1,2]. The reader might find interesting to compare the results of the present note to those in the aforementioned references.

3 The Particular Case: Countable Critical Points

This section is mainly devoted to a sketch of the proof of Theorems 1, 2, and to the analysis of the examples of Proposition 1.

3.1 Two Microlocal Wigner Measures Associated to a Critical Point

Our goal in this section will be to compute the restriction to $\{\xi = \xi_0\}$, with $\xi_0 \in \Lambda$, of the semiclassical measure μ_t associated to sequences of solutions to (2) in terms of quantities that depend only on the sequence of initial data.

The results of the previous section applied to the particular case $\{\xi = \xi_0\}$, $p = d$, ensure the existence of measures $\gamma_t \in \mathcal{M}_+(\mathbf{R}^d \times \mathbf{S}^{d-1})$ and of a positive family of Hermitian operators $M_t \in \mathcal{L}^1(L^2(\mathbf{R}^d))$ such that, for all $(a, \Xi) \in \mathcal{S}^0(d) \times L^1(\mathbf{R})$,

$$\int_{\mathbf{R}} \Xi(t) \int_{\mathbf{R}^d\times\mathbf{S}^{d-1}} a_\infty(x, \xi_0, \omega)\gamma_t(dx, d\omega)dt = \lim_{\delta\to 0}\lim_{R\to\infty}\lim_{\varepsilon\to 0} I^\varepsilon_{u^\varepsilon}(a^{R,\delta}, \Xi) \quad (27)$$

and

$$\int_{\mathbf{R}} \Xi(t)\mathrm{Tr}_{L^2(\mathbf{R}^d)}[a^W(y,\xi_0,D_y)M_t]dt = \lim_{\delta\to 0}\lim_{R\to\infty}\lim_{\varepsilon\to 0} I^\varepsilon_{u^\varepsilon}(a_{R,\delta},\Xi), \tag{28}$$

where $a^{R,\delta}(x,\xi,\eta)$ and $a_{R,\delta}(x,\xi,\eta)$ are defined in (25) and (26) respectively. The localization property (19) and Proposition 2, together with identity (23), then assert that, for every $b \in \mathcal{C}_0^\infty(\mathbf{R}^d \times \mathbf{R}^d)$ and a.e. $t \in \mathbf{R}$:

$$\int_{\{\xi=\xi_0\}} b(x,\xi)\mu_t(dx,d\xi) = \int_{\mathbf{R}^d\times\mathbf{S}^{d-1}} b(x,\xi_0)\gamma_t(dx,d\omega) + \mathrm{Tr}_{L^2(\mathbf{R}^d)}[b^W(\,\cdot\,,\xi_0)M_t]. \tag{29}$$

The fact that u^ε solves Eq. (2) implies that γ_t and M_t enjoy the following additional properties:

Theorem 5. *Let $(u^\varepsilon)_{\varepsilon>0}$ be a sequence of solutions to (2) issued from a $L^2(\mathbf{R}^d)$-bounded sequence of initial data $(u_0^\varepsilon)_{\varepsilon>0}$, then:*

(i) For almost every $t \in \mathbf{R}$, the measure γ_t is invariant through the flow

$$\phi_s^2 : \mathbf{R}^d \times \mathbf{S}^{d-1} \ni (x,\omega) \longmapsto (x + s\,\nabla^2\lambda(\xi_0)\omega,\omega) \in \mathbf{R}^d \times \mathbf{S}^{d-1}.$$

(ii) $M_t = |u_{\xi_0}(t,\cdot)\rangle\langle u_{\xi_0}(t,\cdot)|$, where u_{ξ_0} solves

$$\begin{cases} i\partial_t u_{\xi_0}(t,x) = \frac{1}{2}\nabla^2\lambda(\xi_0)D_x\cdot D_x u_{\xi_0}(t,x) + V(x)u_{\xi_0}(t,x), \\ u_{\xi_0}|_{t=0}(x) = u^0_{\xi_0}(x), \end{cases} \tag{30}$$

and $u^0_{\xi_0}$ is a weak limit in $L^2(\mathbf{R}^d)$ of $(\mathrm{e}^{-\frac{i}{\varepsilon}\xi_0\cdot x}u_0^\varepsilon)_{\varepsilon>0}$ when $\varepsilon \to 0^+$.

The localization property (19) and Corollary 2 below imply together Theorem 1.

Corollary 2. *For every $\xi_0 \in \Lambda$ and almost every $t \in \mathbf{R}$ one has*

$$\mu_t(dx,d\xi)\rceil_{\{\xi=\xi_0\}} \geq |u_{\xi_0}(t,x)|^2 dx\,\delta_{\xi_0}(d\xi),$$

with equality if ξ_0 is a non-degenerate critical point.

Proof. We are going to show that the measure γ_t vanishes identically if ξ_0 is non-degenerate. This is a consequence of the following result, whose proof can be found in [5].

Lemma 1. *Let be $\Phi_s : \mathbf{R}^d\times\mathbf{R}^d \longrightarrow \mathbf{R}^d\times\mathbf{R}^d$ a flow satisfying: for every compact $K \subset \mathbf{R}^d\times\mathbf{R}^d$ containing no stationary points of Φ, there exist constants $\alpha,\beta > 0$ such that:*

$$\alpha|s| - \beta \leq \|\Phi_s(x,\xi)\| \leq \alpha|s| + \beta, \quad \forall (x,\xi) \in K\ \forall s \in \mathbf{R}.$$

Moreover, let μ be a finite, positive Radon measure on $\mathbf{R}^d \times \Omega$ that is invariant by the flow Φ_s. Then μ is supported on the set of stationary points of Φ_s.

When this lemma is applied to the measure γ_t and the flow ϕ_s^2, one finds out that $\gamma_t = 0$.

Remark 3. The formula in Corollary 2 shows in particular that the semiclassical measure μ_t is not uniquely determined by the semiclassical measure μ_0 of the sequence of initial data. Suppose that ξ_0 is a non-degenerate critical point; if $u_0^\varepsilon = \theta(x)e^{\frac{i}{\varepsilon}\xi_0\cdot x}$, $\|\theta\|_{L^2(\mathbf{R}^d)} = 1$, then $u_{\xi_0}^0 = \theta \neq 0$ and $\mu_0 = dx \otimes \delta_{\xi_0}$ and Corollary 2 tells us that:

$$\mu_t(dx, d\xi) = |u_{\xi_0}(t,x)|^2 dx\, \delta_{\xi_0}(d\xi) \neq 0.$$

However, if we choose initial data $v_0^\varepsilon(x) = \theta(x)e^{\frac{i}{\varepsilon}(\xi_0+\varepsilon^\beta\xi_0)\cdot x}$ with $\beta \in (0,1)$, they have the same semiclassical measure μ_0 as $(u_0^\varepsilon)_{\varepsilon>0}$, whereas Corollary 2 now shows that the measure $\mu_t(dx, d\xi)$ is 0, since any weak limit of $e^{-\frac{i}{\varepsilon}\xi_0\cdot x}v_0^\varepsilon$ is 0.

Proof. (Proof of Theorem 5.) Let us start proving part (i), namely the invariance of γ_t. Let $a \in \mathcal{S}_0(d)$ and $a^{R,\delta}$ as defined in (25). We set

$$\widetilde{\phi}_s^2(x,\xi,\eta) := (x + s\nabla^2\lambda(\xi_0)\frac{\eta}{|\eta|}, \xi, \eta), \quad (x,\xi,\eta) \in \mathbf{R}^d \times \mathbf{R}^d \times \mathbf{R}^d$$

and we note that $a^{R,\delta} \circ \widetilde{\phi}_s^2$ also is a symbol of $\mathcal{S}^0(d)$ (in particular, it is smooth because it is supported on $|\eta| > R$). Besides

$$\left(a^{R,\delta} \circ \widetilde{\phi}_s^2\right)_\infty = a_\infty \circ \phi_s^2.$$

Our aim is to prove that for all $\Xi \in \mathcal{C}_0^\infty(\mathbf{R}_t)$, as ε goes to 0, then R to $+\infty$ and finally δ to 0,

$$I_{u^\varepsilon}^\varepsilon\left(a^{R,\delta} \circ \widetilde{\phi}_s^2, \Xi\right) = I_{u^\varepsilon}^\varepsilon\left(a^{R,\delta}, \Xi\right) + o(1). \tag{31}$$

We observe that the quantification of $a^{R,\delta} \circ \widetilde{\phi}_s^2$ has the following property

$$\begin{aligned}\left(a^{R,\delta} \circ \widetilde{\phi}_s^2\right)_\varepsilon^\sharp &= a^{R,\delta}\left(x + \frac{s}{|\xi - \xi_0|}\nabla^2\lambda(\xi_0)(\xi - \xi_0), \xi, \frac{\xi - \xi_0}{\varepsilon}\right)\\ &= a^{R,\delta}\left(x + \frac{s}{|\xi - \xi_0|}\nabla\lambda(\xi), \xi, \frac{\xi - \xi_0}{\varepsilon}\right) + r_{\varepsilon,\delta,R}(x,\xi)\end{aligned}$$

with

$$\|\mathrm{op}_\varepsilon(r_{\varepsilon,\delta,R})\|_{\mathcal{L}(L^2(\mathbf{R}^d))} = O(\delta).$$

For this, we have used that $|\xi - \xi_0| < \delta$ on the support of $a^{R,\delta}$, that $\nabla\lambda(\xi_0) = 0$ and that there exists a smooth bounded tensor of degree 3, Γ, with bounded derivatives such that

$$\nabla\lambda(\xi) = \nabla^2\lambda(\xi_0)(\xi - \xi_0) + \Gamma(\xi)[\xi - \xi_0, \xi - \xi_0].$$

As a consequence, the claim (31) is equivalent to proving that for all $\Xi \in \mathcal{C}_0^\infty(\mathbf{R}_t)$, as ε goes to 0, then δ to 0 and R to $+\infty$,

$$I_{u^\varepsilon}^\varepsilon\left(b_\varepsilon^{R,\delta}(s), \Xi\right) = I_{u^\varepsilon}^\varepsilon\left(a^{R,\delta}, \Xi\right) + o(1)$$

where $b_\varepsilon^{R,\delta}(s)$ is the symbol

$$b_\varepsilon^{R,\delta}(s)(x,\xi) = a^{R,\delta}\left(x + \frac{s}{|\xi-\xi_0|}\nabla^2\lambda(\xi_0)(\xi-\xi_0), \xi, \frac{\xi-\xi_0}{\varepsilon}\right).$$

To this aim, we will show that for all $\Xi \in \mathcal{C}_0^\infty(\mathbf{R}_t)$, as ε goes to 0, then R to $+\infty$ and finally δ to 0,

$$I_{u^\varepsilon}^\varepsilon\left(\partial_s b_\varepsilon^{R,\delta}(s), \Xi\right) = o(1). \tag{32}$$

We observe that

$$\begin{aligned}\partial_s b_\varepsilon^{R,\delta}(s)(x,\xi) &= |\xi-\xi_0|^{-1}\nabla\lambda(\xi)\cdot\nabla_x a^{R,\delta}\left(x + \frac{s}{|\xi-\xi_0|}\nabla^2\lambda(\xi_0)(\xi-\xi_0), \xi, \frac{\xi-\xi_0}{\varepsilon}\right)\\ &= \nabla\lambda(\xi)\cdot\nabla_x\widetilde{b}_\varepsilon^{R,\delta}(s,x,\xi),\end{aligned}$$

with

$$\widetilde{b}_\varepsilon^{R,\delta}(s,x,\xi) := |\xi-\xi_0|^{-1}b_\varepsilon^{R,\delta}(s)(x,\xi).$$

This function satisfies: for all $\alpha \in \mathbb{N}^d$ there exists a constant C_α such that for all $x,\xi \in \mathbf{R}^d$

$$|\widetilde{b}_\varepsilon^{R,\delta}(s,x,\xi)| + |\partial_x^\alpha\widetilde{b}_\varepsilon^{R,\delta}(x,\xi)| \le C_\alpha(R\varepsilon)^{-1}.$$

By the symbolic calculus of semiclassical pseudodifferential operators, we have

$$\begin{aligned}\mathrm{op}_\varepsilon^\sharp\left(\nabla\lambda(\xi)\cdot\nabla_x\widetilde{b}_\varepsilon^{R,\delta}\right) &= \frac{i}{\varepsilon}\left[\lambda(\varepsilon D), \mathrm{op}_\varepsilon\left(\widetilde{b}_\varepsilon^{R,\delta}\right)\right] + O(\varepsilon) + O(1/R)\\ &= \frac{i}{\varepsilon}\left[\lambda(\varepsilon D_x) + \varepsilon^2 V, \mathrm{op}_\varepsilon\left(\widetilde{b}_\varepsilon^{R,\delta}\right)\right] + O(1/R) + O(\varepsilon).\end{aligned}$$

On the other hand,

$$\frac{d}{dt}\left(\mathrm{op}_\varepsilon\left(\widetilde{b}_\varepsilon^{R,\delta}\right)u^\varepsilon(t), u^\varepsilon(t)\right) = \frac{i}{\varepsilon^2}\left(\left[\lambda(\varepsilon D_x) + \varepsilon^2 V, \mathrm{op}_\varepsilon\left(\widetilde{b}_\varepsilon^{R,\delta}\right)\right]u^\varepsilon(t), u^\varepsilon(t)\right) + O(\varepsilon).$$

Therefore

$$\begin{aligned}I_{u^\varepsilon}^\varepsilon\left(\partial_s b_\varepsilon^{R,\delta}(s), \Xi\right) &= \int \Xi(t)\left(\mathrm{op}_\varepsilon^\sharp\left(\nabla\lambda(\xi)\cdot\nabla_x\widetilde{b}_\varepsilon^{R,\delta}\right)u^\varepsilon(t), u^\varepsilon(t)\right)dt\\ &= \frac{i}{\varepsilon}\int \Xi(t)\left(\left[\lambda(\varepsilon D), \mathrm{op}_\varepsilon\left(\widetilde{b}_\varepsilon^{R,\delta}\right)\right]u^\varepsilon(t), u^\varepsilon(t)\right)dt + O(1/R) + O(\varepsilon)\\ &= \varepsilon\int \Xi(t)\frac{d}{dt}\left(\mathrm{op}_\varepsilon\left(\widetilde{b}_\varepsilon^{R,\delta}\right)u^\varepsilon(t), u^\varepsilon(t)\right)dt + O(1/R) + O(\varepsilon)\\ &= -\varepsilon\int \Xi'(t)\left(\mathrm{op}_\varepsilon\left(\widetilde{b}_\varepsilon^{R,\delta}\right)u^\varepsilon(t), u^\varepsilon(t)\right)dt + O(1/R) + O(\varepsilon)\\ &= O(1/R) + O(\varepsilon),\end{aligned}$$

which gives (32), thus (31), and concludes the proof.

To prove part (ii) one starts noticing that, by symbolic calculus and (21),

$$\left(\mathrm{op}_\varepsilon^\sharp\left(a_{R,\delta}\right)u^\varepsilon(t), u^\varepsilon(t)\right) = \left(\mathrm{op}_1(A_{R,\delta}^\varepsilon)\Phi^\varepsilon(t), \Phi^\varepsilon(t)\right),$$

where

$$\Phi^\varepsilon(t,x) := \mathrm{e}^{-\frac{i}{\varepsilon}\xi_0\cdot x} u^\varepsilon(t,x), \quad A^\varepsilon_{R,\delta}(x,\xi) := a_{R,\delta}\left(x,\xi_0+\varepsilon\xi,\xi\right).$$

Since u^ε solves (2), one sees that Φ^ε satisfies

$$i\partial_t\Phi^\varepsilon(t,x) = \frac{1}{\varepsilon^2}\lambda(\xi_0+\varepsilon D_x)\Phi^\varepsilon(t,x) + V(x)\Phi^\varepsilon(t,x) + O(\varepsilon).$$

A Taylor expansion for $\lambda(\xi)$ around ξ_0 shows that setting

$$u^\varepsilon_{\xi_0}(t,x) := \mathrm{e}^{\frac{it}{\varepsilon^2}\lambda(\xi_0)}\Phi^\varepsilon(t,x),$$

then $u^\varepsilon_{\xi_0}$ solves in $L^2(\mathbf{R}^d)$,

$$i\partial_t u^\varepsilon_{\xi_0}(t,x) = \nabla^2\lambda(\xi_0)D_x\cdot D_x\, u^\varepsilon_{\xi_0}(t,x) + V(x)u^\varepsilon_{\xi_0}(t,x) + O(\varepsilon).$$

We still have,

$$\left(\mathrm{op}^\sharp_\varepsilon\left(a_{R,\delta}\right)u^\varepsilon(t), u^\varepsilon(t)\right) = \left(\mathrm{op}_1(A^\varepsilon_{R,\delta})u^\varepsilon_{\xi_0}(t), u^\varepsilon_{\xi_0}(t)\right).$$

On the other hand,

$$\begin{aligned} A^\varepsilon_{R,\delta}(x,\xi) &= a\left(x,\xi_0+\varepsilon\xi,\xi\right)\chi\left(\varepsilon\xi/\delta\right)\chi\left(\xi/R\right) \\ &= a\left(x,\xi_0,\xi\right)\chi\left(\xi/R\right) + O(\varepsilon) = A^0_R(x,\xi) + O(\varepsilon). \end{aligned}$$

Notice that the remainder depends on R and δ, but that this is harmless since we shall first let ε go to 0. Using again the symbolic calculus, we write

$$\mathrm{op}_1(A^\varepsilon_{R,\delta}) = \mathrm{op}_1(A^0_R) + O(\varepsilon).$$

Therefore,

$$\lim_{\varepsilon\to 0^+} I^\varepsilon_{u^\varepsilon}(a_{R,\delta},\Xi) = \lim_{\varepsilon\to 0^+}\int_{\mathbf{R}} \Xi(t)\left(\mathrm{op}_1(A^0_R)u^\varepsilon_{\xi_0}(t), u^\varepsilon_{\xi_0}(t)\right)dt. \tag{33}$$

By [26, Lemma 4.26], $\mathrm{op}_1(A^0_R)$ is a compact operator on $L^2(\mathbf{R}^d)$; therefore, if

$$u^{\varepsilon_k}_{\xi_0}(0,\cdot) \rightharpoonup u^0_{\xi_0}$$

along some subsequence $(\varepsilon_k)_{k\in\mathbb{N}}$, it follows that, for every $t\in\mathbf{R}$,

$$\lim_{k\to\infty}\left(\mathrm{op}_1(A^0_R)u^{\varepsilon_k}_{\xi_0}(t), u^{\varepsilon_k}_{\xi_0}(t)\right) = \left(\mathrm{op}_1(A^0_R)u_{\xi_0}(t), u_{\xi_0}(t)\right),$$

where u_{ξ_0} solves:

$$i\partial_t u_{\xi_0}(t,x) = \nabla^2\lambda(\xi_0)D_x\cdot D_x\, u_{\xi_0}(t,x) + V(x)u_{\xi_0}(t,x), \quad u_{\xi_0}|_{t=0}(x) = u^0_{\xi_0}(x).$$

In particular, if the convergence (33) takes place then the sequence $(u^\varepsilon_{\xi_0}(0,\cdot))$ must have a unique weak accumulation point and:

$$\lim_{\varepsilon\to 0^+} I^\varepsilon_{u^\varepsilon}(a_{R,\delta},\Xi) = \int_{\mathbf{R}} \Xi(t)\mathrm{Tr}_{L^2(\mathbf{R}^d)}[\mathrm{op}_1(A^0_R)|u_{\xi_0}(t)\rangle\langle u_{\xi_0}(t)|]dt.$$

The result follows from (28) by letting R go to $+\infty$.

In order to prove Theorem 2, notice that the assumption that is made in its statement implies that $\gamma_0 = 0$ (by the characterization of γ in (27)), and the result comes from the conservation of mass of γ_t:

Lemma 2. *For all $t \in \mathbf{R}$,*

$$\int_{\mathbf{R}^d \times \mathbf{S}^{d-1}} \gamma_t(dx, d\omega) = \int_{\mathbf{R}^d \times \mathbf{S}^{d-1}} \gamma_0(dx, d\omega).$$

Proof. Using the characterization in (27), we have, for $R, \delta > 0$ and χ as in (25):

$$J^{\varepsilon}_{R,\delta}(t) = \left((1-\chi)\left(\frac{\varepsilon D_x - \xi_0}{R\varepsilon}\right) \chi\left(\frac{\varepsilon D_x - \xi_0}{\delta}\right) u^{\varepsilon}(t,\cdot), u^{\varepsilon}(t,\cdot) \right).$$

Using the dynamical equation (2), we obtain

$$\begin{aligned} \frac{d}{dt} J^{\varepsilon}_{R,\delta}(t) &= -i \left(\left[(1-\chi)\left(\frac{\varepsilon D_x - \xi_0}{R\varepsilon}\right) \chi\left(\frac{\varepsilon D_x - \xi_0}{\delta}\right), V \right] u^{\varepsilon}(t,\cdot), u^{\varepsilon}(t,\cdot) \right) \\ &= O(\varepsilon) + O(1/R) + O(\delta) \end{aligned}$$

by semiclassical symbolic calculus. Therefore, taking limits in all the parameters one concludes.

3.2 Degenerate Critical Points

In this section we focus on the situation of Proposition 1 with the family of initial data given by (12). These concentrate microlocally onto ξ_0, and the two-microlocal measures associated with them depend on the value of α:

(i) If $\alpha = 0$, then $\gamma_0(dx, d\omega) = |\theta(x)|^2 dx\, \delta_{\omega_0}(d\omega)$ and $M_0 = 0$.
(ii) If $\alpha \neq 0$, then $\gamma_0(dx, d\omega) = \delta_0(dx)\, \delta_{\omega_0}(d\omega)$ and $M_0 = 0$.

Comparatively, for the data in (4) (which corresponds to $\alpha = 0$ and $\omega_0 = 0$), we have $\gamma_0 = 0$ and M_0 is the projector on θ. The fact that the direction of oscillations has been shifted by $\varepsilon^{\beta}\omega_0$ yields that all mass concentrating onto ξ_0 comes from the infinity with respect to the scale ε. The contributions that we observe in Proposition 1 are reminiscent of the measure γ_t which happens to be non-zero in this situation. Since $V = 0$, it is possible to calculate everything explicitly and one gets the following description, which implies Proposition 1:

Lemma 3. *Assume $\omega_0 \in \ker \nabla^2\lambda(\xi_0)$, $\beta > \frac{2}{3}$ and $V = 0$.*

(i) If $\alpha = 0$, then $\gamma_t(dx, d\omega) = \left| \mathrm{e}^{it\nabla^2\lambda(\xi_0)D_x \cdot D_x} \theta(x) \right|^2 dx \delta_{\omega_0}(d\omega)$ *and therefore:*

$$\mu_t(dx, d\xi) = \left| \mathrm{e}^{it\nabla^2\lambda(\xi_0)D_x \cdot D_x} \theta(x) \right|^2 dx \delta_{\xi_0}(d\xi).$$

(ii) If $\alpha \neq 0$, then $\gamma_t(dx, d\omega) = \|\theta\|^2_{L^2(\mathbf{R}^d)}\delta_0(dx)\delta_{\omega_0}(d\omega)$ and in particular:

$$\mu_t(dx, d\xi) = \|\theta\|^2_{L^2(\mathbf{R}^d)}\delta_0(dx)\delta_{\xi_0}(d\xi).$$

Proof. The proof relies on the analysis of the product

$$L^\varepsilon := \left(\mathrm{op}_\varepsilon\left(a\left(x,\xi,\frac{\xi-\xi_0}{\varepsilon}\right)\right)u^\varepsilon(t), u^\varepsilon(t)\right), \quad a \in \mathcal{S}^0(d).$$

We observe that L^ε reads:

$$L^\varepsilon = (2\pi)^{-3d}\varepsilon^{-3d-d\alpha}\int_{\mathbf{R}^{7d}} a\left(\frac{x+y}{2},\xi,\frac{\xi-\xi_0}{\varepsilon}\right)\overline{\theta}\left(\frac{x'}{\varepsilon^\alpha}\right)\theta\left(\frac{y'}{\varepsilon^\alpha}\right)$$
$$\times \operatorname{Exp}\left[\frac{i}{\varepsilon}\left((x-y)\cdot\xi-(x'-y')(\xi_0+\varepsilon^\beta\omega_0)+\zeta\cdot(y-y')-\eta\cdot(x-x')\right)\right]$$
$$\times \operatorname{Exp}\left[\frac{it}{\varepsilon^2}\left(\lambda(\eta)-\lambda(\zeta)\right)\right]dx'\,dy'\,dx\,dy\,d\xi\,d\zeta\,d\eta.$$

We perform the change of variables

$$x=\varepsilon^\alpha X,\quad x'=\varepsilon^\alpha X',\quad y=\varepsilon^\alpha Y,\quad y'=\varepsilon^\alpha Y',$$
$$\xi=\xi_0+\varepsilon^\beta\omega_0+\varepsilon^{1-\alpha}\xi',\quad \zeta=\xi_0+\varepsilon^\beta\omega_0+\varepsilon^{1-\alpha}\zeta',\quad \eta=\xi_0+\varepsilon^\beta\omega_0+\varepsilon^{1-\alpha}\eta'$$

in order to obtain

$$L^\varepsilon = (2\pi)^{-3d}\int_{\mathbf{R}^{7d}} a\left(\varepsilon^\alpha\frac{X+Y}{2},\xi_0+\varepsilon^\beta\omega_0+\varepsilon^{1-\alpha}\xi',\varepsilon^{-1+\beta}(\omega_0+\varepsilon^{1-\alpha-\beta}\xi')\right)$$
$$\times \operatorname{Exp}\left[i\left(\xi'\cdot(X-Y)+\zeta'\cdot(Y-Y')-\eta'\cdot(X-X')\right)+\frac{it}{\varepsilon^2}\Gamma_\varepsilon(\zeta',\eta')\right]$$
$$\times\,\overline{\theta}\left(X'\right)\theta\left(Y'\right)dX'\,dY'\,dX\,dY\,d\xi'\,d\zeta'\,d\eta',$$

with

$$\begin{aligned}\Gamma_\varepsilon(\zeta',\eta') &= \lambda\left(\xi_0+\varepsilon^\beta\omega_0+\varepsilon^{1-\alpha}\zeta'\right)-\lambda\left(\xi_0+\varepsilon^\beta\omega_0+\varepsilon^{1-\alpha}\eta'\right)\\ &= \varepsilon^{2(1-\alpha)}\left(\nabla^2\lambda(\xi_0)\eta'\cdot\eta'-\nabla^2\lambda(\xi_0)\zeta'\cdot\zeta'\right)+O(\varepsilon^{3\beta}),\end{aligned}$$

where we have used $\nabla^2\lambda(\xi_0)\omega_0=0$ and $\beta<1-\alpha$. Since $3\beta>2$, the term in $O(\varepsilon^{3\beta})$ will be negligible in the phase. Now, the situation depends on whether $\alpha=0$ or not.

If $\alpha\neq 0$, by use of Taylor expansion and by the definition of a, one easily convinces oneself that

$$L^\varepsilon \sim a_\infty\left(0,\xi_0,\omega_0\right)(2\pi)^{-3d}\int_{\mathbf{R}^{7d}}\overline{\theta}\left(X'\right)\theta\left(Y'\right)\operatorname{Exp}\left[i\left(\xi'\cdot(X-Y)+\zeta'\cdot(Y-Y')-\eta'\cdot(X-X')\right)\right]$$
$$\times \operatorname{Exp}\left[\frac{it}{\varepsilon^{2\alpha}}\left(\nabla^2\lambda(\xi_0)\eta'\cdot\eta'-\nabla^2\lambda(\xi_0)\zeta'\cdot\zeta'\right)\right]dX'\,dY'\,dX\,dY\,d\xi'\,d\zeta'\,d\eta'.$$

The integration in ξ' generates a Dirac mass $\delta(X-Y)$, then the integration in X generates a Dirac mass $\delta(\zeta'-\eta')$, whence

$$L^\varepsilon \sim (2\pi)^{-d} a_\infty(0,\xi_0,\omega_0) \int_{\mathbf{R}^{3d}} \overline{\theta}(X')\,\theta(Y')\,\mathrm{Exp}\left[i\eta'\cdot(X'-Y')\right]\, dX'\, dY'\, d\eta',$$

whence

$$L^\varepsilon \sim a_\infty(0,\xi_0,\omega_0)\, \|\theta\|_{L^2(\mathbf{R}^d)}.$$

If $\alpha=0$, similar arguments give

$$\begin{aligned} L^\varepsilon \sim (2\pi)^{-3d} \int_{\mathbf{R}^{7d}} & a_\infty\left(\frac{X+Y}{2},\xi_0,\omega_0\right) \overline{\theta}(X')\,\theta(Y') \\ & \times \mathrm{Exp}\left[i\left(\xi'\cdot(X-Y)+\zeta'\cdot(Y-Y')-\eta'\cdot(X-X')\right)\right] \\ & \times \mathrm{Exp}\left[it\left(\nabla^2\lambda(\xi_0)\eta'\cdot\eta'-\nabla^2\lambda(\xi_0)\zeta'\cdot\zeta'\right)\right] dX'\, dY'\, dX\, dY\, d\xi'\, d\zeta'\, d\eta'. \end{aligned}$$

Integration in ξ' generates a Dirac mass $\delta(X-Y)$ and integration in Y' and X' give

$$\begin{aligned} L^\varepsilon \sim (2\pi)^{-2d} \int_{\mathbf{R}^{3d}} & a_\infty(X,\xi_0,\omega_0)\, \overline{\widehat{\theta}}(\eta')\, \widehat{\theta}(\zeta')\,\mathrm{Exp}\left[iX\cdot(\zeta'-\eta')\right] \\ & \times \mathrm{Exp}\left[it\left(\nabla^2\lambda(\xi_0)\eta'-\nabla^2\lambda(\xi_0)\zeta'\cdot\zeta'\cdot\eta'\right)\right] dX\, d\zeta'\, d\eta'. \end{aligned}$$

We deduce

$$L^\varepsilon \sim \int_{\mathbf{R}^d} a_\infty(x,\xi_0,\omega_0)\, |\mathrm{e}^{it\nabla^2\lambda(\xi_0)D_x\cdot D_x}\theta(x)|^2 dx,$$

as stated in the Proposition.

4 Some Comments on the Case of a Manifold of Critical Points

The proof of Theorem 3 follows essentially the lines of that of Theorem 1; in particular a result analogous to Theorem 5 holds, based on the two-microlocal semiclassical measures described in Sect. 2.2. For the proof of a more general result, the reader may consult [5].

In this section, we develop the arguments of Remark 1, showing that whenever $\dim \Lambda = p > 0$, then the weak limit ν of the energy densities $|u^\varepsilon(t,\cdot)|^2 dx$ may not be absolutely continuous with respect to dx.

Let us first assume **A0**, **A1**, **A2'** and suppose that the Hessian of λ at its critical points is of maximal rank so that we can use Theorems 3, 4. Suppose that $V=0$ and $\Lambda = \{(\xi',0)\in\mathbf{R}^d\}$, where as before we write $\xi=(\xi',\xi'')$, with $\xi'\in\mathbf{R}^r$, $r=d-p$, and $\xi''\in\mathbf{R}^p$. We consider initial data of the form

$$u_0^\varepsilon(x) = \theta(x'')\, v^\varepsilon(x'),$$

where $\alpha \in [0,1)$, $\|\theta\|_{L^2(\mathbf{R}^p)} = 1$, and v^ε is a uniformly bounded family of $L^2(\mathbf{R}^r)$ admitting only one semiclassical measure $m(dx', d\xi')$.

In view of the proof of and the choice of the initial data, we have (with the notations of Theorem 3):

$$M_0(x',\xi') = |\theta\rangle\langle\theta| \qquad \text{and} \qquad \nu^0(dx',d\xi',d\xi'') = m(dx',d\xi')\,\delta_0(d\xi''),$$

whence, by Theorem 3:

$$M_t(x',\xi') = |\theta(t,\xi',\cdot)\rangle\langle\theta(t,\xi',\cdot)|, \text{ with } \theta(t,\xi',y) = \mathrm{e}^{-\frac{it}{2}\nabla^2_{\xi''}\lambda(\xi',0)D_y\cdot D_y}\theta(y).$$

We deduce:

$$\mu_t(dx,d\xi) = |\theta(t,\xi',x'')|^2 dx'' \otimes m(dx',d\xi') \otimes \delta_0(d\xi'').$$

One sees that, if the projection of m on the position space is not absolutely continuous with respect to dx', then the measure describing the weak limit of the energy density will also be singular.

Corollary 3. *The choice of*

$$v^\varepsilon(x') = \varepsilon^{\frac{p-d}{2}}\varphi\left(\frac{x'-z_0}{\varepsilon}\right)\mathrm{e}^{\frac{i}{\varepsilon}x'\cdot\zeta_0},$$

which is the one of Remark 1, implies

$$m(x',\xi') = \|\varphi\|^2_{L^2(\mathbf{R}^r)}\delta_{z_0}(dx') \otimes \delta_{\zeta_0}(d\xi'),$$

whence Eq. (16).

Of course, in the case where the Hessian of λ is not of maximal rank on Λ, for example at a precise point $\xi_0 = (\xi_0', 0)$, there happens a phenomenon similar to those described in Sect. 3.2. Let us take a family $(v^\varepsilon)_{\varepsilon>0}$ which oscillates along the vector $(\xi_0',0)$ as:

$$v^\varepsilon(x') = \mathrm{e}^{\frac{i}{\varepsilon}\xi_0'\cdot x'}\varphi(x'), \quad \varphi \in \mathcal{C}_0^\infty(\mathbf{R}^r).$$

Besides, as in Sect. 3.2, we add shifted oscillations in x'' along a vector $\omega_0 \in \mathbf{S}^{p-1}$ by setting:

$$u_0^\varepsilon(x) = \varepsilon^{-\frac{\alpha p}{2}}\theta\left(\frac{x''}{\varepsilon^\alpha}\right)\mathrm{e}^{\frac{i}{\varepsilon^{1-\beta}}x''\cdot\omega_0}v^\varepsilon(x').$$

The full picture is described in the following proposition.

Proposition 3. *Assume* $\nabla_{\xi''}\lambda(\xi_0',0)\omega_0 = 0$ *and* $\beta > \frac{2}{3}$.

(i) If $\alpha = 0$, *then:*

$$\mu_t(dx,d\xi) = \left|\varphi(x')e^{it\left(\nabla^2_{\xi''}\lambda(\xi_0)D_{x''}\cdot D_{x''}\right)}\theta(x'')\right|^2 dx \otimes \delta_{\xi_0}(d\xi).$$

(ii) If $\alpha \neq 0$, *then:*

$$\mu_t(dx, d\xi) = |\varphi(x')|^2 dx' \otimes \delta_0(dx'') \otimes \delta_{\xi_0}(d\xi).$$

Here again, we see that both situations may occur when the Hessian is not of maximal rank: absolute continuity with respect to Lebesgue measure or singularity.

Proof. The proof relies on the analysis of the integral

$$\begin{aligned} L^\varepsilon &= \left(\mathrm{op}_\varepsilon\left(a\left(x, \xi, \frac{\xi''}{\varepsilon}\right)\right) u^\varepsilon(t), u^\varepsilon(t)\right) \\ &= (2\pi\varepsilon)^{-3d} \varepsilon^{-p\alpha} \int_{\mathbf{R}^{7d}} a\left(\frac{x+y}{2}, \xi, \frac{\xi''}{\varepsilon}\right) \mathrm{Exp}\left[\frac{i}{\varepsilon}\left(\xi \cdot (x-y) + \zeta \cdot (y-z) - \eta \cdot (x-r) + \xi_0' \cdot (z'-r')\right)\right] \\ &\quad \times \mathrm{Exp}\left[\frac{i}{\varepsilon^{1-\beta}} \omega_0 \cdot (z''-r'')\right] \mathrm{Exp}\left[\frac{it}{\varepsilon^2}\left(\lambda(\eta) - \lambda(\zeta)\right)\right] \varphi(z')\overline{\varphi}(r')\theta\left(\frac{z''}{\varepsilon^\alpha}\right) \overline{\theta}\left(\frac{r''}{\varepsilon^\alpha}\right) d\zeta\, dz\, d\eta\, dr\, d\xi\, dx\, dy. \end{aligned}$$

We perform the change of variables

$$\widetilde{x}'' = \varepsilon^\alpha x'',\ \ \widetilde{y}'' = \varepsilon^\alpha y'',\ \ \widetilde{z}'' = \varepsilon^\alpha z'',\ \ \widetilde{r}'' = \varepsilon^\alpha r'',$$
$$\widetilde{\xi}'' = \varepsilon^\beta \omega_0 + \varepsilon^{1-\alpha} \xi'',\ \ \widetilde{\zeta}'' = \varepsilon^\beta \omega_0 + \varepsilon^{1-\alpha}\zeta'',\ \ \widetilde{\eta}'' = \varepsilon^\beta \omega_0 + \varepsilon^{1-\alpha}\eta'',$$
$$\widetilde{\xi}' = \xi_0' + \varepsilon\xi',\ \ \widetilde{\zeta}' = \xi_0' + \varepsilon\zeta',\ \ \widetilde{\eta}' = \xi_0' + \varepsilon\eta',$$

and obtain (letting the tildas down):

$$\begin{aligned} L^\varepsilon &= (2\pi)^{-3d} \int_{\mathbf{R}^{7d}} a\left(\frac{x'+y'}{2}, \varepsilon^\alpha \frac{x''+y''}{2}, \xi_0' + \varepsilon\xi', \varepsilon^{1-\alpha}\xi'' + \varepsilon^\beta\omega_0, \varepsilon^{\beta-1}(\omega_0 + \varepsilon^{1-\alpha-\beta}\xi'')\right) \\ &\times \mathrm{Exp}\left[i\xi \cdot (x-y) + i\zeta \cdot (y-z) - i\eta \cdot (x-r)\right] \\ &\times \mathrm{Exp}\left[\frac{it}{\varepsilon^2} \Gamma_\varepsilon(\zeta, \eta)\right] \varphi(z')\overline{\varphi}(r')\theta\left(z''\right)\overline{\theta}\left(r''\right) d\zeta\, dz\, d\eta\, dr\, d\xi\, dx\, dy, \end{aligned}$$

where, using the assumptions on ξ_0' and ω_0,

$$\begin{aligned} \Gamma_\varepsilon(\zeta, \eta) &= \lambda_j(\xi_0' + \varepsilon\eta', \varepsilon^\beta\omega_0 + \varepsilon^{1-\alpha}\eta'') - \lambda(\xi_0' + \varepsilon\zeta', \varepsilon^\beta\omega_0 + \varepsilon^{1-\alpha}\zeta'') \\ &= \varepsilon^{2(1-\alpha)}\left(\nabla^2_{\xi''}\lambda(\xi_0', 0)(\eta'', \eta'') - \nabla^2_{\xi''}\lambda(\xi_0', 0)(\zeta'', \zeta'')\right) + O(\varepsilon^{3\beta}). \end{aligned}$$

As a consequence, if $\alpha = 0$,

$$\begin{aligned} L^\varepsilon &\sim (2\pi)^{-3d} \int_{\mathbf{R}^{7d}} a_\infty\left(\frac{x+y}{2}, \xi_0', 0, \frac{\omega_0}{\|\omega_0\|}\right) \mathrm{Exp}\left[i\xi \cdot (x-y) + i\zeta \cdot (y-z) - i\eta \cdot (x-r)\right] \\ &\times \mathrm{Exp}\left[it\left(\nabla^2_{\xi''}\lambda(\xi_0', 0)(\eta'', \eta'') - \nabla^2_{\xi''}\lambda(\xi_0', 0)(\zeta'', \zeta'')\right)\right] \varphi(z')\overline{\varphi}(r')\theta\left(z''\right)\overline{\theta}\left(r''\right) d\zeta\, dz\, d\eta\, dr\, d\xi\, dx\, dy. \end{aligned}$$

Integration in ξ generates a Dirac mass $\delta(x-y)$, whence

$$\begin{aligned} L^\varepsilon &\sim (2\pi)^{-2d} \int_{\mathbf{R}^{5d}} a_\infty\left(x, \xi_0', 0, \frac{\omega_0}{\|\omega_0\|}\right) \mathrm{Exp}\left[i\zeta \cdot (x-z) - i\eta \cdot (x-r)\right] \\ &\quad \times \mathrm{Exp}\left[it\left(\nabla^2_{\xi''}\lambda(\xi_0', 0)(\eta'', \eta'') - \nabla^2_{\xi''}\lambda(\xi_0', 0)(\zeta'', \zeta'')\right)\right] \varphi(z')\overline{\varphi}(r')\theta\left(z''\right)\overline{\theta}\left(r''\right) d\zeta\, dz\, d\eta\, dr\, dx \\ &\sim \int_{\mathbf{R}^d} a_\infty\left(x, \xi_0', 0, \frac{\omega_0}{\|\omega_0\|}\right) |\varphi(x')|^2 \left|\mathrm{Exp}\left[-it\left(\nabla^2_{\xi', \xi'}\lambda(\xi_0', 0)(\nabla_{x''}, \nabla_{x''})\right)\right] \theta(x'')\right|^2 dx. \end{aligned}$$

Similarly, when $\alpha \neq 0$,

$$L^\varepsilon \sim (2\pi)^{-3d} \int_{\mathbf{R}^{7d}} a_\infty \left(\frac{x'+y'}{2}, 0, \xi_0', 0, \frac{\omega_0}{\|\omega_0\|} \right) \mathrm{Exp} \left[\frac{it}{\varepsilon^{2\alpha}} \left(\nabla^2_{\xi''} \lambda(\xi_0', 0)(\eta'', \eta'') - \nabla^2_{\xi''} \lambda(\xi_0', 0)(\zeta'', \zeta'') \right) \right]$$
$$\times \mathrm{Exp}\left[i\xi \cdot (x-y) + i\zeta \cdot (y-z) - i\eta \cdot (x-r) \right] \varphi(z') \overline{\varphi}(r') \theta \left(z'' \right) \overline{\theta} \left(r'' \right) d\zeta \, dz \, d\eta \, dr \, d\xi \, dx \, dy.$$

Integration in ξ generates a Dirac mass $\delta(x-y)$, then integration in x'' generates a Dirac mass $\delta(\zeta'' - \eta'')$ and we obtain

$$L^\varepsilon \sim (2\pi)^{-2d+p} \int_{\mathbf{R}^{5d-2p}} a_\infty \left(x', 0, \xi_0', 0, \frac{\omega_0}{\|\omega_0\|} \right) \mathrm{Exp} \left[i\zeta'(x'-z') - i\eta' \cdot (x'-r') - i\zeta''(z''-r'') \right]$$
$$\times \varphi(z') \overline{\varphi}(r') \theta \left(z'' \right) \overline{\theta} \left(r'' \right) d\zeta' \, dz \, d\eta' \, dr \, dx'$$
$$\sim \|\theta\|^2_{L^2(\mathbf{R}^p)} \int_{\mathbf{R}^{d-p}} a_\infty \left(x', 0, \xi_0', 0, \frac{\omega_0}{\|\omega_0\|} \right) |\varphi(x')|^2 \, dx'.$$

Acknowledgments. F. Macià has been supported by grants StG-2777778 (U.E.) and MTM2013-41780-P, TRA2013-41096-P (MINECO, Spain). Part of this work was done while V. Chabu was visiting ETSI Navales at Universidad Politécnica de Madrid in the fall of 2015. V. Chabu was partly supported by grant 2017/13865-0, São Paulo Research Foundation (FAPESP)

References

1. Anantharaman, N., Macià, F.: Semiclassical measures for the Schrödinger equation on the torus. J. Eur. Math. Soc. (JEMS) **16**(6), 1253–1288 (2014)
2. Anantharaman, N., Fermanian-Kammerer, C., Macià, F.: Semiclassical completely integrable systems: long-time dynamics and observability via two-microlocal Wigner measures. Am. J. Math. **137**(3), 577–638 (2015)
3. Ben-Artzi, M., Devinatz, A.: Local smoothing and convergence properties of Schrödinger type equations. J. Funct. Anal. **101**(2), 231–254 (1991)
4. Chabu, V.: Semiclassical analysis of the Schrödinger equation with irregular potentials. Ph.D. thesis, Université Paris-Est, Créteil (2016)
5. Chabu, V., Fermanian-Kammerer, C., Macià, F.: Wigner measures and effective mass theorems. Preprint arXiv:1803.07319 (2018)
6. Constantin, P., Saut, J.-C.: Local smoothing properties of dispersive equations. J. Am. Math. Soc. **1**(2), 413–439 (1988)
7. Dimassi, M., Sjöstrand, J.: Spectral asymptotics in the semiclassical limit. London Mathematical Society Lecture Note Series, vol. 268. Cambridge University Press, Cambridge (1999)
8. Fermanian-Kammerer, C.: Équation de la chaleur et Mesures semi-classiques. Ph.D. thesis, Université Paris-Sud, Orsay (1995)
9. Fermanian-Kammerer, C.: Analyse à deux échelles d'une suite bornée de L^2 sur une sous-variété du cotangent. C. R. Acad. Sci. Paris Sér. I Math. **340**(4), 269–274 (2005)
10. Fermanian-Kammerer, C., Gérard, P.: Mesures semi-classiques et croisement de modes. Bull. Soc. Math. Fr. **130**(1), 123–168 (2002)
11. Folland, G.B.: Harmonic analysis in phase space. Annals of Mathematics Studies, vol. 122. Princeton University Press, Princeton, NJ (1989)
12. Gérard, P.: Mesures semi-classiques et ondes de Bloch. In: Séminaire sur les Équations aux Dérivées Partielles, 1990–1991, Exp. No. XVI, pp. 1–9. École Polytech, Palaiseau (1991)

13. Gérard, P., Leichtnam, É.: Ergodic properties of eigenfunctions for the Dirichlet problem. Duke Math. J. **71**(2), 559–607 (1993)
14. Gérard, P., Markowich, P.A., Mauser, N.J., Poupaud, F.: Homogenization limits and Wigner transforms. Comm. Pure Appl. Math. **50**(4), 323–379 (1997)
15. Hoshiro, T.: Decay and regularity for dispersive equations with constant coefficients. J. Anal. Math. **91**, 211–230 (2003)
16. Kato, T.: On the Cauchy problem for the (generalized) Korteweg-de Vries equation. Studies in Applied Mathematics. Adv. Math. Suppl. Stud., vol. 8, pp. 93–128. Academic Press, New York (1983)
17. Kenig, C.E., Ponce, G., Vega, L.: Oscillatory integrals and regularity of dispersive equations. Indiana Univ. Math. J. **40**(1), 33–69 (1991)
18. Lions, P.-L., Paul, T.: Sur les mesures de Wigner. Rev. Mat. Iberoamericana **9**(3), 553–618 (1993)
19. Macià, F.: Semiclassical measures and the Schrödinger flow on Riemannian manifolds. Nonlinearity **22**(5), 1003–1020 (2009)
20. Macià, F.: High-frequency propagation for the Schrödinger equation on the torus. J. Funct. Anal. **258**(3), 933–955 (2010)
21. Miller, L.: Propagation d'ondes semi-classiques à travers une interface et mesures 2-microlocales. Ph.D. thesis, École Polythecnique, Palaiseau (1996)
22. Nier, F.: A semiclassical picture of quantum scattering. Ann. Sci. École Norm. Sup. (4) **29**(2), 149–183 (1996)
23. Ruzhansky, M., Sugimoto, M.: Smoothing estimates for non-dispersive equations. Math. Ann. **365**, 241–269 (2016)
24. Sjölin, P.: Regularity of solutions to the Schrödinger equation. Duke Math. J. **55**(3), 699–715 (1987)
25. Vega, L.: Schrödinger equations: pointwise convergence to the initial data. Proc. Am. Math. Soc. **102**(4), 874–878 (1988)
26. Zworski, M.: Semiclassical analysis. Graduate Studies in Mathematics, vol. 138. American Mathematical Society, Providence, RI (2012)

Convergence of Fourier-Walsh Double Series in Weighted $L^p_\mu[0,1)^2$

Martin G. Grigoryan(✉), Tigran M. Grigoryan, and L. S. Simonyan

Yerevan State University, Alex Manukyan 1, 0025 Yerevan, Armenia
{gmarting,t.grigoryan}@ysu.am, lussimonyan@mail.ru

Abstract. In this work we discuss the behavior of Fourier coefficients with respect to the Walsh double system, as well as $L^p_\mu[0,1)^2$-convergence of the spherical partial sums of the double Fourier-Walsh series after modification of functions.

Keywords: Fourier series · Fourier-Walsh double series · Lebesgue spaces

1 Introduction

Let $|E|$ be the Lebesgue measure of a measurable set $E \subseteq [0,1)$ (or $E \subseteq [0,1) \times [0,1) = [0,1)^2$), and let $L^p[0,1)$, $p \geq 1$, be the class of all those measurable functions $f(x)$ on $[0,1)$ such that

$$\int_0^1 |f(x)|^p dx < \infty. \tag{1}$$

Let $\mu(x,y)$ be a positive Lebesgue-measurable function (weight function) defined on $[0,1)^2$. By $L^p_\mu[0,1)^2$ we denote the space of all measurable functions on $[0,1)^2$ with the norm

$$\|.\|_{L^p_\mu} = \left(\int_0^1 \int_0^1 |.|^p \ \mu(x,y)dxdy\right)^{\frac{1}{p}} < \infty : \ p \in [1,\infty).$$

In the sequel, we will accept the terms "measure" and "measurable" in sense of Lebesgue.

Definition 1. The nonzero members of a double sequence $\{b_{k,s}\}_{k,s=0}^{\infty}$ are said to be in a monotonically decreasing order over all rays if $b_{k_2,s_2} < b_{k_1,s_1}$when $k_2 \geq k_1$, $s_2 \geq s_1$, $k_2 + s_2 > k_1 + s_1$ $(b_{k_i,s_i} \neq 0, \ i = 1,2)$.

J. Delgado and M. Ruzhansky (Eds.): *Analysis and Partial Differential Equations: Perspectives from Developing Countries*, PROMS 275, pp. 109–136, 2018.
https://doi.org/10.1007/978-3-030-05657-5_8

Let $\Phi = \{\varphi_k(x)\}_{k=1}^{\infty}$ be the Walsh system (see [5,27,29]). This system forms a basis in the spaces $L^p[0,1)$ for all $p > 1$, that is, any function $f(x) \in L^p[0,1)$ can be uniquely represented by the series $\sum_{k=0}^{\infty} c_k(f)\varphi_k(x)$ which converges to f in the $L^p[0,1)$-norm, where $c_k(f) = \int_0^1 f(x)\varphi_k(x)dx$. One could think of $\{c_k(f)\}_{k=0}^{\infty}$ as the sequence of Fourier coefficients of f with respect to the Walsh system.

Let $f(x,y) \in L^p[0,1)^2$, $p \geq 1$ and $c_{k,n}(f)$ be its Fourier coefficients with respect to the Walsh double system, that is,

$$c_{k,n}(f) = \int_0^1 \int_0^1 f(t,\tau)\varphi_k(t)\varphi_n(\tau)dtd\tau, \ k, \ n = 0,1,2,...$$

We define the spectrum of f by

$$spec(f) = \{(k,s): \ c_{k,s}(f) \neq 0, \ k, \ s \in \mathbb{N}\cup\{0\}\},$$

and the spherical and rectangular partial sums of its Fourier double series in the Walsh double system by

$$S_R(x,y,f) = \sum_{k^2+s^2\leq R^2} c_{k,s}(f)\varphi_k(x)\varphi_s(y),$$

and

$$S_{N,M}(x,y,f) = \sum_{k=0}^{N}\sum_{s=0}^{M} c_{k,s}(f)\varphi_k(x)\varphi_s(y)$$

respectively.

Definition 2. We will say that the Fourier-Walsh double series of a function $f(x,y) \in L^p_\mu[0,1)^2 \cap L^1[0,1)^2$ converges to the function $f(x,y)$ in the $L^p_\mu[0,1)^2$-norm with respect to spheres if

$$\lim_{R\to\infty}\left(\int_0^1\int_0^1 |S_R(x,y,f) - f(x,y)|^p \mu(x,y)dxdy\right)^{\frac{1}{p}} = 0.$$

The definition with respect to rectangles will be given in the same way. For the purpose of this paper the above-mentioned definitions are not given in the most general form; more general statements can be found in literature (see [2–4,30]).

In this work we will discuss the behavior of Fourier coefficients with respect to the Walsh double system, as well as $L^p_\mu[0,1)^2$-convergence of the spherical partial sums of the double Fourier-Walsh series after modification of functions.

Note that classes of partial sums (e.g. spherical, rectangular, square) differ sharply from each other when it comes to convergence in $L^p[0,1)^2$, $p \geq 1$, and convergence almost everywhere. Also, many classical results (for instance, Carleson's [1], Riesz's [28], and Kolmogorov's [23] theorems) cannot be extended from the one-dimensional case to the two-dimensional (see Fefferman [3,4], Getsadze [6]). In [22] Harris constructed a function $f \in L^p[0,1)^2$, for any $1 \leq p < 2$, such that the Fourier-Walsh series of $f(x,y)$ in the Walsh double system diverges almost everywhere and in the $L^p[0,1)^2$-norm with respect to spheres.

From this result it follows that it is impossible to find for every function $f(x,y) \in L^p[0,1)^2$ a double series in the Walsh double system that converges to the function $f(x,y)$ in the $L^p[0,1)^2$-norm or almost everywhere with respect to spheres.

In the present work we prove that for any $\varepsilon > 0$ there exists a measurable set $E \subset [0,1)^2$ with measure $|E| > 1-\varepsilon$, such that for any function $f(x,y) \in L^p(E)$, $p \geq 1$, one can find a series$\sum_{k,n=0}^{\infty} b_{k,n}\varphi_k(x)\varphi_n(y)$ with respect to the Walsh double system which converges to the function $f(x,y)$ in the $L^p(E)$-norm with respect to spheres, that is

$$\lim_{R\to\infty} \int\int_E \left| \sum_{0\leq k^2+s^2\leq R^2} b_{k,s}\varphi_k(x)\varphi_s(y) - f(x,y)\right|^p dxdy = 0.$$

Moreover, in this paper we prove the following theorem:

Theorem 1. *For any $\varepsilon > 0$ there exist a measurable set $E \subset [0,1)^2$ with measure $|E| > 1-\varepsilon$ and a weight function $\mu(x,y)$ with $|\{(x,y) \in [0,1)^2;\ \mu(x,y) = 1\}| > 1-\varepsilon, 0 < \mu(x,y) \leq 1$, $(x,y) \in [0,1)^2$, such that for every function $f(x,y) \in L^p_\mu[0,1)^2$ one can find a function $g(x,y) \in L^p_\mu[0,1)^2 \cap L^1[0,1)^2$ coinciding with $f(x,y)$ on E, whose Fourier series$\sum_{k,s=0}^{\infty} c_{k,s}(g)\varphi_k(x)\varphi_s(y)$ in the Walsh double system$\{\varphi_k(x)\varphi_s(y)\}_{k,s=0}^{\infty}$ converges to the function $g(x,y)$ in the $L^p_\mu[0,1)^2$-norm with respect to spheres and the non-zero coefficients in $\{|c_{k,s}(g)|\}_{k,s=0}^{\infty}$ are in decreasing order over all rays.*

This is a continuation of papers [12–19] on the convergence of Fourier series and on the behavior of Fourier coefficients in classical systems. Underlying the first author's studies in this area is the so-called "Luzin's C-property of measurable functions", which reads as follows: for every measurable, almost everywhere finite function f on $[0,1]$ and every $\varepsilon > 0$, there exists a measurable set $E \subset [0,1]$ with $|E| > 1-\varepsilon$ and a continuous function $g(x), x \in [0,1]$, that coincides with $f(x)$ on E. This famous result of Luzin [24] dates back to 1912. Luzin's idea of modifying a function to improve its properties was substantially developed later on.

In 1939 Men'shov obtained the following fundamental result, which is known as "Men'shov's C-strong property" (see Men'shov [25]).

For every measurable, almost everywhere finite function f on $[0,2\pi]$ and every $\varepsilon > 0$, there is a continuous function f_ε such that $|\{x \in [0,2\pi],\ f_\varepsilon(x) \neq f(x)\}| < \varepsilon$ and the Fourier series of the function f_ε converges uniformly in $[0,2\pi]$.

In 1988 M. Grigoryan showed that the trigonometric system possesses the L-strong property for integrable functions: for each $\varepsilon > 0$ there exists a (measurable) set $E \subset [0,2\pi]$ of measure $|E| > 2\pi - \varepsilon$ such that for each function $f \in L^1[0,2\pi]$ there exists a function $g \in L^1[0,2\pi]$ equal to $f(x)$ on E and with Fourier series with respect to the trigonometric system convergent to $g(x)$ in the

$L^1[0, 2\pi]$-norm (see [9,10]). A similar result has been obtained also for the Walsh system (see [21],p. 110).

After Men'shov's proof of the C-strong property, many correction-type theorems were proved for different systems.

We are not going to give a complete survey of all research done in this area.

A number of papers (see [8,11–17,20,26]) have been devoted to the correction theorems, in which the absolute values of non-zero Fourier coefficients (in the Haar, Walsh and Faber-Schauder systems) of the corrected function monotonically decrease.

Below we present results that will be heavily used in this paper.

Theorem 2. *For any $0 < \varepsilon < 1$ and each function $f \in L^\infty[0,1]$ one can find a function $g \in L^\infty[0,1), |\{x \in [0,1);\ g \neq f\}| < \varepsilon$, such that the sequence $\{|c_k(g)|\ ,\ k \in spec(g)\}$ is monotonically decreasing (see [14]).*

Theorem 3. *For any $\varepsilon > 0$ there exist a measurable set $E \subset [0,1)$ with measure $|E| > 1-\varepsilon$ and a weight function $\mu(x)$ with $|\{x \in [0,1);\ \mu(x) = 1\}| > 1-\varepsilon, 0 < \mu(x) \leq 1,\ x \in [0,1)$, such that for every function $f(x) \in L^p_\mu[0,1)$ one can find a function $g(x) \in L^p_\mu[0,1) \cap L^1[0,1)$ coinciding with $f(x)$ on E whose Fourier series $\sum_{k=0}^{\infty} c_k(g)\varphi_k(x)$ in the Walsh system $\{\varphi_k(x)\}_{k=0}^{\infty}$ converges to the function $g(x)$ in the $L^p_\mu[0,1)$-norm and the non-zero coefficients in $\{|c_k(g)|\}_{k=0}^{\infty}$ are in decreasing order (see [13]).*

Theorem 4. *For any $\varepsilon > 0$ there exist a measurable set $G \subset [0,1)$ with measure $|G| > 1-\varepsilon$, such that for each function $f \in L^1[0,1]$ there exists a function $g \in L^1[0,1]$ equal to $f(x)$ on G and with Fourier series with respect to the Walsh system convergent almost everywhere (see [13]).*

Theorems 1 and 2 follow from the more general Theorem 5.

Theorem 5. *Let $\Phi = \{\varphi_k(x)\}$ be the Walsh system and let $\varepsilon \in (0,1)$. Then, there exist*

· a double sequence $\{b_{k,s}\}_{k,s=0}^{\infty}$ with $\sum_{k,s=1}^{\infty} |b_{k,s}|^r < \infty$, for all $r > 2$ and non-zero members in $\{|b_{k,s}|\}_{k,s=0}^{\infty}$, which are in decreasing order over all rays,

· a measurable set $G \subset [0,1)^2$ with measure $|G| > 1-\varepsilon$,

· a measurable function $\mu(x,y)$ with $|\{(x,y) \in [0,1)^2;\ \mu(x,y) = 1\}| > 1-\varepsilon, 0 < \mu(x,y) \leq 1,\ (x,y) \in [0,1)^2$,

with the following property: for each $p \in [1,\infty)$ and for every function $f(x,y) \in L^p_\mu[0,1)^2$ one can find a function $g(x,y) \in L^p_\mu[0,1)^2 \cap L^1[0,1)^2$ coinciding with $f(x,y)$ on G, whose Fourier series $\sum_{k,s=0}^{\infty} c_{k,s}(g)\varphi_k(x)\varphi_s(y)$ in the Walsh double system $\{\varphi_k(x)\varphi_s(y)\}_{k,s=0}^{\infty}$ converges to the function $g(x,y)$ both in the $L^p_\mu[0,1)^2$-norm and $L^1[0,1)^2$-norm with respect to spheres and

$$c_{k,s}(g) = b_{k,s},\ \textit{for all } (k,s) \in spec(g).$$

It should be mentioned that Theorem 5 is new also in the one-dimensional case.

The following general problem remains open: which of above formulated Theorems 1–5 hold for the trigonometric system?

2 Proofs of the Main Lemmas

The Walsh system, an extension of the Rademacher system, may be obtained in the following manner. Let $r(x)$ be the periodic function, of the least period 1, defined on $[0,1)$ by

$$r = \chi_{[0,1/2)} - \chi_{[1/2,1)},$$

where by $\chi_E(x)$ we denote the characteristic function of the set E, that is,

$$\chi_E(x) = \begin{cases} 1, \text{if } x \in E, \\ 0, \ \text{if } x \notin E. \end{cases}$$

The Rademacher system $R = \{r_n : \ n = 0, 1, ...\}$ is defined by the conditions

$$r_n(x) = r(2^n x), \ \textit{for all } x \in R, \ n = 0, 1, ..., \tag{2}$$

and, in the ordering employed by Paley (see [1,10]), the nth element of the Walsh system $\{\varphi_n\}$ is given by

$$\varphi_n(x) = \prod_{k=0}^{\infty} r_k^{n_k}(x), \tag{3}$$

where $\sum_{k=0}^{\infty} n_k 2^k$ is the unique binary expansion of n, with each n_k either 0 or 1.

Introduce some designations. We separate the semiinterval $[0,1)$ to 2^m equal parts and denote those semiintervals by $\Delta_m^{(i)}$, which later will be named dyadic intervals or simply intervals of the form $\Delta_m^{(i)} = [\frac{i-1}{2^m}, \frac{i}{2^m})$, $1 \leq i \leq 2^m$.

Let $\Phi = \{\varphi_k(x)\}$ be the Walsh system and let $f \in L^p[0,1)$, $p \in [1, \infty)$.

Denote by $c_k(f)$ the Fourier-Walsh coefficients of f, that is,

$$c_k(f) = \int_0^1 f(x)\varphi_k(x)dx \ ,$$

and by $S_N(x, f)$ the Nth partial sum of Fourier-Walsh series of the function $f(x)$, that is,

$$S_N(x, f) = \sum_{k=0}^{N} c_k(f)\varphi_k(x).$$

Below we will apply the following inequality (see [7], p. 121).

$$\left(\int_0^1 |S_n(x, f)|^p \, dx\right)^{\frac{1}{p}} \leq C_p \left(\int_0^1 |f(x)|^p \, dx\right)^{\frac{1}{p}}, \tag{4}$$

for all $f(x) \in L^p[0,1]$, $1 < p < \infty$, and $n \geq 0$, where C_p depends only on p.

By Δ we denote the dyadic rectangle of the form

$$\Delta = \Delta_{\lambda,\nu}^{(j,i)} = \Delta_\lambda^{(j)} \times \Delta_\nu^{(i)},$$

where $\Delta_\lambda^{(j)}$, $\Delta_\nu^{(i)}$ are dyadic intervals.

Put for $\lambda = 1, 2, ...$ and $1 \le j \le 2^\lambda$

$$I_\lambda^{(j)}(x) = \begin{cases} 1 \ , \ \text{if } x \in [0,1) \setminus \Delta_\lambda^{(j)} \\ 1 - 2^\lambda \ , \ \text{if } x \in \Delta_\lambda^{(j)} = [\frac{j-1}{2^\lambda}, \frac{j}{2^\lambda}); \end{cases} \quad \lambda = 1, 2, ..., \ 1 \le j \le 2^\lambda, \tag{5}$$

and periodically extend these functions on R^1 with period 1.

Then, clearly,

$$I_\lambda^{(j)}(x) = \varphi_0(x) - 2^\lambda \chi_{\Delta_\lambda^{(j)}}(x),$$

and let for the natural numbers $\lambda \ge 1$, and $j \in [1, 2^\lambda]$

$$a_i(I_\lambda^{(j)}) = \int_0^1 I_\lambda^{(j)}(x)\varphi_i(x)dx = \begin{cases} 0 \ , & \text{when } \ i = 0 \ \text{ or } \ i \ge 2^\lambda \\ \pm 1, & \text{when } \ 1 \le i < 2^\lambda. \end{cases}$$

$$b_i(\chi_{\Delta_N^{(j)}}) = \int_0^1 \chi_{\Delta_N^{(j)}}(x)\varphi_i(x)dx = \ \pm\frac{1}{2^N} \ , \ 0 \le i < 2^N \ .$$

Hence

$$I_\lambda^{(j)}(x) = \sum_{i=1}^{2^\lambda - 1} a_i(I_\lambda^{(j)})\varphi_i(x) \ .$$

$$\chi_{\Delta_N^{(j)}}(x) = \sum_{i=0}^{2^N - 1} b_i(\chi_{\Delta_N^{(j)}})\varphi_i(x). \tag{6}$$

Lemma 1. *Let the dyadic interval* $\Delta_0 = [\frac{i_0 - 1}{2^{m_0}}, \frac{i_0}{2^{m_0}})$, $i_0 \in [1, 2^{m_0}]$, *and numbers* $\gamma_0 \ne 0$, $p_0 > 2$, δ_0, η_0, ε_0, $\varepsilon_1 \in (0,1)$, $N_0 \in \mathbb{N}$ *be given. Then there exist a measurable set* $E \subset \Delta_0$, *a bounded function* $g(x)$, *and a polynomial* $Q\ (x)$ *in the Walsh system* $\{\varphi_k(x)\}$ *of the following form*

$$Q(x) = \sum_{n=N_0}^{N} c_n\varphi_n(x),$$

which satisfy the following conditions:

(1) all nonzero coefficients of the sequence $\{|c_n|\}$ *are in decreasing order,*
(2) $\sum_{n=N_0}^{N} |c_n|^{2+\varepsilon_0} < \varepsilon_1$,
(3) $|E| > (1 - \delta_0)|\Delta_0|$,
(4) $g(x) = \gamma_0\chi_{\Delta_0}(x)$, *for all* $x \in E \cup ([0,1] \setminus \Delta_0)$,
(5) $\int_0^1 |Q(x) - g(x)|^{p_0}\, dx < \eta_0^{p_0}$,
(6) $\int_0^1 |g(x)|\, dx \le 2|\gamma_0||\Delta_0|$
(7) $\left(\int_0^1 |g(x)|^p\, dx\right)^{\frac{1}{p}} \le \frac{2|\gamma_0||\Delta_0|^{\frac{1}{p}}}{\delta_0^{1-\frac{1}{p}}}$, *for all* $p > 1$,
(8) $\max_{N_0 \le m \le N} \left(\int_0^1 \left|\sum_{n=N_0}^{m} c_n\varphi_n(x)\right|^p dx\right)^{\frac{1}{p}} \le 3\frac{|\gamma_0||\Delta_0|^{1/p}}{\delta_0^{1-1/p}}$ *for all* $p \in [1, p_0)$.

Proof. We choose a natural number $m > 4m_0$ such that

$$\left(2^{-m}|\gamma_0|\right)^{\varepsilon_0}\frac{4}{\delta_0}\gamma_0^2|\Delta_0| + \frac{2C_{p_0}|\gamma_0|2^{-\frac{m}{p_0}}}{\delta_0} < \min\{\left(\frac{1}{2}\varepsilon_1\right)^{2+\varepsilon_0};\ \frac{|\gamma_0||\Delta_0|}{2}\}. \quad (7)$$

where C_{p_0} is the constant in (4). Set

$$\nu_0 = 2^{m-m_0}.$$

We represent the interval Δ_0 in the form of a union of dyadic intervals $\{\Delta_\nu,\ 1 \le \nu \le \nu_0\}$:

$$\Delta_0 = \bigcup_{\nu=1}^{\nu_0}\Delta_\nu,\ \textit{with}\ \ |\Delta_\nu| = 2^{-m}, 1 \le \nu \le \nu_0. \quad (8)$$

Let

$$\lambda = \left[\log_2\frac{1}{\delta_0}\right] + 1, \quad (9)$$

$$s_1 = [\log_2 N_0] + m.$$

We define the function (polynomial) $g_1(x)$ and the numbers d_n, a_i , b_i and $a_k^{(1)}$, in the following way:

$$g_1(x) = \gamma_0\chi_{\Delta_1}(x)I_\lambda^{(1)}(2^{s_1}x),\quad x \in [0,1], \quad (10)$$

$$d_n(g_1) = \int_0^1 g_1(x)\varphi_n(x)dx\ ,\ \textit{for all}\ n \ge 0,$$

$$b_i = b_i(\chi_{\Delta_1}),\ 0 \le i < 2^m,\ a_i = a_i(I_\lambda^{(1)})\ ,$$

Taking into consideration the following identity (see (2) and (3))

$$\varphi_i(x)\varphi_j(2^{s_1}x) = \varphi_{j\cdot 2^{s_1}+i}(x)\ ,\ \text{when}\ 0 \le i\ ,\ j < 2^{s_1}, \quad (11)$$

and having the relations (5)–(6) and (8)–(11), we obtain that the polynomial $g_1(x)$ has the form:

$$g_1(x) = \gamma_0\sum_{i=0}^{2^m-1}b_i\varphi_i(x)\sum_{j=1}^{2^\lambda-1}a_j\varphi_j(2^{s_1}x) =$$

$$= \gamma_0\sum_{j=1}^{2^\lambda-1}a_j\sum_{i=0}^{2^m-1}b_i\varphi_{j\cdot 2^{s_1}+i}(x) = \sum_{n=N_0}^{N_1-1}d_n(g_1)\varphi_n(x) = \sum_{k=k_0}^{k_1-1}a_k^{(1)}\varphi_{n_k}(x)\ ,$$

where

$$d_n(g_1) = \begin{cases} \pm\frac{\gamma_0}{2^m} \text{ or } 0, \text{ if } n \in [N_0, N_1] \\ 0, \text{ if } n \notin [N_0, N_1], \end{cases} \quad N_1 = 2^{s_1+\lambda} + 2^m - 2^{s_1} - 1,$$

$$a_k^{(1)} = d_{n_k}(g_1) = \pm\frac{\gamma_0}{2^m}, \quad \{n_k\}_{k=k_0}^{k_1-1} = spec(g_1) = \{k \in N; c_k(g_1) \neq 0\}.$$

We set

$$E_1 = \{x \in [0,1);\ g_1(x) = \gamma_0\}\ .$$

From this, (5), (9) and (10) we obtain

$$|E_1| = |\Delta_1|(1 - 2^{-\lambda})\ ,$$

$$g_1(x) = \begin{cases} \gamma_0, \text{ when } x \in E_1 \\ \gamma_0(1 - 2^\lambda), \text{ when } x \in \Delta_1 \setminus E_1 \\ 0 \text{ out of } \Delta_1\,, \end{cases}$$

$$\left(\int_0^1 |g_1(x)|^p dx\right)^{\frac{1}{p}} \leq (2^{\lambda(p-1)} + 1)^{\frac{1}{p}} |\Delta_1|^{1/p} |\gamma_{\ 0}| \leq$$

$$\leq \left(\frac{2^p}{\delta_0^{p-1}} |\gamma_0|^p |\Delta_1|\right)^{\frac{1}{p}}, \text{ for all } p \geq 1. \tag{12}$$

By (4) and (12) we have

$$\max_{k_0 \leq m < k_1} \left(\int_0^1 \left|\sum_{k=k_0}^m a_k^{(1)} \varphi_{n_k}(x)\right|^p dx\right)^{\frac{1}{p}} \leq \frac{2C_{p_0}|\gamma_0||\Delta_1|^{\frac{1}{p_0}}}{\delta_0^{1-\frac{1}{p_0}}}, 1 \leq p \leq p_0.$$

So, we can use induction to determine sets $E_1, ..., E_{\nu_0}$, numbers $s_1 < ... < s_\nu < ... < s_{\nu_0}$, $k_0 < k_1 < ... < k_\nu < ... < k_{\nu_0}$, and functions $g_1(x),\ g_\nu(x) ...,\ g_{\nu_0}(x)$ of the form

$$g_\nu(x) = \gamma_0 \chi_{\Delta_\nu}(x) I_\lambda^{(1)}(2^{s_\nu} x) = \sum_{n=N_{\nu-1}}^{N_\nu - 1} d_n(g_\nu)\varphi_n(x) = \sum_{k=k_{\nu-1}}^{k_\nu - 1} a_k^{(\nu)} \varphi_{n_k}(x), \nu \in [1, \nu_0],$$

satisfying the conditions (2.27)-(2.32) for all $\nu \in [1, \nu_0]$:

$$s_\nu = [\log_2 N_{\nu-1}] + m, \quad \{n_k\}_{k=k_{\nu-1}}^{k_\nu - 1} = spec(g_\nu), \tag{13}$$

$$a_k^{(\nu)} = d_{n_k}(g_\nu) = \int_0^1 g_\nu(x)\varphi_{n_k}(x) dx \ = \pm\frac{\gamma_0}{2^m}, \tag{14}$$

$$g_\nu(x) = \begin{cases} \gamma_0, \quad x \in E_\nu\,; \\ 0, \ x \notin \Delta_\nu\,. \end{cases} \tag{15}$$

$$|E_\nu| > (1 - \delta_0)|\Delta_\nu|, \tag{16}$$

$$\int_0^1 |g_\nu(x)|^p dx < \frac{2^p}{\delta_0^{p-1}}|\gamma_0|^p|\Delta_\nu|, p \geq 1, \tag{17}$$

$$\max_{k_{\nu-1}\leq m<k_\nu}\left(\int_0^1\left|\sum_{k=k_{\nu-1}}^{m} a_k^{(\nu)}\varphi_{n_k}(x)\right|^p dx\right)^{\frac{1}{p}} \leq$$

$$\leq \frac{2C_{p_0}|\gamma_0||\Delta_\nu|^{\frac{1}{p_0}}}{\delta_0^{1-\frac{1}{p_0}}} \leq \frac{2C_{p_0}|\gamma_0|2^{-\frac{m}{p_0}}}{\delta_0}, \ 1\leq p\leq p_0. \tag{18}$$

We put

$$E=\bigcup_{\nu=1}^{\nu_0} E_\nu, \tag{19}$$

$$g(x)=\sum_{\nu=1}^{\nu_0} g_\nu(x)=\sum_{\nu=1}^{\nu_0}\sum_{k=k_{\nu-1}}^{k_\nu-1} a_k^{(\nu)}\varphi_{n_k}(x)=\sum_{k=k_0}^{\bar{k}} a_k\varphi_{n_k}(x), \bar{k}=k_{\nu_0}-1, \tag{20}$$

$$Q(x)=\sum_{n=N_0}^{N} c_n\varphi_n(x)=\sum_{k=k_0}^{\bar{k}} h_k\varphi_{n_k}(x), \bar{k}=k_{\nu_0}-1, \ N=N_{\nu_0}-1, \tag{21}$$

where

$$c_n=\begin{cases} h_k, \ when \ n=n_k; \\ 0, \ in \ other \ cases,\end{cases} \tag{22}$$

$$h_k=a_k+\beta_k, \ \beta_k=2^{-k}\min\{\frac{\varepsilon_1\eta_0}{2};\frac{|\gamma_0||\Delta_0|}{2}\}sgn\ a_k,$$

$$a_k=a_k^{(\nu)}, \quad k\in[k_{\nu-1},k_\nu), \quad 1\leq\nu\leq\nu_0. \tag{23}$$

Taking into account the relations (15), (16) and (19)–(21) and (23) we get

$$g(x)=\gamma_0\chi_{\Delta_0}(x), \ for \ all \ x\in E\cup([0,1]\setminus\Delta_0),$$

$$|E|>(1-\delta_0)|\Delta_0|,$$

$$\left(\int_0^1|Q(x)-g(x)|^{p_0}dx\right)^{1/p_0}=\left(\int_0^1|\sum_{k=k_0}^{\bar{k}}\beta_k\varphi_k(x)|^{p_0}dx\right)^{\frac{1}{p_0}}\leq\sum_{k=k_0}^{\bar{k}}|\beta_k|<\eta_0.$$

From (2), (14) and (17) for all $\nu\in[1,\nu_0]$ we have

$$\sum_{k=k_{\nu-1}}^{k_\nu-1}|a_k^{(\nu)}|^2=$$

$$=\int_0^1|g_\nu(x)|^2dx\leq\frac{4}{\delta_0}|\gamma_0|^2|\Delta_\nu|.$$

By (23) we have

$$\sum_{k=k_0}^{\bar{k}} |a_k|^2 = \sum_{\nu=1}^{\nu_0} \left[\sum_{k=k_{\nu-1}}^{k_\nu - 1} |a_k^{(\nu)}|^2 \right] \le \frac{4}{\delta_0} \gamma_0^2 |\Delta_0|,$$

Hence from (7) and (14) we have

$$\sum_{k=k_0}^{\bar{k}} |a_k|^{2+\epsilon_0} < \left(\max_{k_0 \le k \le \bar{k}} |a_k|^{\varepsilon_0} \right) \cdot \sum_{k=k_0}^{\bar{k}} |a_k|^2 \le \left(2^{-m}|\gamma_0|\right)^{\varepsilon_0} \frac{4}{\delta_0} \gamma_0^2 |\Delta_0| < \left(\frac{1}{2} \varepsilon_1 \right)^{2+\varepsilon_0}.$$

From this and (23) we have

$$\left(\sum_{k=k_0}^{\bar{k}} |h_k|^{2+\varepsilon_0} \right)^{\frac{1}{2+\varepsilon_0}} \le \left(\sum_{k=k_0}^{\bar{k}} |a_k|^{2+\varepsilon_0} \right)^{\frac{1}{2+\varepsilon_0}} + \left(\sum_{k=k_0}^{\bar{k}} (2^{-k} \frac{\varepsilon_1}{2})^{2+\varepsilon_0} \right)^{\frac{1}{2+\varepsilon_0}} < \varepsilon_1.$$

$$|h_k| > |h_{k+1}| > 0, \; k \in [k_0, \bar{k}).$$

We verified that the conditions (1)–(6) of Lemma 1 hold.

Now verify the statement (7) of Lemma 1.

By virtue of (2), (15), (17) and (20) for all $\nu \in [1, \nu_0]$, $p \ge 1$ we have

$$\int_0^1 |\sum_{n=1}^{\nu} g_n(x)|^p dx \le \int_0^1 |g(x)|^p dx = \sum_{i=1}^{\nu_0} \int_{\Delta_i} |\sum_{n=1}^{\nu_0} g_n(x)|^p dx =$$

$$= \sum_{i=1}^{\nu_0} \int_{\Delta_i} |g_i(x)|^p dx \le \sum_{i=1}^{\nu_0} \frac{2^p |\gamma_0|^p |\Delta_i|}{\delta_0{}^{p-1}} = \frac{2^p}{\delta_0^{p-1}} |\gamma_0|^p |\Delta_0|. \tag{24}$$

Now verify condition (8) of Lemma 1.

Let $M \in [k_0, \bar{k}]$, then for some $\nu \in [1, \nu_0]$ we have $k_{\nu-1} \le M < k_\nu$, so that from (23) we obtain

$$\sum_{k=k_0}^{M} h_k \varphi_{n_k}(x) = \sum_{k=k_0}^{M} (a_k + \beta_k) \varphi_{n_k}(x) = \sum_{j=1}^{\nu-1} \sum_{k=k_{j-1}}^{k_j - 1} a_k^{(j)} \varphi_{n_k}(x) +$$

$$+ \sum_{k=k_{\nu-1}}^{M} a_k^{(\nu)} \varphi_{n_k}(x) + \sum_{k=k_0+1}^{M} \beta_k \varphi_{n_k}(x).$$

From this, (7), (22) and (24) for all $p \in [1;\, p_0]$ we obtain

$$\left(\int_0^1 \left| \sum_{k=k_0}^{M} h_k \varphi_{n_k}(x) \right|^p dx \right)^{\frac{1}{p}} \le \left(\int_0^1 \left| \sum_{j=1}^{\nu-1} g_j(x) \right|^p dx \right)^{\frac{1}{p}} + \left(\int_0^1 \left| \sum_{k=k_{\nu-1}}^{M} a_k^{(\nu)} \varphi_{n_k}(x) \right|^p dx \right)^{\frac{1}{p}} +$$

$$+ \sum_{k=k_0}^{\bar{k}} |\beta_k| \le \left(\int_0^1 |\, g(x) \,|^p \, dx \right)^{1/p} + \frac{2C_{p_0} |\gamma_0| 2^{-\frac{M}{p_0}}}{\delta_0} + \frac{|\gamma_0| |\Delta_0|}{2} \le 3 \frac{|\gamma_0| |\Delta_0|^{1/p}}{\delta_0^{1-1/p}}.$$

□

Lemma 2. *Let the numbers* $p_0 > 2,\ \gamma \neq 0,\ \beta,\ \delta,\ \varepsilon,\ \eta \in (0,1),\ N > 1,\ N \in \mathbb{N}$ *and the dyadic rectangle* $\Delta = \Delta_1 \times \Delta_2 \subset [0,1)^2$ *be given. Then one can find a set* $E \subset [0,1)^2$*, a bounded function* $g(x,y)$ *and a double polynomial in the Walsh system of the form*

$$Q(x,y) = \sum_{k,s=N,M_0}^{N_1,M} c_{k,s}\varphi_k(x)\varphi_s(y),$$

satisfying the following conditions:

(1) the nonzero coefficients in $\{|c_{k,s}|,\ N \le k \le N_1 < M_0 \le s \le M\}$ *are in decreasing order over all rays and*

$$\sum_{k,s=N,M_0}^{N_1,M} |c_{k,s}|^{2+\varepsilon} < \beta,$$

(2) $|E| > (1-\delta)|\Delta|$,
(3) $g(x,y) = \gamma\chi_E(x,y),$ *for all* $(x,y) \in E \cup ([0,1] \setminus \Delta)$,
(4) $\int_0^1\int_0^1 |g(x,y)|dxdy \le 2|\gamma||\Delta|$,
(5) $\int_0^1\int_0^1 |Q(x,y) - g(x,y)|^{p_0}dxdy < \eta^{p_0}$,
(6) $\max_{N^2+M_0^2 \le R^2 \le N_1^2+M^2}\left(\int_0^1\int_0^1 \left|\sum_{N^2+M_0^2 \le k^2+s^2 \le R^2} c_{k,s}\varphi_k(x)\varphi_s(y)\right|^p dxdy\right)^{\frac{1}{p}} \le$

$$\le 108\frac{|\gamma||\Delta|^{1/p}}{\delta^{2(1-1/p)}},\ \textit{for all}\ p \in [1,p_0).$$

Proof. We use Lemma 1, with

$$\gamma_0 = \gamma,\ \Delta_0 = \Delta_1,\ \ N_0 = N,\ \delta_0 = \frac{\delta}{2},\ \varepsilon_0 = \varepsilon,\ \varepsilon_1 = \sqrt{\beta},\ \eta_0 = \frac{\eta\delta^{1+\frac{1}{p_0}}}{8|\Delta_2|^{\frac{1}{p_0}}},\ (p_0 = p_0)$$

in its formulation. Then there exist a measurable set $E_1 \subset [0,1]$, a function $g_1(x)$, a polynomial $Q_1(x)$ of the form

$$Q_1(x) = \sum_{k=N}^{N_1} a_k\varphi_k(x), \tag{25}$$

satisfying the following conditions:

$$|E_1| > (1-\frac{\delta}{2})|\Delta_1|, \tag{26}$$

$$g_1(x) = \gamma\chi_{\Delta_1}(x),\ \textit{for all}\ x \in E_1 \cup ([0,1] \setminus \Delta_1), \tag{27}$$

$$\max_{N_0 < m \le N_1}\left(\int_0^1 \left|\sum_{k=N_0}^{m} a_k\varphi_k(x)\right|^p dx\right) \le \frac{6^p}{\delta^{p-1}}|\gamma|^p|\Delta_1|,\ \textit{for all}\ p \in [1,p_0), \tag{28}$$

$$\int_0^1 |g_1(x)|\, dx \le 2|\gamma||\Delta_1|, \tag{29}$$

$$\left(\int_0^1 |Q_1(x) - g_1(x)|^{p_0}\, dx\right)^{\frac{1}{p_0}} < \frac{\eta\delta^{1-\frac{1}{p_0}}}{8|\Delta_2|^{\frac{1}{p_0}}}, \tag{30}$$

$$\left(\int_0^1 |g_1(x)|^p\, dx\right)^{\frac{1}{p}} < 4\frac{|\gamma||\Delta_1|^{\frac{1}{p}}}{\delta^{1-\frac{1}{p}}}, \tag{31}$$

$$\{|a_k| \searrow,\ k \in spec(Q_1)\}, \tag{32}$$

that is, the nonzero coefficients in $\{|a_k|,\ k \in [N, N_1]\}$ are in decreasing order.

$$\sum_{k=N}^{N_1} |a_k|^{2+\varepsilon} < \sqrt{\beta}. \tag{33}$$

We set

$$M_0 = 2\left(N_1^2 + 1\right). \tag{34}$$

Again we use Lemma 1 with

$$\gamma_0 = 1,\ \Delta_0 = \Delta_2,\ N_0 = M_0,\ \delta_0 = \frac{\delta}{2},\ \varepsilon_0 = \varepsilon,\ \varepsilon_1 = \sqrt{\beta},\ \eta_0 = \frac{\eta}{2(\int_0^1 |Q_1(x)|^{p_0}\, dx)^{\frac{1}{p_0}}}\ (p_0 = p_0),$$

in its formulation.

Then we determine a measurable set $E_2 \subset \Delta_2$, a function $g_2(x)$, a polynomial $Q_2(y)$ of the form

$$Q_2(y) = \sum_{s=M_0}^{M} b_s\varphi_s(y), \tag{35}$$

satisfying the following conditions:

$$g_2(y) = \chi_{\Delta_2}(y),\quad \forall y \in E_2 \cup ([0,1] \setminus \Delta_2),\quad |E_2| > (1 - \frac{\delta}{2})|\Delta_2|, \tag{36}$$

$$\max_{M_0 < m \le M}\left(\int_0^1 \left|\sum_{s=M_0}^{m} b_s\varphi_s(y)\right|^p dy\right) \le \frac{6^p}{\delta^{p-1}}|\Delta_2|,\ \textit{for all}\ p \in [1, p_0) \tag{37}$$

$$(\int_0^1 |Q_2(y) - g_2(y)|^{p_0}\, dy)^{\frac{1}{p_0}} < \frac{\eta}{2(\int_0^1 |Q_1(x)|^{p_0}\, dx)^{\frac{1}{p_0}}}, \tag{38}$$

$$\int_0^1 |g_2(y)|\, dy \le 2|\Delta_2|, \tag{39}$$

$$\left(\int_0^1 |g_2(y)|^{p_0} dy\right)^{\frac{1}{p_0}} \le \frac{4|\Delta_2|^{\frac{1}{p_0}}}{\delta^{1-\frac{1}{p_0}}}, \tag{40}$$

and

$$\{|b_s|\searrow,\ s\in spec(Q_2)\},\ \sum_{k=M_0}^{M}|b_s|^{2+\varepsilon}<\sqrt{\beta}. \tag{41}$$

Define a set E, a function $g(x,y)$, a polynomial $Q(x,y)$ as follows:

$$g(x,y)=g_1(x)g_2(y),\quad E=E_1\times E_2, \tag{42}$$

$$Q(x,y)=Q_1(x)Q_2(y)=\sum_{k,s=N,M_0}^{N_1,M}c_{k,s}\varphi_k(x)\varphi_s(y)=\sum_{k=N}^{N_1}a_k\varphi_k(x)\sum_{s=M_0}^{M}b_s\varphi_s(y), \tag{43}$$

where

$$c_{k,s}=\begin{cases}a_kb_s, & N\le k\le N_1,\quad M_0\le s\le M;\\ 0, & k\notin[N,N_1],\ or\ n\notin[M_0,M].\end{cases} \tag{44}$$

Taking into account the relations (25)–(27), (32), (33), (35), (36) and (40)–(44) we get

$$g(x,y)=\gamma\chi_\Delta(x,y),\quad (x,y)\in E\cup([0,1)^2\setminus\Delta),\ |E|>(1-\delta)|\Delta|,$$

nonzero members in the sequence $\{|c_{k,s}|\ ,\ N\le k\le N_1,\quad M_0\le s\le M\}$ are located in decreasing order over all rays, and

$$\sum_{k,s=N,M_0}^{N_1,M}|c_{k,s}|^{2+\varepsilon}<\beta.$$

From conditions (30), (38) and (40) it follows that

$$\int_0^1\int_0^1|g(x,y)|dxdy\le 4|\gamma||\Delta|,$$

$$\left(\int_0^1\int_0^1|Q(x,y)-g(x,y)|^{p_0}dxdy\right)^{1/p_0}\le\left(\int_0^1|Q_2(y)-g_2(y)|^{p_0}dy\right)^{1/p_0}\left(\int_0^1|Q_1(x)|^{p_0}dx\right)^{1/p_0}+$$
$$+\left(\int_0^1|Q_1(x)-g_1(x)|^{p_0}dx\right)^{1/p_0}\left(\int_0^1|g_2(y)|^{p_0}dy\right)^{1/p_0}\le\eta.$$

Now we will verify implementation of statement (6) of Lemma 2. Let $N^2+M_0^2<R^2<N_1^2+M^2$, then for some $m_0>M_0$ we have $m_0\le R<m_0+1$.

From (34) it follows $R^2-N_1^2\ge m_0^2-N_1^2\ge(m_0-1)^2$ and hence, from (28), (37), (43) and (44) for all $p\in[1,p_0)$ we obtain

$$\max_{N^2+M_0^2\le R^2\le N_1^2+M^2}\left(\int_0^1\int_0^1\left|\sum_{N^2+M_0^2\le k^2+s^2\le R^2}c_{k,s}\varphi_k(x)\varphi_s(y)\right|^p dxdy\right)^{1/p}\le$$

$$\le\left(\int_0^1\int_0^1\left|\sum_{k=N}^{N_1}\sum_{s=M_0}^{m_0-1}c_{k,s}\varphi_k(x)\varphi_s(y)\right|^p dxdy\right)^{1/p}+$$

$$+\left(\int_0^1\int_0^1\left|\sum_{k=N}^{s}c_{k,m_0}\varphi_k(x)\varphi_{m_0}(y)\right|^p dxdy\right)^{1/p}=$$

$$=\left(\int_0^1\left|\sum_{k=N}^{N_1}a_k\varphi_k(x)\right|^p dx\right)^{1/p}\left(\int_0^1\left|\sum_{s=M_0}^{m_0-1}b_s\varphi_s(y)\right|^p dy\right)^{1/p}+$$

$$+\left(\int_0^1|b_{m_0}\varphi_{m_0}(y)|^p dy\right)^{1/p}\cdot\left(\max_{N\le m\le N_1}\int_0^1\left|\sum_{k=N}^{m}a_k\varphi_k(x)\right|^p dx\right)^{1/p}<108\frac{|\gamma||\Delta|^{1/p}}{\delta^{2(1-1/p)}}.$$

□

Lemma 3. *Let numbers $p_0>2$, $\delta,\ \varepsilon\in(0,1)$, $N_0>1$, $N_0\in\mathbb{N}$ and function $f(x,y)\in L^p[0,1)^2$ are given. Then one can find a set $E\subset[0,1)^2$, a bounded function $g(x,y)$ and a double polynomial with respect to the Walsh double system of the form*

$$Q(x,y)=\sum_{j,m=N_0}^{M}c_{j,m}\varphi_j(x)\varphi_m(y),$$

satisfying the following conditions:

(1) the nonzero coefficients in $\{|c_{j,m}(Q)|,\ j,m\in[N_0,M]\}$ are in decreasing order over all rays, and $\sum_{k,n=N_0}^{M}|c_{k,n}|^{2+\varepsilon}<\varepsilon$,
(2) $g(x,y)=f(x,y)$, for all $(x,y)\in E$, $|E|>1-\delta$,
(3) $\int_0^1\int_0^1|Q(x,y)-g(x,y)|^{p_0}dxdy<\varepsilon^{p_0}$,
(4) $\int_0^1\int_0^1|g(x,y)|\,dxdy\le 3\int_0^1\int_0^1|f(x,y)|\,dxdy$,
(5) $\max_{\sqrt{2}N_0\le R\le\sqrt{2}M}\int_0^1\int_0^1\left|\sum_{2N_0^2\le j^2+m^2\le R^2}c_{j,m}\varphi_j(x)\varphi_m(y)\right|dxdy\le$

$$\le 2\int_0^1\int_0^1|f(x,y)|\,dxdy+\varepsilon,$$

(6) $\max_{\sqrt{2}N_0\le R\le\sqrt{2}M}\left(\int\int_e\left|\sum_{2N_0^2\le j^2+m^2\le R^2}c_{j,m}\varphi_j(x)\varphi_m(y)\right|^p dxdy\right)^{\frac{1}{p}}\le$

$$\le\left(\int\int_e|f(x,y)|^p dxdy\right)^{\frac{1}{p}}+\varepsilon,\ \textit{for all } p\in[1,p_0],$$

for each measurable set $e\subset E$.

Proof. We take a step function

$$\varphi(x,y)=\sum_{\nu=1}^{\nu_0}\gamma_\nu\chi_{\Delta_\nu}(x,y)\tag{45}$$

such that

$$\max_{1\le\nu\le\nu_0}\left(108\frac{|\gamma_\nu|}{\delta^2}|\Delta_\nu|^{\frac{1}{p_0}}\right) < \min\{\frac{\varepsilon}{4}; \int_0^1\int_0^1 |f(x,y)|dxdy\}, \tag{46}$$

$$\left(\int_0^1\int_0^1 |f(x,y)-\varphi(x,y)|^{p_0}\,dxdy\right)^{\frac{1}{p_0}} < \min\{\frac{\varepsilon}{4}; \frac{\int_0^1\int_0^1 |f(x,y)|dxdy}{3}\}, \tag{47}$$

where Δ_ν-are dyadic squares, $\nu = 1, ..., \nu_0$.

Successively applying Lemma 2 one can define functions $g_1(x,y),\dots,$ $g_{\nu_0}(x,y)$, sets $E_\nu \subset \Delta_\nu$, $\nu = 1,2,\dots,\nu_0$, and polynomials

$$Q_\nu(x,y) = \sum_{k=N_{\nu-1}}^{N_\nu-1}\sum_{s=M_{\nu-1}}^{M_\nu-1} c_{k,s}^{(\nu)}\varphi_k(x)\varphi_s(y),\ \nu=1,2,\dots\nu_0, \tag{48}$$

where the nonzero members of the sequence $\{|c_{k,s}^{(\nu)}|,\ k\in[N_{\nu-1},N_\nu),\ s\in[M_{\nu-1},M_\nu)\}$ $(N_{\nu-1}<N_\nu<M_{\nu-1}<M_\nu)$ are in decreasing order over all rays for a fixed ν and

$$\max_{N_\nu\le k<N_{\nu+1},\ M_\nu\le s<M_{\nu+1}} |c_{k,s}^{(\nu+1)}| <$$

$$< \min_{(k,s)\in specQ_\nu} |c_{k,s}^{(\nu)}|,\ \nu=1,2,\dots\nu_0-1. \tag{49}$$

$$\sum_{k=N_{\nu-1}}^{N_\nu-1}\sum_{s=M_{\nu-1}}^{M_\nu-1} |c_{k,s}^{(\nu)}|^{2+\varepsilon} < 2^{-\nu_0}\varepsilon, \tag{50}$$

$$|E_\nu| > (1-\delta)|\Delta_\nu|, \tag{51}$$

$$g_\nu(x,y) = \gamma_\nu\chi_{E_\nu}(x,y),\ for\ all (x,y)\in E_\nu\cup([0,1)^2\setminus\Delta_\nu), \tag{52}$$

$$\int_0^1\int_0^1 |g_\nu(x,y)|dxdy \le 2|\gamma_\nu||\Delta_\nu|, \tag{53}$$

$$\int_0^1\int_0^1 |Q_\nu(x,y)-g_\nu(x,y)|^{p_0}dxdy < (\frac{\varepsilon}{2^{\nu+2}})^{p_0}, \tag{54}$$

$$\max_{N_{\nu-1}^2+M_{\nu-1}^2\le R^2\le(N_\nu-1)^2+(M_\nu-1)^2}\int_0^1\int_0^1\left|\sum_{N_{\nu-1}^2+M_{\nu-1}^2\le k^2+s^2\le R^2} c_{k,s}^{(\nu)}\varphi_k(x)\varphi_s(y)\right|^p dxdy \le$$

$$\le \frac{108^p}{\delta^{2(p-1)}}|\gamma_\nu|^p|\Delta_\nu|,\ for\ all\ p\in(1,p_0), \tag{55}$$

$$\max_{N_{\nu-1}^2+M_{\nu-1}^2\le R^2\le(N_\nu-1)^2+(M_\nu-1)^2}\int_0^1\int_0^1\left|\sum_{N_{\nu-1}^2+M_{\nu-1}^2\le k^2+s^2\le R^2} c_{k,s}^{(\nu)}\varphi_k(x)\varphi_s(y)\right| dxdy \le$$

$$\le 108|\gamma_\nu||\Delta_\nu|. \tag{56}$$

Define a set E, a function $g(x,y)$, a polynomial $Q(x,y)$ as follows:

$$E=\bigcup_{\nu=1}^{\nu_0} E_\nu, \tag{57}$$

$$g(x,y)=\sum_{\nu=1}^{\nu_0} g_\nu(x,y)+[f(x,y)-\varphi(x,y)], \tag{58}$$

$$Q(x,y)=\sum_{\nu=1}^{\nu_0} Q_\nu(x,y)=\sum_{\nu=1}^{\nu_0}\sum_{k=N_{\nu-1}}^{N_\nu-1}\sum_{s=M_{\nu-1}}^{M_\nu-1} c_{k,s}^{(\nu)}\varphi_k(x)\varphi_s(y)=$$

$$=\sum_{j,m=N_0,}^{M} c_{j,m}\varphi_j(x)\varphi_m(y),\ M=M_{\nu_0}-1, \tag{59}$$

where

$$c_{k,s}=\begin{cases} c_{k,s}^{(\nu)},\ k\in[N_{\nu-1},N_\nu),\ s\in[M_{\nu-1},M_\nu),\ \nu=1,2,...,\nu_0 \\ 0,\ \text{in other cases.}\end{cases} \tag{60}$$

Taking into account the relations (45)–(54) and (57)–(60), we get

$$g(x,y)=f(x,y),\ \textit{for all}\ (x,y)\in E,\ |E|>1-\delta,$$

$$\int_0^1\int_0^1 |Q(x,y)-g(x,y)|^{p_0}\,dxdy<\varepsilon^{p_0},$$

$$\int_0^1\int_0^1 |g(x,y)|\,dxdy\le 3\int_0^1\int_0^1 |f(x,y)|\,dxdy,$$

all nonzero members of the sequence $\{|c_{k,s}(Q)|,\ k,s\in[N_O,M]\}$ are in decreasing order over all rays, and

$$\sum_{k,s=N_0}^{M} |c_{k,s}|^{2+\varepsilon}<\varepsilon,$$

that is, the statements (1)–(4) of Lemma 3 hold.

Now we will verify the implementation of statements (5) and (6) of Lemma 3.

Taking into account the equality $g_\nu(x,y)=\gamma_\nu\chi_{E_\nu}(x,y)$ when $(x,y)\in E$ (see (2.66)), for all $p\in[1,p_0)$, $\nu\in[1,\nu_0]$, and for any measurable set $e\subset E$ we will have

$$\int\int_e |\sum_{s=1}^{\nu-1} g_s(x,y)|^p dxdy=\sum_{\nu=1}^{\nu_0}\int\int_{e\cap\Delta_\nu} |\sum_{s=1}^{\nu-1} g_s(x,y)|^p dxdy\le\sum_{\nu=1}^{\nu_0}\int\int_{e\cap\Delta_\nu} |g_\nu(x,y)|^p dxdy\le$$

$$\le\sum_{\nu=1}^{\nu_0}|\gamma_\nu|^p|e\cap\Delta_\nu|=\int\int_e|\varphi(x,y)|^p dxdy. \tag{61}$$

Let $R \in [\sqrt{2}N_0, \sqrt{2}M]$, then for some $\nu' \in [1, \nu_0]$,

$$\sum_{2N_0^2 \le j^2+m^2 \le R^2} c_{j,m}\varphi_j(x)\varphi_m(y) =$$

$$= \sum_{\nu=1}^{\nu'-1} Q_\nu(x,y) + \sum_{N_{\nu'}^2+M_{\nu'}^2 \le k^2+s^2 \le R^2} c_{k,s}^{(\nu')}\varphi_k(x)\varphi_s(y). \tag{62}$$

From this and from the relations (45)–(47), (54), (55) for all $p \in [1, p_0)$, and for any measurable set $e \subset E$ we get

$$\left(\int\!\!\int_e \left|\sum_{2N_0^2 \le j^2+m^2 \le R^2} c_{j,m}\varphi_j(x)\varphi_m(y)\right|^p dxdy\right)^{\frac{1}{p}} \le \sum_{s=1}^{\nu'-1}\left(\int_0^1\int_0^1 |Q_s(x,y) - g_s(x,y)|^p\, dxdy\right)^{\frac{1}{p}} +$$

$$+\left(\int\!\!\int_e \left|\sum_{s=1}^{\nu'-1} g_s(x,y)\right|^p dxdy\right)^{\frac{1}{p}} + \left(\int_0^1\int_0^1 \left|\sum_{N_{\nu'}^2+M_{\nu'}^2 \le k^2+s^2 \le R^2} c_{k,s}^{(\nu')}\varphi_k(x)\varphi_s(y)\right|^p dxdy\right)^{\frac{1}{p}} \le$$

$$\le \frac{\varepsilon}{4} + \left(\int\!\!\int_e |\varphi(x,y)|^p dxdy\right)^{\frac{1}{p}} + 108\frac{|\gamma_{\nu'}|}{\delta^{2(1-1/p_0)}}|\Delta_{\nu'}|^{\frac{1}{p_0}} \le \left(\int\!\!\int_e |f(x,y)|^p dxdy\right)^{\frac{1}{p}} + \varepsilon.$$

Taking into account the relations (45)–(47), (53), (54)–(56) and (62) it can be proved similarly that

$$\max_{\sqrt{2}N_0 \le R \le \sqrt{2}M} \int_0^1\int_0^1 \left|\sum_{2N_0^2 \le j^2+m^2 \le R^2} c_{j,m}\varphi_j(x)\varphi_m(y)\right| dxdy \le 2\int_0^1\int_0^1 |f(x,y)|\, dxdy + \varepsilon.$$

□

Lemma 4. *Let $\{\varphi_k\}$ be the Walsh system. Then for each $0 < \delta < 1$ there exists a measurable positive function $\mu(x,y)$ with $|\{(x,y) \in [0,1)^2; \mu(x,y) = 1\}| > 1-\delta$ and a measurable set $E \subset [0,1)^2$, such that for any numbers $\varepsilon \in (0,1)$, $N \in \mathbb{N}$, $p_0 > 1$ and for each function $f \in L^{p_0}[0,1)^2$, $\|f\|_{p_0} > 0$, one can find a function $g(x,y)$, a polynomial $Q(x,y)$ of the form*

$$Q(x,y) = \sum_{k,s=N}^{M} c_{k,s}\varphi_k(x)\varphi_s(y),$$

satisfying the following conditions:

(1) the nonzero coefficients in $\{|c_{k,n}|,\ k,n = N,...,M\}$, are in decreasing order over all rays, and

$$\sum_{k,n=N}^{M} |c_{k,n}|^{2+\varepsilon} < \varepsilon,$$

(2) $g(x,y) = f(x,y)$, for all $(x,y) \in E$, $|E| > 1-\delta\varepsilon$,

(3) $\int_0^1 \int_0^1 |Q(x,y) - g(x,y)|^{p_0}\, dxdy < \varepsilon^{p_0}$,

(4) $\int_0^1 \int_0^1 |g(x,y)| dxdy \le 3 \int_0^1 \int_0^1 |f(x,y)|\, dxdy$,

(5) $\int_0^1 \int_0^1 |g(x,y)|^p \mu(x,y) dxdy \le 3 \int_0^1 \int_0^1 |f(x,y)|^p\, \mu(x,y) dxdy$, *for all* $p \in [1, p_0]$,

(6) $\max_{\sqrt{2}N \le R \le \sqrt{2}M} \int_0^1 \int_0^1 \left| \sum_{2N^2 \le k^2+n^2 \le R^2} c_{k,n} \varphi_k(x) \varphi_n(y) \right| dxdy \le$

$$\le 2 \int_0^1 \int_0^1 |f(x,y)|\, dxdy + \varepsilon,$$

(7)

$$\max_{\sqrt{2}N \le R \le \sqrt{2}M} \left(\int_0^1 \int_0^1 \left| \sum_{2N^2 \le k^2+s^2 \le R^2} c_{k,s} \varphi_k(x) \varphi_s(y) \right|^p \mu(x,y) dxdy \right)^{\frac{1}{p}} \le$$

$$\le 2 \left(\left(\int_0^1 \int_0^1 |f(x,y)|^p\, \mu(x,y) dxdy \right)^{\frac{1}{p}} + \varepsilon \right), \ \textit{for all } p \in [1, p_0].$$

Proof. Let $\{ p_k \}_{k=1}^{\infty} \nearrow \infty$, $p_1 > 1$ and let

$$F = \{ f_k(x,y) \}_{k=1}^{\infty} \tag{63}$$

be a sequence of all Walsh polynomials with rational coefficients. Then using Lemma 3 repeatedly we can find a sequence $\{g_n(x,y)\}$ of functions, a sequence $\{E_k\}$ of sets and a sequence of polynomials

$$Q_n(x,y) = \sum_{k,s=m_{n-1}}^{m_n - 1} c_{k,s}^{(n)} \varphi_k(x) \varphi_s(y), \tag{64}$$

satisfying the following conditions: for each $n \ge 1$ the nonzero coefficients in $\{|c_{k,s}^{(n)})|\ ,\ k,s \in [m_{n-1}, m_n)\}$ are in decreasing order over all rays,

$$\max_{m_n \le\ k,s < m_{n+1}} |c_{k,s}^{(n+1)}| < \min_{(k,s) \in specQ_n} |c_{k,s}^{(n)}|, \text{ for all } n = 1, 2...., \tag{65}$$

$$\sum_{k,s=m_{n-1}}^{m_n - 1} |c_{k,s}^{(n)}|^{2+2^{-n}} < 2^{-n}, \tag{66}$$

$$g_n(x,y) = f_n(x,y), \ \textit{for all } (x,y) \in E_n, \ |E_n| > 1 - 2^{-n}, \tag{67}$$

$$\int_0^1 \int_0^1 |g_n(x,y)| dxdy \le 3 \int_0^1 \int_0^1 |f_n(x,y)|\, dxdy, \tag{68}$$

$$\int_0^1 \int_0^1 |Q_n(x,y) - g_n(x,y)|^{p_n}\, dxdy < 2^{-np_n}, \tag{69}$$

$$\max_{\sqrt{2}m_{n-1}\le R\le\sqrt{2}M}\int_0^1\int_0^1\left|\sum_{2{m_{n-1}}^2\le k^2+s^2\le R^2} c_{k,s}^{(n)}\varphi_k(x)\varphi_s(y)\right|dxdy\le$$

$$\le 2\int_0^1\int_0^1|f_n(x,y)|\,dxdy+2^{-n},\tag{70}$$

$$\max_{\sqrt{2}m_{n-1}\le R\le\sqrt{2}m_n}\left(\int\int_e\left|\sum_{2{m_{n-1}}^2\le k^2+s^2\le R^2} c_{k,s}^{(n)}\varphi_k(x)\varphi_s(y)\right|^p dxdy\right)^{\frac{1}{p}}\le$$

$$\le\left(\int\int_e|f_n(x,y)|^p dxdy\right)^{\frac{1}{p}}+2^{-n},\tag{71}$$

for all $p\in[1,p_n]$ and for any measurable subset $e\subset E_n$.

Let

$$\begin{cases}\Omega_n=\bigcap_{s=n}^{\infty}(E_s),\quad n=1,2,...;\\ E=\Omega_{n_0}=\bigcap_{s=n_0}^{\infty}(E_s),\quad n_0=[\log_{1/2}(\delta\varepsilon)]+4;\\ B=\Omega_{n_0}\bigcup\left(\bigcup_{n=n_0+1}^{\infty}\Omega_n\setminus\Omega_{n-1}\right).\end{cases}\tag{72}$$

It is clear that (see (42) and (48))

$$|B|=1,\quad |E|>1-\delta\varepsilon.$$

We define a function $\mu(x,y)$ as follows:

$$\mu(x,y)=\begin{cases}1,\ x\in E\cup([0,1)^2\setminus B);\\ \mu_n\ ,\ (x,y)\in\Omega_n\setminus\Omega_{n-1},\ n\ge n_0+1,\end{cases}\tag{73}$$

where

$$\mu_n=\left[2^{2n\cdot p_n}\cdot\prod_{s=1}^{n}h_s\right]^{-1};\tag{74}$$

$$h_k=\sup_{1\le p\le p_k}\left(1+\int_0^1\int_0^1|g_k(x,y)|^p dxdy\right.$$

$$\left.+\max_{m_{k-1}\le R<m_k}\int_0^1\int_0^1\left|\sum_{2m_{k-1}^2\le j^2+s^2\le R^2}c_{j,s}^{(k)}\varphi_j(x)\varphi_s(y)\right|^p dxdy\right).$$

By (72)–(74) for all $k > n_0$ and $p \in [1, p_k]$ we have

$$\int\int_{[0,1)^2\setminus\Omega_k} |g_k(x,y)|^p \mu(x,y)dxdy = \sum_{n=k+1}^{\infty} \left(\int\int_{\Omega_n\setminus\Omega_{n-1}} |g_k(x,y)|^p \mu_n dxdy \right) \leq$$

$$\leq \sum_{n=k+1}^{\infty} 2^{-2np_n} \left(\int_0^1 \int_0^1 |g_k(x,y)|^p dxdy \right) h_k^{-1} < \frac{1}{3} 2^{-2k\cdot p}. \tag{75}$$

Analogously, for all $k > n_0$ we will get

$$\max_{m_{k-1}\leq R<m_k} \left(\int\int_{[0,1)^2\setminus\Omega_k} \left| \sum_{2m_{k-1}^2 \leq j^2+s^2 \leq R^2} c_{j,s}^{(k)} \varphi_j(x)\varphi_s(y) \right|^{p_k} \mu(x,y)dxdy \right)^{\frac{1}{p_k}} <$$

$$< \frac{1}{3} 2^{-2k}. \tag{76}$$

Taking into account the relations (67), (69), (72) and (76) for all $k > n_0$ we obtain

$$\int_0^1 \int_0^1 |g_k(x,y)|^p \mu(x,y)dxdy$$

$$= \int\int_{\Omega_k} |f_k(x,y)|^p \mu(x,y)dxdy + \int\int_{[0,1)^2\setminus\Omega_k} |g_k(x,y)|^p \mu(x,y)dxdy \leq$$

$$\leq \int_0^1 \int_0^1 |f_k(x,y)|^p \mu(x,y)dxdy + 2^{-2kp}, \ \textit{for all } p \in [1, p_k]. \tag{77}$$

By the conditions (72), (73), (74) and (76) for $R \in [\sqrt{2}m_{k-1}, \sqrt{2}m_k)$, for each $k > n_0$ and for any $p \leq p_k$ we have

$$\int_0^1 \int_0^1 \left| \sum_{2m_{k-1}^2 \leq j^2+s^2 \leq R^2} c_{j,s}^{(k)} \varphi_j(x)\varphi_s(y) \right|^p \mu(x,y)dxdy$$

$$= \int\int_{[0,1)^2\setminus\Omega_k} \left| \sum_{2m_{k-1}^2 \leq j^2+s^2 \leq R^2} c_{j,s}^{(k)} \varphi_j(x)\varphi_s(y) \right|^p \mu(x,y)dxdy +$$

$$+ \int\int_{\Omega_k} \left| \sum_{2m_{k-1}^2 \leq j^2+s^2 \leq R^2} c_{j,s}^{(k)} \varphi_j(x)\varphi_s(y) \right|^p \mu(x,y)dxdy \leq$$

$$\leq \frac{1}{3} \cdot 2^{-2k} + \int\int_{\Omega_{n_0}} \left| \sum_{2m_{k-1}^2 \leq j^2+s^2 \leq R^2} c_{j,s}^{(k)} \varphi_j(x)\varphi_s(y) \right|^p dxdy +$$

$$+\sum_{n=n_0+1}^{k}\left[\int\int_{\Omega_n\setminus\Omega_{n-1}}\left|\sum_{2m_{k-1}^2\le j^2+s^2\le R^2} c_{j,s}^{(k)}\varphi_j(x)\varphi_s(y)\right|^p dxdy\right]\mu_n \le$$

$$\le \frac{1}{3}\cdot 2^{-2k}+2^p\left(\int\int_{\Omega_{n_0}}|f_k(x,y)|^p\,dxdy+2^{-pk}\right)$$

$$+\sum_{n=n_0+1}^{k}\left[2^{-2(k+1)}+\left(\int\int_{\Omega_n\setminus\Omega_{n-1}}|f_k(x,y)|^p dxdy\right)^{\frac{1}{p}}\right]^p\cdot\mu_n =$$

$$=\frac{1}{3}\cdot 2^{-2k}+2^p\left(\int\int_{\Omega_{n_0}}|f_k(x,y)|^p\,dxdy+2^{-pk}\right)$$

$$+\sum_{n=n_0+1}^{k}\left[\frac{\mu_n^{\frac{1}{p}}}{2^{2(k+1)}}+\left(\int\int_{\Omega_n\setminus\Omega_{n-1}}|f_k(x,y)|^p\cdot\mu_n dxdy\right)^{\frac{1}{p}}\right]^p \le$$

$$\le \frac{1}{3}\cdot 2^{-2kp}+2^p\left(\int\int_{\Omega_{n_0}}|f_k(x,y)|^p\,dxdy+2^{-pk}\right)$$

$$+\sum_{n=n_0+1}^{k}2^p\cdot\left[\frac{\mu_n}{2^{2p(k+1)}}+\int\int_{\Omega_n\setminus\Omega_{n-1}}|f_k(x,y)|^p\mu_n dxdy\right]\le$$

$$\le \frac{1}{3}\cdot 2^{-2kp}+2^p\left(\int\int_{\Omega_{n_0}}|f_k(x,y)|^p\,dxdy+2^{-pk}\right)$$

$$+2^p\cdot 2^{-2p(k+1)}\cdot\sum_{n=n_0+1}^{k}\mu_n+2^p\int\int_{\Omega_k}|f_k(x,y)|^p\cdot\mu(x,y)dxdy\le$$

$$\le 2^p\left(2^{-k+1}+\int_0^1\int_0^1|f_k(x,y)|^p\mu(x,y)dxdy\right). \tag{78}$$

Take a function $f_{k_0}(x,y)$ from the sequence (63) such that

$$\left(\int_0^1\int_0^1|f(x,y)-f_{k_0}(x,y)|^{p_0}\,dxdy\right)^{\frac{1}{p_0}}<$$

$$<\min\{\frac{\varepsilon\delta}{4};\frac{\int_0^1\int_0^1|f(x,y)|\,dxdy}{2}\}, \tag{79}$$

$$k_0>n_0+p_0[\log_2\frac{2}{\varepsilon}]+2,\ p_{k_0}>p_0;\ m_{k_0}>N. \tag{80}$$

Denote

$$g(x,y)=f(x,y)+g_{k_0}(x,y)-f_{k_0}(x,y)$$

and

$$Q(x,y)=Q_{k_0}(x,y)=\sum_{k,s=m_{k_0-1}}^{m_{k_0}-1}c_{k,s}^{(k_0)}\varphi_k(x)\varphi_s(y))=\sum_{k,s=N}^{M}c_{k,s}\varphi_k(x)\varphi_s(y)$$

Taking into account the relations (64)–(71), (73) and (77)–(80), we get that the functions $g(x, y)$, $\mu(x, y)$, the set E and the polynomial $Q(x, y)$ satisfy the demands of Lemma 4. □

3 Proof of Theorem 5

Proof. Let $0 < \varepsilon < 1$, $p_n \nearrow \infty$ $(p_1 > 1)$ and let

$$\{f_k(x, y)\}_{k=1}^{\infty} \tag{81}$$

be a sequence of all polynomials in the Walsh system with rational coefficients.

Successively applying Lemma 4 we can find a measurable weight function $\mu(x, y)$ with

$$\mid \{(x, y) \in [0, 1)^2;\ \mu(x, y) = 1\} \mid > 1 - \epsilon/2, \tag{82}$$

and sequences of functions $\{\bar{g}_n(x, y)\}_{n=1}^{\infty}$, sets $\{G_n\}$ and polynomials

$$\overline{Q}_n(x, y) = \sum_{k,s=m_{n-1}}^{m_n - 1} b_{k,s}^{(n)} \varphi_k(x)\varphi_s(y),\ m_n \nearrow, \tag{83}$$

which satisfy the following conditions for every $n \geq 1$:

$$\overline{g}_n(x, y) = f_n(x, y), \quad x \in G_n, \tag{84}$$

$$|G_n| > 1 - 2^{-n-2}, \tag{85}$$

$$\int_0^1 \int_0^1 |\overline{g}_n(x, y)| dx dy < 3 \int_0^1 \int_0^1 |f_n(x, y)| dx dy, \tag{86}$$

$$\int_0^1 \int_0^1 |\overline{Q}_n(x, y) - \overline{g}_n(x, y)|^{p_n} dx dy < 2^{-8p_n(n+1)}, \tag{87}$$

$$\int_0^1 \int_0^1 |\overline{g}_k(x, y)|^p \mu(x, y) dx dy \leq 3 \int_0^1 \int_0^1 |f_k(x, y)|^p \mu(x, y) dx dy,\ for\ all\ p \in [1, p_k]. \tag{88}$$

All nonzero members in the sequence $\left\{\left|b_{k,s}^{(n)}\right|,\ k, s \in [m_{n-1};\ m_n)\right\}$ are in decreasing order over all rays for any fixed $n \geq 1$ and

$$\max_{k,s \in [m_{n-1}, m_n)} \left|b_{k,s}^{(n)}\right| < \min_{(k,s) \in spec \overline{Q}_{n-1}} \left|b_{k,s}^{(n)}\right|, \text{ for all } n = 1, 2..., \tag{89}$$

$$\sum_{k,s=m_{n-1}}^{m_n - 1} \left|b_{k,s}^{(n)}\right|^{2+2^{-n}} < \frac{1}{2^{8(n+1)}},\ n \geq 1 \tag{90}$$

$$\max_{\sqrt{2}m_{n-1} \leq R < \sqrt{2}m_n} \left(\int_0^1 \int_0^1 \left| \sum_{2m_{n-1}^2 \leq k^2 + s^2 \leq R^2} b_{k,s}^{(n)} \varphi_k(x) \varphi_s(y) \right|^p \mu(x, y) dx dy \right)^{\frac{1}{p}} \leq$$

$$\leq 2\left(\int_0^1\int_0^1 |f_n(x,y)|^p\mu(x,y)dxdy\right)^{\frac{1}{p}} + 2^{-2n}, \ for\ all\ p\in[1,p_n], \tag{91}$$

$$\max_{\sqrt{2}m_{n-1}\leq R<\sqrt{2}m_n}\int_0^1\int_0^1\left|\sum_{2m^2_{n-1}\leq k^2+s^2\leq R^2} b^{(n)}_{k,s}\varphi_k(x)\varphi_s(y)\right|dxdy\leq$$

$$\leq 2\left(\int_0^1\int_0^1 |f_n(x,y)|dx\right) + 2^{-2n}. \tag{92}$$

We put

$$G=(\bigcap_{n=n_0}^{\infty} G_n)\bigcap\{(x,y)\in[0,1)]^2;\ \mu(x,y)=1\},\ where\ n_0=[\log_{\frac{1}{2}}\epsilon]+2, \tag{93}$$

and

$$b_{k,s}=\begin{cases} b^{(n)}_{k,s},\ k,s\in[m_{n-1},m_n), n\geq 1\\ \qquad 0,\ \text{in other cases.}\end{cases} \tag{94}$$

From this and (82), (85), (93) we obtain that $|G|>1-\varepsilon$.

The nonzero coefficients in $\{|b_{k,s}|\}_{k,s=1}^{\infty}$ are in decreasing order over all rays and $\sum_{k,s=1}^{\infty}|b_{k,s}|^r<\infty$, for all $r>2$.

Let $p\geq 1$ and let $f(x,y)\in L^p_\mu[0,1)^2$, i. e. $\int_0^1\int_0^1|f(x,y)|^p\mu(x,y)dx<\infty$.

We set

$$\overline{f}(x,y)=\begin{cases} f(x,y), & (x,y)\in G;\\ \quad 0, & (x,y)\notin G.\end{cases} \tag{95}$$

It is not hard to see that there is a subsequence $\{f_{k_n}(x,y)\}_{n=1}^{\infty}$ of the sequence (81) such that

$$\lim_{N\to\infty}\int_0^1\int_0^1\left|\sum_{n=1}^{N} f_{k_n}(x,y)-\overline{f}(x,y)\right|^p dxdy=0,\ p_{k_1}>p,\ k_1>n_0, \tag{96}$$

$$\int_0^1\int_0^1 |f_{k_n}(x,y)|^p\,dxdy<2^{-8n\cdot p},\ n\geq 2. \tag{97}$$

Assume that we have already determined the integers $k_1=\nu_1<...<\nu_{q-1}$, the functions $g_n(x,y)$, $f_{\nu_n}(x,y)$, $1<n\leq q-1$ and the polynomials

$$\overline{Q}_{\nu_n}(x,y)=\sum_{k,s=m_{\nu_n-1}}^{m_{\nu_n}-1} b^{(n)}_{k,s}\varphi_k(x)\varphi_s(y)$$

satisfying the conditions:

$$g_n(x,y)=f_{k_n}(x,y),\ \ x\in G,\ 1\leq\ n\leq q-1,$$

$$\int_0^1\int_0^1|g_n(x)|dxdy<2^{-(n+1)},\ \left(\int_0^1\int_0^1|g_n(x)|^p\mu(x,y)dxdy\right)^{\frac{1}{p}}<2^{-(n+1)}, 1<\ n\leq q-1,$$

$$\int_0^1\int_0^1\left|\sum_{k=1}^{n}\left[\overline{Q}_{\nu_k}(x,y)-g_k(x,y)\right]\right|dxdy<2^{-(n+1)},\ 1<\ n\le q-1, \tag{98}$$

$$\int_0^1\left|\sum_{k=1}^{n}\left[\overline{Q}_{\nu_k}(x,y)-g_k(x,y)\right]\right|^p\mu(x,y)dxdy<2^{-2n\cdot p},\ 1<\ n\le q-1\ , \tag{99}$$

$$\max_{\sqrt{2}m_{\nu_n-1}\le R<\sqrt{2}m_{\nu_n}}\int_0^1\int_0^1\left|\sum_{2m_{\nu_n-1}^2\le k^2+s^2\le R^2}b_{k,s}^{(n)}\varphi_k(x)\varphi_s(y)\right|dxdy<2^{-n},\ 1<\ n\le q-1,$$

$$\max_{\sqrt{2}m_{\nu_n-1}\le R<\sqrt{2}m_{\nu_n}}\int_0^1\int_0^1\left|\sum_{2m_{\nu_n-1}^2\le k^2+s^2\le R^2}b_{k,s}^{(n)}\varphi_k(x)\varphi_s(y)\right|^p\mu(x,y)dxdy<2^{-n\cdot p},$$

for all $1<\ n\le q-1$.

It is not hard to see that one can choose a natural number $\nu_q>\nu_{q-1}$ (function $f_{\nu_q}(x,y)$ is from the sequence (81)), such that

$$\int_0^1\int_0^1\left|f_{\nu_q}(x,y)-\left(f_{k_q}(x,y)-\sum_{i=1}^{q-1}\left[\overline{Q}_{\nu_i}(x,y)-g_i(x,y)\right]\right)\right|^p\mu(x,y)dxdy<$$
$$<2^{-8\cdot q\cdot p}. \tag{100}$$

$$\int_0^1\int_0^1\left|f_{\nu_q}(x,y)-\left(f_{k_q}(x,y)-\sum_{i=1}^{q-1}\left[\overline{Q}_{\nu_i}(x,y)-g_i(x,y)\right]\right)\right|dxdy<2^{-8\cdot q}. \tag{101}$$

By (97)–(99) we have

$$\left(\int_0^1\int_0^1\left|f_{k_q}(x,y)-\sum_{i=1}^{q-1}\left[\overline{Q}_{\nu_i}(x,y)-g_i(x,y)\right]\right|^p\mu(x,y)dxdy\right)^{\frac{1}{p}}<2\cdot2^{-2(q-1)},$$

$$\int_0^1\int_0^1\left|f_{k_q}(x,y)-\sum_{i=1}^{q-1}\left[\overline{Q}_{\nu_i}(x,y)-g_i(x,y)\right]\right|dxdy<2\cdot2^{-(q-1)}.$$

From this and (100) and (101) we obtain

$$\left(\int_0^1\int_0^1\left|f_{\nu_q}(x,y)\right|^p\mu(x,y)dxdy\right)^{\frac{1}{p}}<3\cdot2^{-2(q-1)}, \tag{102}$$

$$\int_0^1\int_0^1\left|f_{\nu_q}(x,y)\right|dxdy<3\cdot2^{-(q-1)}. \tag{103}$$

We put

$$g_q(x,y)=f_{k_q}(x,y)+\left[\overline{g}_{\nu_q}(x,y)-f_{\nu_q}(x,y)\right]. \tag{104}$$

Taking into account the relations (84), (88) and (98)–(104) we get

$$g_q(x,y) = f_{k_q}(x,y),\ (x,y) \in G. \tag{105}$$

$$\int_0^1\int_0^1 |g_q(x,y)|dxdy \le \int_0^1\int_0^1 \left| f_{\nu_q}(x,y) - \left(f_{k_q}(x,y) - \sum_{i=1}^{q-1}\left[\overline{Q}_{\nu_i}(x,y) - g_i(x,y)\right]\right)\right| dx+$$

$$+\int_0^1\int_0^1 |\overline{g}_{\nu_q}(x,y)|dxdy + \int_0^1\int_0^1 \left|\sum_{i=1}^{q-1}\left[\overline{Q}_{\nu_i}(x,y) - g_i(x,y)\right]\right| dxdy < 2^{-(q-5)}, \tag{106}$$

$$\left(\int_0^1\int_0^1 |g_q(x,y)|^p\mu(x,y)dxdy\right)^{\frac{1}{p}} \le \left(\int_0^1\int_0^1 |\overline{g}_{\nu_q}(x,y)|^p\mu(x,y)dxdy\right)^{\frac{1}{p}} +$$

$$+\left(\int_0^1\int_0^1 \left|\sum_{i=1}^{q-1}\left[\overline{Q}_{\nu_i}(x,y) - g_i(x,y)\right]\right|^p \mu(x,y)dxdy\right)^{\frac{1}{p}} +$$

$$+\left(\int_0^1\int_0^1 \left| f_{\nu_q}(x,y) - \left(f_{k_q}(x,y) - \sum_{i=1}^{q-1}\left[\overline{Q}_{\nu_i}(x,y) - g_i(x,y)\right]\right)\right|^p \mu(x,y)dxdy\right)^{\frac{1}{p}} <$$

$$< 2^{-(q-5)}, \tag{107}$$

Taking into account the relations $\|.\|_{L^p_\mu} \le \|.\|_{L^p}$, $\nu_q \ge 1$ and (87), (100), (104) one can obtain:

$$\left(\int_0^1\int_0^1 \left|\sum_{i=1}^{q}\left[\overline{Q}_{\nu_i}(x,y) - g_i(x,y)\right]\right|^p \mu(x,y)dxdy\right)^{\frac{1}{p}}$$

$$\le \left(\int_0^1\int_0^1 \left|\overline{Q}_{\nu_q}(x,y) - \overline{g}_{\nu_q}(x,y)\right|^p \mu(x,y)dxdy\right)^{\frac{1}{p}} +$$

$$+\left(\int_0^1\int_0^1 \left| f_{\nu_q}(x,y) - \left(f_{k_q}(x,y) - \sum_{i=1}^{q-1}\left[\overline{Q}_{\nu_i}(x,y) - g_i(x,y)\right]\right)\right|^p \mu(x,y)dxdy\right)^{\frac{1}{p}} <$$

$$< 2^{-2q}. \tag{108}$$

By (91), (102) we have

$$\max_{\sqrt{2}m_{\nu_q-1}\le R<\sqrt{2}m_{\nu_q}} \left(\int_0^1\int_0^1 \left|\sum_{2m^2_{\nu_q-1}\le k^2+s^2\le R^2} b^{(\nu_q)}_{k,s}\varphi_k(x)\varphi_s(y)\right|^p \mu(x,y)dxdy\right)^{\frac{1}{p}} \le$$

$$\le 2\left(\int_0^1\int_0^1 |f_{\nu_q}(x,y)|^p\mu(x,y)dxdy\right)^{\frac{1}{p}} + 2^{-2(\nu_q+1)} < 2^{-(q-5)}. \tag{109}$$

Analogously for all $q>1$

$$\int_0^1\int_0^1\left|\sum_{i=1}^{q}|\overline{Q}_{\nu_i}(x,y)-g_i(x,y)\right|dxdy<2^{-2q}, \tag{110}$$

$$\max_{\sqrt{2}m_{\nu_q-1}\le R<\sqrt{2}m_{\nu_q}}\left(\int_0^1\int_0^1\left|\sum_{2m^2_{\nu_q-1}\le k^2+s^2\le R^2} b^{(\nu_q)}_{k,s}\varphi_k(x)\varphi_s(y)\right|dxdy\right)<2^{-(q-5)}. \tag{111}$$

So, we can use induction to determine a sequence $\{g_q(x,y)\}_{q=1}^{\infty}$ $(g_1(x,y)=f_{k_1}(x,y))$ of functions, numbers $\{\nu_q\}\nearrow$ and a sequence $\{\overline{Q}_{\nu_q}(x,y)\}$ of polynomials satisfying the conditions (105)–(111) for all $q\ge 1$.

From (106) it follows that

$$\sum_{q=1}^{\infty}\int_0^1\int_0^1|g_q(x,y)|dxdy<\infty. \tag{112}$$

Let the function $g(x,y)$ and the sequence $\{\varepsilon_{k,s}\}_{k,s=0}^{\infty}$ be defined as follows

$$g(x,y)=\sum_{q=1}^{\infty}g_q(x,y), \tag{113}$$

$$\varepsilon_{k,s}=\begin{cases}1, & k,s\in\bigcup_{q=1}^{\infty}[m_{\nu_q-1},m_{\nu_q})\\ 0, & \text{in other cases.}\end{cases} \tag{114}$$

From (95), (96), (105), (112) and (113) it follows

$$g(x,y)\in L^1[0,1),\ g(x,y)=f(x,y),\ x\in G.$$

Let $R>\sqrt{2}m_{\nu_1}$, then for some q, $q>1$, we have $\sqrt{2}m_{\nu_q-1}\le R<\sqrt{2}m_{\nu_q}$, so that from (94), (107)–(109) and (114) we obtain

$$\left(\int_0^1\int_0^1\left|\sum_{0\le k^2+s^2\le R^2}\varepsilon_{k,s}b_{k,s}\varphi_k(x)\varphi_s(y)-g(x,y)\right|^p\mu(x,y)dxdy\right)^{\frac{1}{p}}\le$$

$$\le\left(\int_0^1\int_0^1\left|\sum_{i=1}^{q-1}[\overline{Q}_{\nu_i}(x,y)-g_i(x,y)]\right|^p\mu(x,y)dxdy\right)^{\frac{1}{p}}+$$

$$+\sum_{i=q}^{\infty}\left(\int_0^1\int_0^1|g_i(x,y)|^p\mu(x,y)dxdy\right)^{\frac{1}{p}}+$$

$$+\max_{\sqrt{2}m_{\nu_q-1}\le R<\sqrt{2}m_{\nu_q}}\left(\int_0^1\int_0^1\left|\sum_{2m_{\nu_q-1}{}^2\le k^2+s^2\le R^2}b^{(\nu_q)}_{k,s}\varphi_k(x)\varphi_s(y)\right|^p\mu(x,y)dxdy\right)^{\frac{1}{p}}<2^{-(q-5)}.$$

From this follows that the series $\sum_{k,s=0}^{\infty} \varepsilon_{k,s} b_{k,s} \varphi_k(x)\varphi_s(y)$ converges to the function $g(x,y)$ in the $L^p_\mu[0,1)^2$-norm with respect to spheres. Analogously we have

$$\int_0^1 \int_0^1 \left| \sum_{0\le k^2+s^2\le R^2} \varepsilon_{k,s} b_{k,s} \varphi_k(x)\varphi_s(y) - g(x,y) \right| dxdy < 2^{-(q-5)} \to 0,\ R \to \infty (q \to \infty)$$

and therefore

$$\varepsilon_{k,s} b_{k,s} = \int_0^1 \int_0^1 g(x,y)\varphi_k(x)\varphi_s(y) dxdy, \quad i = 1, 2, ...$$

□

References

1. Carleson, L.: On convergence and growth of partial sums of Fourier series. Acta Math. **116**, 135–157 (1966)
2. D'yachenko, M.I.: Some problems in the theory of multiple trigonometric series. Uspekhi Mat. Nauk **47**(5) (16992), 97–162 (1992) (English transl. In Russian Math. Surveys **47**)
3. Fefferman, C.: On the divergence of multiple Fourier series. Bull Am. Math. Soc. **77**(2), 191–195 (1971)
4. Fefferman, C.: The multiple problem for the ball. Ann. Math. **94**(2), 330–336 (1971)
5. Galoyan, L.N., Grigoryan, M.G., Kobelyan, AKh: Convergence of Fourier series in classical systems. Mat. Sb. **206**(7), 55–94 (2015)
6. Getsadze, R.D.: On divergence in measure of general multiple orthogonal Furier series. Dokl. Akad. Nauk SSSR **306**, 24–25 (1989) (English transl. in Soviet Math. Dokl. **39** 1989)
7. Golubov, B.I., Efimov, A.F., Skvortsov, V.A.: Series and Transformations of Walsh. Moskow (1987) (in Russian)
8. Golubov, B.I.: Multiple Furier series and integrals. Itogi Nauki i Tekhniki Ser. Mat. Anal. **47**(5), 97–162 (1992) (English transl. in J. Soviet Math. **24** 1984)
9. Grigorian, M.G.: On the convergence of Fourier series in the metric of L. Anal. Math. **17**, 211–237 (1991)
10. Grigorian, M.G.: On the L^p-strong property of orthonormal systems. Math. sb. **194**(10), 77–106 (2012) (in Russ., English transl. Sbornik: Math. **194**(10), 1503–1532)
11. Grigorian, M.G., Sargsyan, A.A.: On the coefficients of expansion of elements from C[0,1] space by the Faber-Schauder system. J. Funct. Spaces Appl. **2**, 34–42 (2011)
12. Grigorian, M.G., Zink, R.E.: Greedy approximation with respect to certain subsystems of the Walsh orthonormal system. Proc. Am. Math. Soc. **134**(12), 3495–3505 (2006)
13. Grigoryan, M.G.: Modifications of functions, Fourier coefficients and nonlinear approximation. Mat. Sb. **203**(3), 49–78 (2012)
14. Grigoryan, M.G.: Uniform convergence of the greedy algorithm with respect to the Walsh system. Studia Math. **198**(2), 197–206 (2010)
15. Grigoryan, M.G., Gogyan, S.L.: On nonlinear approximation with respect to the Haar system and modifications of functions. An. Math. **32**, 49–80 (2006)

16. Grigoryan, M.G., Krotov, V.G.: Luzin's correction theorem and the coefficients of Fourier expansions in the faber-schauder system. Mat. Zametki **93**(2), 172–178 (2013)
17. Grigoryan, M.G., Navasardyan, K.A.: On behavior of Fourier coefficients by Walsh systems. J. Contemp. Math. Anal. (Armen. Acad. Sci.) **51**(1), 1–13 (2016)
18. Grigoryan, M.G.: On some properties of orthogonal systems. Izv. Ross. Akad. Nauk, Ser. Mat. **57**(5), 75–105 (1993)
19. Grigoryan, M.G., Sargsyan, A.A.: On the universal functions fr the class Lp[0,1]. J. Funct. Anal. **270**, 3111–3133 (2016)
20. Grigoryan, M.G., Sargsyan, S.A.: On the Fourier-Vilenkin coefficients. Acta Math. Sci. **37**(B)(2), 293–300 (2017)
21. Grigoryan, M.G.: Series in the classical systems. LAP LAMBERT Academic Publishing, Saarbruken (2017) (in Russian)
22. Harris, D.C: Almost everywhere divergence of multiple Walsh-Fourier series. Am. Math. Soc. **101**(4) (1987)
23. Kolmogorov, A.H.: Sur les fonctions harmoniques conjugees et les series. de Fourier, FM **7**, 23–28 (1925)
24. Luzin, N.N.: On the fundamental theorem of the integral calculus. Mat. Sb. **28**, 266–294 (1912). (in Russian)
25. Men'shov, D.E.: Sur la convergence uniforme des series de Fourier. Mat. Sb. **11**(53), 67–96 (1942) (French; Russian)
26. Men'shov, D.E.: On Fourier series of integrable functions. Trudy Moskov. Mat. Obshch. **1**, 5–38 (1952)
27. Paley, R.E.A.C.: A remarkable set of orthogonal functions. Proc. Lond. Math. Soc. **34**, 241–279 (1932)
28. Riss, M.: Sur les fonctions conjugees. Math. Zeit. **27**, 214–244 (1927)
29. Walsh, J.L.: A closed set of normal orthogonal functions. Am. J. Math. **45**, 5–24 (1923)
30. Zhizhiashvili, L.V.: Some problems in the theory of simple and multiple trigonometric and orthogonal series. Uspekhi Mat. Nauk **28**(2), 65–119 (1973) (English transl. In Russian, Math. Surveys **28** 1973)

"Strong" Turing-Hopf Instability for Reaction-Diffusion Systems

Giani Egaña Fernández[1], J Sarría González[2], and Mariano Rodríguez Ricard[1](✉)

[1] Facultad de Matemática y Computación, San Lázaro y L, Vedado, Universidad de La Habana, Havana, Cuba
gegana@matcom.uh.cu, rricard@matcom.uh.cu

[2] Facultad 4, Universidad de Ciencias Informáticas, Carretera a San Antonio Km 2 $\frac{1}{2}$, Havana, Cuba
jsarria@uci.cu

Abstract. Turing-Hopf instabilities for reaction-diffusion systems provide spatially inhomogeneous time-periodic patterns of chemical concentrations. In this presentation, it is shown the parameter space in which the reaction-diffusion system modelling glycolysis and the Lengyel-Epstein model could show twinkling patterns. To do so, we follow the Ricard-Mischler procedure in Ricard and Mischler (J Nonlinear Sci 19(5):467–496, 2009, [18]), i.e., considering this phenomenom as a consequence of the instability generated by diffusion on the limit cycle which appears due to a Hopf bifurcation about the spatially homogeneous steady state.

Keywords: Turing instability · Hopf bifurcation · Reaction diffusion · Twinkling pattern

1 Introduction

The twinkling pattern phenomenon is connected with both diffusive instability, say Turing patterns, and Hopf bifurcation, say the emergence of limit cycles surrounding the monodromic steady state in a 2D reaction diffusion system. In accordance to [18], it is featured by an intermittent temporal behavior of the ultimate spatial pattern. The study of Turing-Hopf (TH) instabilities is frequently done by determining the region of the parameter space at which bifurcations and instabilities coexist. We recall that some conditions are required in addition for the emergence of such time-periodic spatially heterogeneous patterns. In [18] the authors propose a set of conditions and asymptotic estimates of the solutions corresponding to this phenomenon. In that paper were considered the parameters lying on a neighborhood of a codimension-two TH point, at which

J. Delgado and M. Ruzhansky (Eds.): *Analysis and Partial Differential Equations: Perspectives from Developing Countries*, PROMS 275, pp. 137–158, 2018.
https://doi.org/10.1007/978-3-030-05657-5_9

the asymptotic expansion of the limit cycle and the Turing analysis of diffusive instabilities are valid. Diffusive linear instability analysis is done through modified Fourier normal modes, there called the *extended* normal modes.

The extended modes are valid only in the presence of the limit cycle, i.e. close to a Hopf bifurcation point. These modes, being unstable or not, show interactions between a spatial pattern (the spatial eigenfunction) and the oscillations due to the cycle. The knowledge of such modes would serve to get a better representation of the ultimate oscillatory pattern and the way in which Hopf bifurcation interact with diffusive instability at the onset. Such phenomenon was called *strong* diffusive instability, which is the terminology we shall use here. Strong instabilities are featured by an intermittent switching between the spatially inhomogeneous pattern, represented by the set of positiveness of the spatial eigenfunction, with its "complementary pattern", represented by the set of negativeness of the proper eigenfunction. The frequency of these oscillations are different in general from the frequency of the cycle.

Turing instability is caused by the destabilization of the stable steady state due to different enough diffusion coefficients of the reactives, in the 2D reaction-diffusion system. On the contrary, strong diffusive instability is a consequence of the destabilization of the asymptotically stable limit cycle provided the diffusion coefficients are close enough, or even equal. In order that diffusion could destabilize the limit cycle, a couple of inequalities which involve invariants of the Jacobian matrix at the steady state, the spatial eigenvalues (the geometry of the spatial domain) and the diffusion coefficients, must hold.

The limit cycle emerges surrounding the unstable monodromic steady state due to a supercritical Hopf bifurcation. Additionally, for the strong one it is required that the lowest positive spatial eigenvalue (eigenvalues of the operator minus Laplacian) is small enough. Naturally, the last condition involves the shape of the region. We assume that the ultimate pattern emerges (see [15]) due to the boundedness of the unstable modes by the nonlinear reaction terms in Eq. 1. We expect that spatiotemporal oscillations could be observed at least for a while, so that the resulting solution moves closely to some pattern, being time-oscillatory or not. We hypothesize that, between all destabilizing extended modes, those that grow fastest will have a prevailing influence, relaying their spatiotemporal structure on the resulting time-periodic inhomogeneous pattern.

For instance, in [1] the behavior of a predator-prey system showing oscillatory patterns as a consequence of TH instabilities was studied and it was concluded that these instabilities can be considered as one important mechanism for the appearance of complex spatiotemporal phenomena in population dynamics models.

The expected applications of the procedure in [18] are in mathematical modelling of chemical systems [29]-[28]-[25]-[12], semiconductor physics [6]-[14], population (mainly prey-predator) systems [1]-[3]. More recent applications of this procedure appear in morphochemical electrodeposition [8]-[22]-[24]-[23].

Our purpose in this paper is to show conditions for the emergence of twinkling patterns in two special widely studied reaction-diffusion systems. Namely, a type of glycolysis model and the Lengyel-Epstein system. The plan of the paper is as follows. Section 2 is devoted to the mathematical preliminaries mainly following [18], and Sections 3 and 4 concern to the glycolysis system and Lengyel-Epstein, respectively. Some general comments are included in the Conclusions.

2 Preliminaries

We shall study the onset of diffusive instabilities of the spatially homogeneous periodic solution $(\overline{u}(t), \overline{v}(t))$ to the following system:

$$u_t = D_u \, \Delta u + f(u, v; a) \tag{1}$$

$$v_t = D_v \, \Delta v + g(u, v; a) \tag{2}$$

with Neumann boundary conditions

$$\partial_n u = \partial_n v = 0 \text{ on } \partial\Omega \,. \tag{3}$$

Functions f and g shall been taken in correspondence to each system in Sections 3 and 4. Consider that the system has an isolated spatially uniform steady state (u_0, v_0). For simplicity, one may assume that the steady state do not depend on a. We denote by $J_a = \left(j_{ij}^a\right)$ the Jacobian matrix of the reaction term at the steady state, and

$$\delta_a = \det(J_a) \tag{4}$$

$$\tau_a = \operatorname{trace}(J_a) \,. \tag{5}$$

We assume that the eigenvalues of J_a are complex conjugate numbers for all values of a, so

$$\tau_a^2 - 4\delta_a < 0 \tag{6}$$

and further, we assume the existence of a bifurcation value a_* in which the trace τ_a vanishes to change in sign. Then, for any value a in some open neighborhood of a_* we have

$$\delta_a > 0 \,. \tag{7}$$

We derive first an asymptotic expansion of the limit cycle solution $(\overline{u}(t), \overline{v}(t))$ to the nonlinear reaction system which appears via a Hopf bifurcation. To do so, we use an asymptotic expansion and the averaging method. This cycle solution appears while the parameter a in the reaction part of the system varies, considering now the fact that reactants can diffuse within the spatial (time-independent) bounded region Ω with regular boundary. We denote the Laplacian operator by Δ. It is well-known that patterns will depend not only on the reaction dynamics, but also on geometrical features of Ω including their spatial dimension. In general, there is no restriction to the shape or to the spatial dimension of the domain Ω , but in this paper we shall consider that the spatial region is a parallelogram, say $\Omega = [0, L_1] \times [0, L_2]$; $x = (x_1, x_2)$, the eigenfunctions of $-\Delta$ can be taken as: $U_k(x) = \prod_{j=1}^{2} \cos k_j x_j$ where $k = (k_1, k_2)$ and $k_j = \left(\frac{\pi}{L_j} p_j\right)$, for such $p_j \in \mathbb{N}$. The corresponding spatial eigenvalue to such an eigenfunction will be $\lambda_k = |k|^2 = \left(k_1^2 + k_2^2\right)$.

Here u, v are the profiles of component concentrations under diffusion and a is a scalar parameter. All parameters and variables in this paper are considered dimensionless. For bounded spatial domains and natural boundary conditions it is known from [5] and [9] that the non-constant spatially homogeneous periodic solution to Eqs. 1 and 3 is orbitally stable if (D_u, D_v) belongs to a certain open neighborhood of the bisectrix of the first quadrant in the Cartesian product of diffusion coefficients while the non-zero Flocquet's exponent of the linearized system is negative. Unfortunately, the sign of this exponent using such asymptotic expansions cannot be properly determined.

We recall that Prop. 9.1 provides an algorithm allowing the reduction of the reaction system into a second order differential equation representing a weakly nonlinear oscillator in normal form. This change of variables is, in general, nonlinear, but it was proved [18] that it is enough to consider the linear part of this map to obtain the equation of the oscillator, preserving the required accuracy. The main idea of this procedure is that the change of variables can be taken "close" to the appropriate linear map in a neighborhood of the origin. Then, applying the Krylov-Bogoliubov-Mitropolski averaging method [2] to the oscillator, in Theorem 9.1 we use a discriminant function to determine the appearance of an HB, sub or supercritical, as in the Andronov-Hopf normal form [7]. It can be shown that, the existence of at least one nonzero term with even degree in the expansion of the discriminant function is a necessary and sufficient condition to determine subcritical or supercritical bifurcations, in accordance with the sign of the coefficient in the lowest even degree nonzero term. It was concluded in [18] that terms with even degree in the Taylor expansion of f and g in the right hand of Eq. 1 do not have any contribution to the appearance of a Hopf bifurcation. Only appropriate interactions within the terms having odd degree in the expansions of f and g can contribute to the appearance of such bifurcation.

Using the asymptotic expansion to the limit cycle we shall build appropriate normal modes in Proposition 9.3. The extended modes, which can be unstable or not, allow the study of the central matter in this paper: the appearance of diffusive (Turing) instabilities of the stable limit cycle in the reaction diffusion system. In the classic theory, the steady patterns are associated with the region of positiveness of the spatial eigenfunction participating in the unstable Fourier mode. The extended modes in the case of strong instability, suggest an intermittent switching between the unstable Turing pattern and its "complementary". The frequency of these oscillations would not necessarily the same as the frequency of the cycle solution.

For a dynamical system, an isolated non-punctual closed orbit is called *limit cycle*. The appearance of such solutions to a dynamical system depending on a real parameter a, would happen when the parameter takes values slightly beyond a certain threshold value a_0, called the *bifurcation value* [13]. The situation would be sketched as follows: for $a < a_0$ the steady state Y_0 is a stable focus, but for $a > a_0$ this steady state is an unstable focus. So, the behavior of the solutions changes abruptly at the bifurcation value. The reader can find a remarkable study of this bifurcation in [13], or in [7].

Assume that the reaction system has a limit cycle solution $\Theta(t)$ due to a Hopf bifurcation as the parameter τ_a varies, then $\Theta(t)$ is also a spatially homogeneous solution to Eqs. 1-3. The concept of an orbitally stable solution to this system results as a natural extension of the precedent concept to an infinite dimensional dynamical system defined in an appropriate Hilbert space. More exactly, given a bounded domain $\Omega \subset \mathbb{R}^n$, let $\mathfrak{A}$ be the non-negative self-adjoint linear operator in $L_2(\Omega)$ defined in the dense subset $\mathfrak{D}(\mathfrak{A}) = \{\phi \in H^1(\Omega) \cap H^2(\Omega) \,|\, \partial_{\mathbf{n}}\phi = 0 \text{ on } \partial\Omega\}$ by the correspondence $\mathfrak{A}u = -\Delta u$. Here $H^k(\Omega)$ is the usual Sobolev space and the boundary condition is understood in the sense of traces. Let us consider in $\mathfrak{D}(\mathfrak{A})$ the abstract (infinite dimensional) dynamical system

$$\eta_t = -D\,\mathfrak{A}\eta + F(\eta) \tag{8}$$

where D represents the diagonal matrix of diffusion coefficients, and the function F represents the reaction without diffusion. A solution to Eq. 8 is a continuous function defined in an open subset of $\mathbb{R}^+$ with values in $[\mathfrak{D}(\mathfrak{A})]^2 \subset \mathbf{H} = [L_2(\Omega)]^2$, which satisfy the equation in a mild sense [9]. In what follows, we will identify a solution to Eq. 8 as a solution to the boundary value problem Eqs. 1-3. Let us take an appropriate norm in $\mathfrak{D}(\mathfrak{A})$, say $\|\cdot\|$.

Definition 9.1. The spatially homogeneous periodic solution $\Theta(t)$ to Eq. 8 is *orbitally asymptotically stable* if there are positive constants ρ, δ, μ such that, for every solution $\Psi(t)$ to Eq. 8 satisfying

$$\min_t \|\Psi(t_0) - \Theta(t)\| \leq \rho$$

for some t_0, then exists $h \in \mathbb{R}$ such that

$$\|\Psi(t+h) - \Theta(t)\| \leq \delta \exp(-\mu t)$$

for all $t \geq t_0$. If $\mu = 0$ the solution $\Theta(t)$ is said to be *neutrally stable*. In any other case the solution is said to be *unstable*.

In Turing's seminal paper [26] a spatially homogeneous stable steady state was assumed, allowing a separate analysis of the destabilizing influence of diffusion. Turing showed that dissimilar diffusion coefficients of the participating reactants would destabilize the steady state of the reaction kinetics. Such instabilities lead to steady spatially varying profiles in the reactant concentrations which are called *patterns*. The mathematical basis for these assertions is the stability analysis of spatially homogeneous solutions with compact orbit, so leading to spectral analysis. The standard Turing procedure for the stability analysis is done for systems with two chemical reactants considering *normal modes* of the type

$$Z(x,t) = \exp(\sigma t)\ U_k(x)\ R \tag{9}$$

as non-trivial solutions to the linearization of Eq. 8,

$$\frac{\partial Z}{\partial t} = D\ \Delta Z + J_a\ Z \tag{10}$$

where $U_k(x)$ are eigenfunctions associated to the spatial eigenvalue λ_k ($k \in \mathbb{N}$) of the unbounded non-negative linear operator $(-\Delta)$ with Neumann boundary conditions at $\partial\Omega$. We call these λ_k the *spatial eigenvalues*, and the $U_k(x)$ are called the *spatial eigenfunctions*. Perturbations with the form in Eq. 9 allow us to study the stability on bounded domains, due to the fact that any small perturbation Z can be expanded in terms of such basic functions

$$Z(x,t) = \sum_{k=1}^{\infty} \exp(\sigma_k\ t)\ U_k(x)\ R_k \tag{11}$$

that is, in Fourier series. In this series, eigenvalues are eventually repeated several times according with their (finite) multiplicity. The R_k are determined by the Fourier development of the initial condition. In Eq. 9 the temporal eigenvalues σ_k are determined by the second degree equation

$$\det(J_a - \lambda_k\ D - \sigma I) = 0 \tag{12}$$

with R_k being a non-zero σ-eigenvector of the matrix

$$E_k = J_a - \lambda_k\ D\ . \tag{13}$$

Here J_a, D and I are 2×2 matrixes which are the Jacobian matrix at the spatially homogeneous steady state solution to Eq. 1, the diagonal matrix of diffusion coefficients and the identity matrix respectively. Notice that varying k and a, the eigenvalues σ of the matrix E_k are the eigenvalues of the linear part operator in Eq. 8.

Let us denote

$$\tau_T = \mathrm{trace}(J_a - \lambda_k\ D) = \tau_a - \lambda_k(D_u + D_v) \tag{14}$$

$$\delta_T = \det(J_a - \lambda_k\ D) = \delta_a - \lambda_k(D_u\ j_{22}^a + D_v\ j_{11}^a) + \lambda_k^2\ D_u D_v \tag{15}$$

which are functions of the eigenvalue λ_k. If for some critical value of a parameter, there is a change in the sign of τ_T, then a Hopf bifurcation in Eq. 8 occurs. We will refer to a *subcritical* bifurcation if the limit cycle appears for close to zero but with negative values of τ_T, and a *supercritical* bifurcation if it appears for positive small values of τ_T. To fix ideas, throughout this paper any reference to a Hopf bifurcation will be related to the case $k = (0,0)$, corresponding to the bifurcation of spatially homogeneous steady solutions. We also recall that, the emergence of a limit cycle for Eq. 8 could be seen in the frame of the Centre Manifold Theorem, reducing the calculus to a 2D manifold.

Let us rewrite the reaction system near the steady state as

$$\dot{X} = \mathfrak{F}(X) = J_a X + \Psi(X) \tag{16}$$

where $X = (U(t), V(t))^T$ are the new reaction variables representing the shift of the steady state to the origin, and the vector function $\Psi(X)$ contains all nonlinearities. Let us do, in Eq. 16, an invertible analytical change of coordinates of a neighborhood of the origin onto another

$$Y = \mathfrak{H}(X) = \Gamma X + \mathfrak{G}(X) \tag{17}$$

driven by a non-singular matrix Γ, and the analytic vector function $\mathfrak{G}(X)$ which is assumed to have a positive radius of convergence. Note that, in the Inverse Function Theorem [19], the existence of the inverse $\mathfrak{H}^{-1}$ is warranted only being Γ non-singular and $\mathfrak{H}$ with smooth continuous derivatives. If $\mathfrak{H}$ is such that

$$Y = (z, \dot{z})^{\mathrm{T}} \tag{18}$$

being $z(t)$ an unknown function, the integration of the reaction system can be reduced to the integration of a second order differential equation in the variable z.

Proposition 9.1. *Let us assume Eq. 6 holds. Then, there exists an invertible change of variables Eq. 17 such that Eq. 18 holds. The matrix Γ is any non-trivial linear combination of*

$$\left\{ \begin{pmatrix} 1 & 0 \\ j_{11}^a & j_{12}^a \end{pmatrix}, \begin{pmatrix} 0 & 1 \\ j_{21}^a & j_{22}^a \end{pmatrix} \right\} . \tag{19}$$

The function z in Eq. 18 satisfies the following second order equation:

$$\ddot{z} - \tau_a \dot{z} + \delta_a z = G(z, \dot{z}) \tag{20}$$

where the right hand side in Eq. 20 does not involve linear terms in z, $\dot{z}$.

The expression of G and details of the proof can be found in the cited paper. For any non-trivial matrix Γ which is a linear combination of matrixes in Eq. 19 we get

$$\Gamma\, J_a\, \Gamma^{-1} = \begin{pmatrix} 0 & 1 \\ -\delta_a & \tau_a \end{pmatrix} . \tag{21}$$

So, in a neighborhood of the origin the transform $\mathfrak{H}$ is "close enough" to the matrix transform Γ and the function G in Eq. 20 can be approximately represented as

$$G(z, \dot{z}) = \Pi_2 \left\{ \Gamma \left(\Psi \left(\Gamma^{-1} Y \right) \right) \right\} + O\left(\|Y\|^{M+1} \right) . \tag{22}$$

For instance, a suitable transform is that engendered by $\overline{\varphi} = U^{M+1}$.

From Eq. 20 we look for an oscillation with positive and small, but finite, amplitude ε. Taking in Eq. 20 the change of variables

$$z(t) = \varepsilon \varsigma (t) \tag{23}$$

follows the equation of a weakly nonlinear oscillator in *normal form*:

$$\ddot{\varsigma} - \tau_a \dot{\varsigma} + \delta_a \varsigma = \varepsilon G(\varsigma, \dot{\varsigma}; \varepsilon) . \tag{24}$$

Then, the cycle solution to Eq. 8 will correspond to the non-trivial periodic solution to Eq. 24. We shall apply the Krylov-Bogoliubov-Mitropolski averaging method [20] in order to find an asymptotic expansion to the solution. By a linear shift of coordinates we may assume that $\delta_{a_0} = 1$; if this equality do not take place, then the following formulae have to change a little bit (see for instance [17]). So, let us consider the new variables $r = r(t)$ and $\theta = \theta(t)$ defined as follows

$$\varsigma = r \cos(t + \theta) \tag{25}$$

$$\dot{\varsigma} = -r \sin(t + \theta) \tag{26}$$

then, the corresponding averaged equations are

$$\dot{r} = -\frac{1}{2\pi} \int_0^{2\pi} \sin\phi \, \{-\tau_a \, r \sin\phi + \varepsilon G(r\cos\phi, -r\sin\phi; \varepsilon)\} \, d\phi \tag{27}$$

$$\dot{\theta} = -\frac{1}{2\pi r} \int_0^{2\pi} \cos\phi \, \{-\tau_a \, r \sin\phi + \varepsilon G(r\cos\phi, -r\sin\phi; \varepsilon)\} \, d\phi \tag{28}$$

so,

$$\dot{r} = \frac{r}{2} \{\tau_a - p(r; \varepsilon)\} \tag{29}$$

$$\dot{\theta} = q(r; \varepsilon) \tag{30}$$

in which

$$p(r; \varepsilon) = \frac{\varepsilon}{\pi r} \int_0^{2\pi} \sin\phi \, G(r\cos\phi, -r\sin\phi; \varepsilon) \, d\phi \tag{31}$$

$$q(r; \varepsilon) = -\frac{\varepsilon}{2\pi r} \int_0^{2\pi} \cos\phi \, G(r\cos\phi, -r\sin\phi; \varepsilon) \, d\phi . \tag{32}$$

Functions $p(r; \varepsilon)$ and $q(r; \varepsilon)$ have at least order $O(\varepsilon^2)$. Moreover, the Taylor expansions of $p(r; \varepsilon)$ and $q(r; \varepsilon)$ must not contain odd powers of r.

Proposition 9.2. *If the function $p(r; \varepsilon)$ is not identically zero, there must exist a positive integer N and a positive real number r_0 such that $p(r, \varepsilon)$ has the non-trivial Taylor expansion:*

$$p(r; \varepsilon) = \omega \, \varepsilon^{2N} \, r_0^{-2N} \, r^{2N} + O\left(\varepsilon^{2N+2} \, r^{2N+2}\right) \tag{33}$$

where $\omega = +1$ *or* -1. *Let us assume the existence of a positive root* r *to the equation*

$$p(r;\varepsilon) - \tau_a = 0 \tag{34}$$

for positive (respect. negative) values τ_a *sufficiently close to zero. Then, up to the leading term, the root to Eq. 34 has the form*

$$r = r_0 \left(\frac{|\tau_a|}{\varepsilon^{2N}}\right)^{\frac{1}{2N}} + O\left(\varepsilon^2\right) . \tag{35}$$

It can be concluded that the expansion Eq. 33 has the form

$$p(r;\varepsilon) = p_3\ \varepsilon^2 r^2 + p_5\ \varepsilon^4 r^4 + \cdots \tag{36}$$

in which $p_s = p_s(\tau_a)$. In the following, we shall take

$$\varepsilon = |\tau_a|^{1/2N} \tag{37}$$

as the small parameter. From Eq. 37 it follows that the root in Eq. 35 can now be written as

$$r = r_0 + O\left(|\tau_a|^{\frac{1}{N}}\right) . \tag{38}$$

Due to Proposition 9.2 the function p in Eq. 31 will play the role of a "discriminant" with respect to the emergence of a cycle, as we shall see in the next Theorem.

Theorem 9.1. *(averaging in Hopf bifurcation) Let us assume that Eq. 6 holds and that Eq. 34 has the root Eq. 35 for positive (respectively, negative) but sufficiently close to zero values of the bifurcation parameter* τ_a. *Then, a spatially homogeneous limit cycle to the system Eq. 8 appears. This bifurcation is supercritical if the root exists for* $0 < \tau_a \ll 1$, *and subcritical if the root exists for* $0 < -\tau_a \ll 1$. *Furthermore, the limit cycle is orbitally asymptotically stable if and only if the bifurcation is supercritical.*

For the proof, see [18]. Notice that, the existence of a positive root to Eq. 34 is equivalent to the existence of a limit cycle to the averaged system Eqs. 29 and 30. If the steady state is stable (unstable) then the cycle must be unstable (stable). In other words, the sign of τ_a must be equal to the sign of $p(r;\varepsilon)$ being $0 < r < r_0$ and $|\tau_a|$ small enough.

From Eq. 30 we will obtain the angular speed of the oscillation. Finally, going back to the substitutions given in Eq. 23 and Eq. 17 we shall obtain the uniform asymptotic expansion of the solution to Eq. 20. We develop the periodic solution $\Theta(t) = (\overline{u}(t), \overline{v}(t))$ to Eq. 8 the orbit of which is the limit cycle, by

$$\overline{u}(t) = u_0(a) + \ u_1(t)\ \ (|\tau_a|)^{\frac{1}{2N}} + O\left(|\tau_a|^{\frac{1}{N}}\right) \tag{39}$$

$$\overline{v}(t) = v_0(a) + v_1(t)\ \ (|\tau_a|)^{\frac{1}{2N}} \ + O\left(|\tau_a|^{\frac{1}{N}}\right) \tag{40}$$

where $(u_0(a); v_0(a))$ are appropriate constants, and

$$\begin{pmatrix} u_1(t) \\ v_1(t) \end{pmatrix} = r_0 \, \Gamma^{-1} \begin{pmatrix} \cos(\varpi \, t) \\ -\sin(\varpi \, t) \end{pmatrix} \tag{41}$$

with frequency given

$$\varpi = 1 + q\left(r_0, (|\tau_a|)^{\frac{1}{2N}}\right) \tag{42}$$

and period $T = \frac{2\pi}{\varpi}$. Note that, up to the leading terms, the expansions of the cycle solution in Eqs. 39 and 40 are uniform, as the $O\left(|\tau_a|^{\frac{1}{2N}}\right)$-terms are bounded functions.

By studying the appearance of diffusive instabilities for the stable limit cycle solution to the reaction system due to a supercritical bifurcation, we implicitly are considering an unstable steady state. In this situation, fortunately, for any initial condition close to the steady state the corresponding outgoing solution is bounded for all t and so, the loss of stability is non-catastrophic. So, it is reasonable to study diffusive instabilities for an unstable steady state in the presence of a supercritical Hopf bifurcation, because wavenumber-zero instabilities will not be a source of unbounded solutions.

In the presence of a linearly stable limit cycle we have $\tau_a > 0$ so, according to the formula Eq. 14 the sign of τ_T would be positive only if $\lambda_k > 0$ is small enough, say $0 < \lambda_1 \ll 1$. Now we shall study small perturbations of the spatially homogeneous periodic solution

$$\Theta(t) = (\overline{u}(t), \overline{v}(t)) \tag{43}$$

to Eqs. 1 and 3. Denoting the corresponding perturbations by capital letters we get,

$$u(t,x) = \overline{u}(t) + U(t,x)$$
$$v(t,x) = \overline{v}(t) + V(t,x)$$

substituting into Eq. 1 and linearizing we get the following system with periodic coefficients for the perturbations

$$\frac{\partial Z}{\partial t} = D \, \Delta Z + J_\Theta(t) \; Z \tag{44}$$

where $Z(x,t)$ is the column vector with components U and V satisfying Eq. 3. Substituting the development of Θ into Eqs. 39 and 40 into the Jacobian $J_\Theta(t)$ we get

$$J_\Theta(t) = J_a + \tau_a^{\frac{1}{2N}} \; J_{1/2N}(t) + O\left(\tau_a^{\frac{1}{N}}\right)$$

where

$$J_{1/2N}(t) = (\kappa_{ij}) \tag{45}$$

is a time-periodic matrix.

Let us assume that the solutions to Eq. 44 could be asymptotically developed in the small parameter as follows,

$$Z = Z_0(t,x) + \tau_a^{\frac{1}{2N}} \; Z_1(t,x) + O\left(\tau_a^{\frac{1}{N}}\right) . \tag{46}$$

Then, substituting Eq. 46 into the system Eq. 44, we get a hierarchy of equations determining Z_j:

$$\frac{\partial Z_0}{\partial t} = D \; \Delta Z_0 + J_a Z_0 \tag{47}$$

which corresponds to $O(1)$ terms, and

$$\frac{\partial Z_1}{\partial t} = D \; \Delta Z_1 + J_a Z_1 + J_{1/2N}(t) \; Z_0 \tag{48}$$

corresponding to the $O\left(\tau_a^{\frac{1}{2N}}\right)$ terms. The functions Z_0, Z_1 are determined from Eq. 47 and Eq. 48 considering homogeneous Neumann boundary conditions in both cases. As the solution Z_0 can be expanded in a Fourier series, we would focus only on terms with the form Eq. 9. Corresponding to such Z_0 we expect the solution to Eq. 48 to be in the form

$$Z_1(t,x) = \exp(\sigma t) \; U_k(x) \; W_k(t) \; R \tag{49}$$

for some specific matrix-valued function $W_k(t)$ satisfying certain requirements. Finally, we have,

Proposition 9.3. *Let us assume the existence of a supercritical Hopf bifurcation, and let* $0 < \tau_a \ll 1$. *Then, the* extended normal modes*:*

$$Z(x,t) = \exp(\sigma t) \, U_k(x) \; \{I + O(\varepsilon)\} \; R \tag{50}$$

are asymptotic expansions of solutions to Eq. 44, or more exactly, they are normal modes disturbances corresponding to the spatial eigenvalue λ_k *in the stability analysis of* $\Theta(t)$ *as a spatially homogeneous solution to Eqs. 1 and 3.*

As in the former Turing analysis, the appearance of instabilities depends on the eigenvalue σ which is connected with the remaining parameters and the spatial eigenvalues through Eq. 12. It is implicitly assumed the existence of a bounding domain (see [15]) in the phase space of the reaction system which simultaneously contains the steady state, the limit cycle and the neighboring orbits when diffusion is included. Furthermore, if TH instabilities associated with the spatial eigenvalue λ_k appear, we shall consider that the linearly unstable extended modes are bounded by the nonlinear terms and from these interactions emerge the ultimate steady or spatiotemporal pattern.

The sign of τ_T (see Eq. 14) becomes relevant in the study of these instabilities, because $\tau_a > 0$ holds. For instance, complex roots with positive real part to Eq. 12 appear if $\tau_T > 0$, and it would be interesting to study the way in which the oscillations due to the limit cycle are transferred to the resulting diffusive instabilities. We shall address this question by an analysis using the extended modes.

Definition 9.2. Let us assume that the reaction part in Eq. 9.1 admits a supercritical HB and let $0 < \tau_a \ll 1$. If the roots to Eq. 9.12 are complex conjugated $\sigma = \sigma_r \pm i\ \sigma_i$ with $\sigma_r > 0$, then we shall call it *strong TH instability*.

Notice that

$$\tau_T^2 - 4\delta_T = \left(\tau_a^2 - 4\delta_a\right) + 2\lambda_k\left(\, j_{11}^a - \ j_{22}^a\right)\left(D_v - D_u\right) + \lambda_k^2\ \left(D_v - D_u\right)^2 \quad (51)$$

so, from Eq. 6, it follows that the existence of complex eigenvalues σ as in the definition 9.2, could be expected only if $(D_v - D_u)$ is small enough. In the following Theorem (see [18]) are gathered the conditions for the appearance of twinkling patterns.

Theorem 9.2. *Let λ_k be a given positive spatial eigenvalue ; τ_T and δ_T are given in Eq. 14 and Eq. 15 respectively. Assume further that the reaction system has a limit cycle via a Hopf bifurcation. If $\tau_T > 0$ and $\tau_T^2 - 4\delta_T < 0$ strong instabilities appear.*

Consequently, if the diffusion coefficients are equal or close enough to each other, only strong instabilities would appear. Under strong instabilities the extended modes show reinforced time-periodic oscillations with frequency σ_i, like an alternate switching between the steady pattern associated with the set of positiveness of the eigenfunction $U_k\,(x)$ and its "complementary" pattern, associated with the set of negativeness of the same eigenfunction. In this situation, the real part of the Eq. 50 can be represented asymptotically as

$$Z\,(x,t) = \exp\left(\sigma_r t\right)\ \left\{\cos\left(\sigma_i t\right)\ U_k\,(x) + O\left(\tau_a^{\frac{1}{2N}}\right)\right\}\ R\ . \quad (52)$$

For strong TH instabilities we would depict the oscillatory pattern by the set of positiveness of

$$\cos\left(\sigma_i t\right)\ U_k\,(x) \quad (53)$$

oscillating with the frequency $\sigma_i = \sqrt{\delta_T - \tau_T^2\ /4}$ which is different from the frequency of the limit cycle.

If we have a large enough nondimensional characteristic length in Ω, the lowest positive spatial eigenvalue λ_1 would be so small that $\tau_T > 0$. If in addition $\tau_T^2 - 4\delta_T < 0$ holds (see Eq. 51), then instabilities associated with this eigenvalue must appear inducing a "twinkling" pattern.

3 Strong Turing-Hopf instabilities to the glycolysis system

An important example of two-species biological model describing oscillatory behavior is the glycolysis system. Glycolysis (or glucolysis) is the metabolic pathway entrusted to rust the glucose with the purpose to obtain energy for the cell. We shall represent the situation assuming that the chemicals concentrations

u and v correspond to the glycolysis model for two species in reaction and diffusion in the plane. In this Section we resume the main results in [4], and show the way in which the emerging cycle destabilizes due to the diffusion coefficients, leading to a time-periodic oscillations of a spatially inhomogeneous solution.

Here we shall focus on the formation of TH patterns. Interesting inhomogeneous patterns, consisting either of squares, hexagons, stripes or combinations, have been studied in easy-shaped domains. We can find important contributions in which the main question is the appearance of instabilities, which are caused by the imposition of simultaneous conditions for Turing instabilities and Hopf bifurcations. In [18] the authors proposed a new approach to the problem of TH (TH) instability, seen it as the instability of the limit cycle generated by the Hopf bifurcation in a 2-D reaction diffusion system. Particularly, it was introduced there the concept of strong TH instability. The matter in this Section is the study of such instabilities in the glycolysis system

$$\begin{aligned} u_t &= D_u \Delta_x u + \delta - \kappa u - uv^2, \; x \in \Omega, \; t > 0 \\ v_t &= D_v \Delta_x v + \kappa u + uv^2 - v, \; x \in \Omega, \; t > 0 \end{aligned} \tag{54}$$

with non-negative initial conditions. The reaction occurs in a bounded domain $\Omega \subset \mathbb{R}^2$ with an impermeable, for the reactants, smooth boundary $\partial\Omega$. Here, $\delta > 0$ is the dimensionless input flux and $\kappa > 0$ is the dimensionless constant rate for the low activity rate. This system has a unique positive spatially homogeneous steady-state solution,

$$(u, v) = (\delta/\alpha, \delta), \tag{55}$$

where $\alpha := \delta^2 + \kappa$.

One of the main results in [31] about the glycolysis system Eq. 54 is that the constant steady-state solution Eq. 55 is globally asymptotically stable when $\kappa > 4\delta^2$ and $\kappa \geq (\max_{x\in\Omega} v_0(x))^2$. In this case, we obtain more accurate condition that guarantee the orbital asymptotic stability of the steady-state solution, so we expect a supercritical bifurcation when this condition fails.

Our result can be summarized as follows,

Theorem 9.3. *Let $\delta > 0$ and $0 < \kappa \leq 1/8$, then for the unique equilibrium Eq. 55 there exists two Hopf bifurcation curves in the parameter space (δ, κ). Moreover, both bifurcations are generics and the reaction system has a orbitally asymptotically stable periodic orbit provided*

$$\left|\delta^2 + \kappa - 1/2\right| < \sqrt{1 - 8\kappa} \qquad \text{and} \qquad 0 < \kappa < 1/8.$$

To do so, we shall start with the Hopf bifurcation analysis via the standard procedure of Lyapunov coefficients [7]. Let us denote by $J_0(\alpha, \kappa)$ the Jacobian matrix of the reaction system at the steady state. Let

$$\mathbf{tr}(\alpha, \kappa) := \text{trace } J_0(\alpha, \kappa) = -\alpha + \frac{\alpha - 2\kappa}{\alpha},$$

so the neutral stability curve in the parameter space is given by

$$\alpha^2 - \alpha + 2\kappa = 0.$$

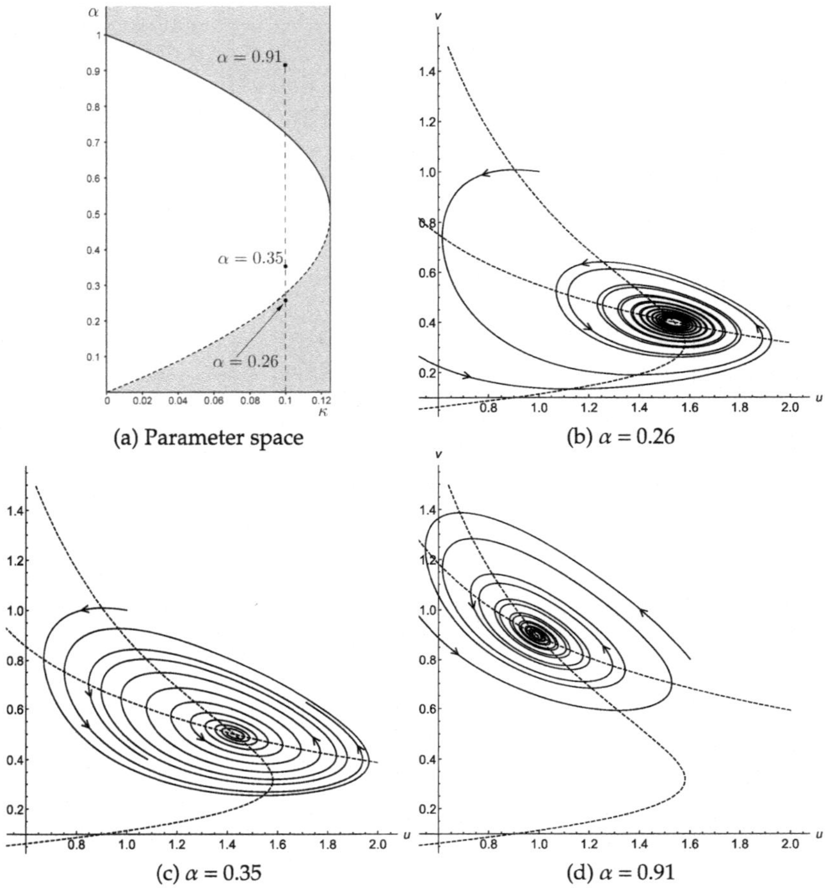

Fig. 1. Hopf bifurcation in glycolysis model with $\kappa = 0.1$. The bifurcations points are the abscissas of the intersection of the bifurcation curves and the line $\kappa = 0.1$. In graphic (a), $a_0^- \approx 0,276$, $a_0^+ \approx 0,724$. The graphics (b) and (d) represent stable focuses, while we can observe an unstable focus in graphic (c), where the orbits have to approximate a limit cycle

We shall take α as the bifurcation parameter for each fixed value of κ, say

$$\alpha_0^{\pm} = \alpha_0^{\pm}(\kappa) = \frac{1}{2}\left(1 \pm \sqrt{1-8\kappa}\right) > 0 \tag{56}$$

provided $0 < \kappa \leq 1/8$. In addition, the determinant is

$$\omega^2\left(\alpha_0^{\pm}\right) = \det J_0\left(\alpha_0^{\pm}, \kappa\right) = \alpha_0^{\pm} > 0. \tag{57}$$

For α close enough to $\alpha_0^{\pm}$, the eigenvalues of $J_0\left(\alpha_0^{\pm}, \kappa\right)$ are complex and have the form $\lambda_1(\alpha) = \lambda(\alpha) = \overline{\lambda_2(\alpha)}$, where

$$\lambda(\alpha) := \frac{1}{2}\,\mathbf{tr}(\alpha) + i\,\omega(\alpha) \tag{58}$$

and $\mathbf{tr}\left(\alpha_0^{\pm}\right) = 0$, $\omega\left(\alpha_0^{\pm}\right) > 0$. Then, if $0 < \kappa < 1/8$, we have two positive values $\alpha_0^{\pm}$ of α at which the unique steady state may shows a Hopf bifurcation. Obviously, the transversality condition $\partial\mathbf{tr}/\partial\alpha\left(\alpha_0^{\pm}\right) \neq 0$ is satisfied. After a shift of variables the system can be represented in the vectorial form

$$\xi_t = \mathbf{A}\xi + \frac{1}{2}\,\mathbf{B}\left(\xi,\xi\right) + \frac{1}{6}\,\mathbf{C}\left(\xi,\xi,\xi\right)$$

where, $\mathbf{A} = J_0\left(\alpha_0,\kappa\right)$, $\mathbf{B}$ and $\mathbf{C}$ are appropriate multilinear functions, $\xi\left(t\right) \in \mathbb{R}^2$. After large calculations, we compute the first Lyapunov coefficient by

$$l_1\left(\alpha_0\right) = \frac{2\omega^4 - 2\omega^2 - 1}{8\omega^3}$$

being ω as in Eq. 57.

The interesting observation is that, the limit cycle emerging at the supercritical Hopf bifurcation value α_0^- persists for greater values and finally collapses at $\alpha = \alpha_0^+$. This fact prepare the system Eq. 54 for the emergence of strong TH instabilities in the sense of [18], provided $\alpha \in \left(\alpha_0^-, \alpha_0^+\right)$.

The next step is the study of conditions for the emergence of strong instabilities, as in Theorem 9.2. For all values $\alpha \in \left(\alpha_0^-, \alpha_0^+\right)$, the emerging cycle is orbitally aymptotically stable, leading to the formation of twinkling patterns provided the first positive spatial eigenvalue is so small that $\tau_T > 0$ (Eq. 14). It is well-known that, the first eigenvalue strongly depends on the spatial region. In addition, the diffusion coefficients should be so close in order that $\tau_T^2 - 4\delta_T < 0$ (Eq. 51).

To avoid a repetition in this presentation, we shall present in Section 4 more details about the calculations, including the value of p_3 in the expansion Eq. 36 for the Hopf bifurcation, and further, the explicit inequalities giving the conditions for the appearance of strong instability.

4 Strong Turing-Hopf instability to the Lengyel-Epstein system

The goal of this Section is the study of strong TH instabilities in the sense of [18] for the Lengyel-Epstein CIMA reaction diffusion model [10]-[11]. These are instabilities of the limit cycle which emerges due to a Hopf bifurcation. Our intention is to find conditions for the formation of periodic spatially heterogeneous patterns induced by diffusive instabilities of the emerging limit cycle at bifurcation, extrapolating the Turing procedure for the steady point. This system is featured by the presence of two limit cycles at the Bautin-Hopf bifurcation [7]. We shall consider here the Bautin codimension one bifurcation, characterized by the emergence of one limit cycle at the bifurcation value while the other one persists throughout the bifurcation. As before, we are interested in the emerging orbitally asymptotically stable limit cycle due to the supercritical bifurcation, consequently, the greater cycle will be unstable.

Several papers on this subject for the Lengyel-Epstein reaction-diffusion system can be found in literature. For instance, in [30] the authors derive conditions on the parameters for diffusive instability of the spatially homogeneous steady state and for the homogeneous limit cycle. In [28] the authors studied Bautin-Hopf bifurcation for the Lengyel-Epstein model without diffusion. In [27] some sufficient conditions for Turing instability are obtained. Conditions for strong TH instabilities do not require different diffusion coefficients, on the contrary, it is required equal or very close diffusion coefficients.

The corresponding dimensionless reaction-diffusion system takes the following form ([27])

$$\begin{cases} u_t = \Delta u + \left(a - u - \frac{4uv}{1+u^2}\right) \\ v_t = \rho\left(d\Delta v + b(u - \frac{uv}{1+u^2})\right) \end{cases} \tag{59}$$

$x \in \Omega$, $t > 0$, where $u = u(x,t)$, $v = v(x,t)$ are normalized dimensionless concentrations of the two reactants iodide I^- and chlorite CIO_2^- respectively, verifying appropriate initial conditions; a and b are the parameters related to the concentration of acid, d is the diffusion coefficient ratio, $\rho > 1$ is an adjustable parameter which is bound up with starch concentration. No flux boundary conditions on $\partial\Omega$ are imposed to the reactants.

To simplify the calculations, we shall considering in Eq. 59 the following new parameters,

$$\alpha := 1/5\ a;\ \eta^2 := \rho \cdot d;\ \beta := \rho \cdot b. \tag{60}$$

If, in addition, we realize a shift of coordinates $u' = u - \alpha$, $v' = v - (1 + \alpha^2)$, rejecting the primes in the new variables, the system Eq. 59 can be rewritten in the form :

$$\begin{cases} u_t = \Delta u - \frac{4(u+\alpha)(v+1+\alpha^2)}{1+(u+\alpha)^2}\ ;\ x \in \Omega,\ t > 0, \\ v_t = \eta^2 \Delta v + \beta\left(u + \alpha - \frac{(u+\alpha)(v+1+\alpha^2)}{1+(u+\alpha)^2}\right)\ ;\ x \in \Omega,\ t > 0, \end{cases} \tag{61}$$

where Ω is a bounded spatial domain in $\mathbb{R}^p$, $(p \geq 1)$ with sufficiently smooth boundary $\partial\Omega$, thus homogeneous Neumann boundary condition are imposed in addition to the initial values of the unknowns.

In the absence of diffusion, the system Eq. 61 has a unique constant steady state in $P_\alpha(\alpha, 1+\alpha^2)$. Let us denote by Σ the part of the neutral stability curve $\tau = 0$ located in the first quadrant, with

$$\Sigma = \left\{(\alpha,\beta)|\ \beta = \beta^H(\alpha) := \frac{3\alpha^2 - 5}{\alpha};\ \sqrt{\frac{5}{3}} < \alpha < \alpha^H\right\}, \tag{62}$$

as follows from Eq. 66. As we shall see in Fig.2, the condition for the emergence of a supercritical non-degenerate Hopf bifurcation (in the sense of [17]) provided the parameters vary throughout the curve Σ is given by the condition $p_3(\tau = 0) > 0$ (see Eq. 36). This condition corresponds to the standard inequality for the non-degeneracy of the bifurcation, using Lyapunov coefficients. The point $Q\left(\alpha^H, \beta^H\left(\alpha^H\right)\right)$ is determined as the intersection point of the curves $\tau = 0$ and $p_3(\alpha, \beta) = 0$, i.e. α^H is determined by

$$p_3\left(\alpha^H, \beta^H\left(\alpha^H\right)\right) = 0 \; .$$

Notice that the steady state P_α is independent of β, so we may choose β as the Hopf bifurcation parameter for a fixed α. Hence, a non-degenerate supercritical Hopf bifurcation takes place when β decreases from above across the bifurcation value $\beta^H(\alpha)$. We underline the domain of parametric dependence in the (α, β)-space, which we shall called the *parameter space.* For all pairs of parameters in this region, the steady state P_α is linear unstable and an asymptotically stable limit cycle emerging at the critical value, exists.

The Jacobian matrix on the steady state is

$$J = \begin{pmatrix} \frac{3\alpha^2-5}{1+\alpha^2} & -\frac{4\alpha}{1+\alpha^2} \\ \frac{2\beta\alpha^2}{1+\alpha^2} & -\frac{\beta\alpha}{1+\alpha^2} \end{pmatrix} \; . \tag{63}$$

Let us denote $\Psi(X) = (f, g)^T$ in the Eq. 16. Then, the fifth order Taylor expansion:

$$\begin{aligned} f(Z,\alpha) = & \frac{4\left(-1+\alpha^2\right)}{\left(1+\alpha^2\right)^2} VU - \frac{4\alpha\left(-3+\alpha^2\right)}{\left(1+\alpha^2\right)^2} U^2 \\ & - \frac{4\left(\alpha\left(-3+\alpha^2\right)\right)}{\left(1+\alpha^2\right)^3} VU^2 + \frac{4\left(1-6\alpha^2+\alpha^4\right)}{\left(1+\alpha^2\right)^3} U^3 \\ & + \frac{4\left(1-6\alpha^2+\alpha^4\right)}{\left(1+\alpha^2\right)^4} U^3 V - \frac{4\left(5\alpha-10\alpha^3+\alpha^5\right)}{\left(1+\alpha^2\right)^4} U^4 \\ & - \frac{4\left(5\alpha-10\alpha^3+\alpha^5\right)}{\left(1+\alpha^2\right)^5} U^4 V \\ & + \frac{4\left(-1+15\alpha^2-15\alpha^4+\alpha^6\right)}{\left(1+\alpha^2\right)^5} U^5 + O\left(\|Z\|^6\right) \; , \end{aligned} \tag{64}$$

$$g(Z,\alpha) = \frac{\beta}{4} f(Z,\alpha) \; . \tag{65}$$

Let us denote by τ and δ the trace and determinant of J respectively. Then,

$$\tau = \frac{3\alpha^2 - \alpha\beta - 5}{1+\alpha^2} \; ; \; \omega^2 := \delta = \frac{5\alpha\beta}{1+\alpha^2} \; . \tag{66}$$

The eigenvalues of the matrix J are $\mu = \mu_r \pm i\mu_i$, provided that (α, β) belongs to the parameter space. More precisely,

$$\mu_r = \mu_r(\alpha,\beta) = \frac{1}{2}\cdot\frac{3\alpha^2-\alpha\beta-5}{(1+\alpha^2)} \; ; \; \mu_i = \mu_i(\alpha,\beta) = \frac{1}{2}\cdot\frac{\sqrt{\Lambda}}{1+\alpha^2} \tag{67}$$

where we have denoted

$$\Lambda = 4\delta - \tau^2 = 4\left(\frac{5\alpha\beta}{1+\alpha^2}\right) - \left(\frac{3\alpha^2-\alpha\beta-5}{1+\alpha^2}\right)^2 > 0, \tag{68}$$

which is a necessary condition for the emergence of an HB. Hence,

$$\mu_r^0(\alpha, \beta^H(\alpha)) = 0 \; ; \; \mu_i^0 := \mu_i(\alpha, \beta^H(\alpha)) = \sqrt{5}\sqrt{\frac{3\alpha^2-5}{(1+\alpha^2)^3}}. \tag{69}$$

We have taken

$$\Gamma = \frac{\beta}{4}\Gamma_1 - \Gamma_2 = \begin{pmatrix} \frac{\beta}{4} & -1 \\ -\frac{5\beta}{4} & 0 \end{pmatrix}, \tag{70}$$

from Eq. 19. Then, following the procedure in Section 2, we get

$$\begin{aligned} p(r;\varepsilon) = &-12r^2\omega(\alpha,\beta)^2\varepsilon^2\frac{\left(4\alpha^4-\beta\alpha^3-24\alpha^2+3\beta\alpha+4\right)}{25\beta^2\left(\alpha^2+1\right)^3} + \\ &32r^4\omega(\alpha,\beta)^4\varepsilon^4\frac{\left(-4\alpha^6+\beta\alpha^5+60\alpha^4-10\beta\alpha^3-60\alpha^2+5\beta\alpha+4\right)}{125\beta^4\left(\alpha^2+1\right)^5} \\ &+\cdots \end{aligned}$$

and

$$q(r;\varepsilon) = 2r^2\omega(\alpha,\beta)^2\alpha\varepsilon^2\frac{\alpha^2-3}{5\beta\left(\alpha^2+1\right)^3} + 16r^4\omega(\alpha,\beta)^4\alpha\varepsilon^4\frac{\alpha^4-10\alpha^2+5}{125\beta^3\left(\alpha^2+1\right)^5} + \cdots \tag{71}$$

up to the fifth order terms. Here $\varepsilon = \tau^{1/2}$. So, it can be concluded the emergence of a unique stable limit cycle at the supercritical Hopf bifurcation. However, there is an unstable limit cycle throughout the monoparametric bifurcation (Bautin), as $p_5 \neq 0$. This situation can be found for instance in [17, Subsection 4.1].

The following assertion takes place,

Theorem 9.4. *If α, β are positive and close to the parametric curve S, then:*

i) The equilibrium P_α is locally asymptotically stable when $\beta > \beta^H(\alpha)$ and unstable when $\beta < \beta^H(\alpha)$;

ii) The system Eq. 61 in the absence of diffusion undergoes a supercritical (respect. subcritical) Hopf bifurcation at P_α when $\beta = \beta^H(\alpha)$ provided:

$$\sqrt{\frac{5}{3}} < \alpha < \sqrt{\frac{27+\sqrt{769}}{4}},$$

(respect. $\alpha > \sqrt{\frac{27+\sqrt{769}}{4}}$).

The parameter domain of our interest is given in Fig.2 . Let us consider the rectangle $\Omega = [0, l] \times [0, m]$ in which the non-dimensional lengths are taken to satisfy $l > m$, in order that the smallest positive spatial eigenvalue is $\lambda_{(1,0)} = \pi^2/l^2$.

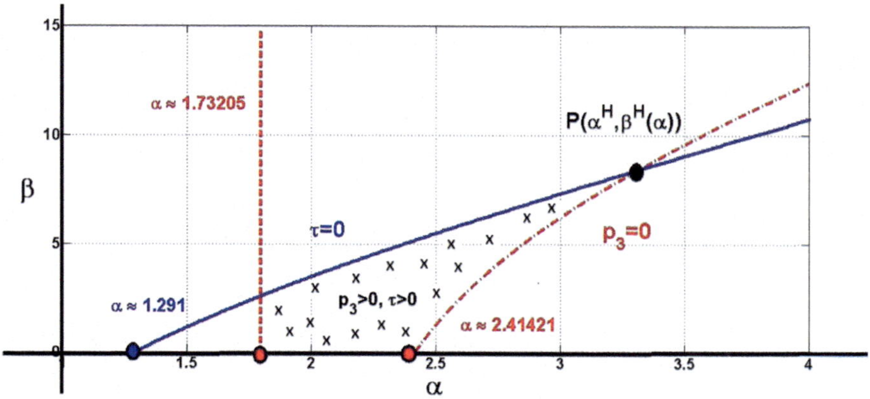

Fig. 2. Neutral stability curve (blue) in the $\alpha\beta$-plane. A single limit cycle surrounding the steady state can be found when (α, β) belongs to the shadowed domain

For instance, let $\alpha = 2$, $\beta = 3.2$, $D_u = D_v = D$, so $\tau = 0.12$, $\delta = 6.4$, and

$$\tau_T = 0.12 - 2D\pi^2/l^2$$

$$\tau_T^2 - 4\delta_T = \tau^2 - 4\delta < 0.$$

Hence, taking $l^2 > 164.33\ D$, twinkling patterns should appear associated with the eigenfunction corresponding to $\lambda_{(1,0)}$, in accordance to the theory. We would depict the oscillatory pattern by the set of positiveness of the function in Eq. 53, say

$$\cos(\sigma_i t)\ U_{(1,0)}(x_1, x_2)$$

oscillating with the frequency $\sigma_i = \sqrt{\delta_T - \tau_T^2 4}$. We are gathering numerical results in this direction in [21].

5 Conclusions

In the study of twinkling patterns for a reaction-diffusion system, we first check the conditions for the emergence of a limit cycle, for instance, due to a non-degenerate Hopf bifurcation of the spatially homogeneous steady state. The next step is the study of conditions for the emergence of strong instability, as in Theorem 9.2. So we need to check that the first positive spatial eigenvalue is so small that $\tau_T > 0$ (Eq. 14). In addition, the diffusion coefficients should be so close in order that $\tau_T^2 - 4\delta_T < 0$ (Eq. 51).

The conditions for Hopf bifurcation based on the extended modes require that the system moves close enough the manifold $\tau_a = 0$ in the parameter space, so for the stability of the limit cycle. However, for a limit cycle with finite radius ($\tau_a = O(1)$) the diffusive instability analysis could be done by the Flocquet theory [9]. In addition, in the finite radius case, more greater spatial eigenvalues would satisfy Eq. 14 and, a superposition of twinkling patterns would appear. Strong Turing Hopf instabilities presumably lead to well defined intermittent spatiotemporal patterns with a frequency different from the one of the limit cycle.

Acknowledgments. Mariano Rodríguez Ricard would like to express his gratitude to the organizers of the ODA Week at Imperial College London (Nov 23-28, 2016), Prof. Michael Ruzhansky and Dr. Julio Delgado, for their support and cordiality during the workshop.

References

1. Baurmann, M., Gross, T., Feudel, U.: Instabilities in spatially extended predator-prey systems: spatio-temporal patterns in the neighborhood of Turing-Hopf bifurcations. J. Theor. Biol. **245**(2), 220–229 (2007)
2. Bogoliubov, N.N., Mitropolski, Y.A.: Asymptotic Methods in the Theory of Non-linear Oscillations. Gordon and Breach Science Publishers, New York (1961)
3. Díaz Rodrígues, L.A., Mistro, D.C., Petrovskii, S.: Pattern formation, long-term transients, and the Turing-Hopf Bifurcation in a space- and time-discrete predator-prey system. Bull. Math. Biol. (2010). https://doi.org/10.1007/s11538-010-9593-5
4. Egaña Fernández, G., Rodríguez Ricard, M.: Emergence and collapse of limit cycles in the glycolysis model. Revista Investigación Operacional **39**(1), 23–32 (2018)
5. Henry, D.: Geometric Theory of Semilinear Parabolic Equations. Springer, New York (1981)
6. Just, W., Bose, M., Bose, S., Engel, H., Schöll, E.: Spatiotemporal dynamics near a supercritical Turing-Hopf bifurcation in a two-dimensional reaction-diffusion system. Phys. Rev. E. **64**(026219), 1–12 (2001). https://doi.org/10.1103/PhysRevE.64.026219
7. Kuznetsov, YuA: Elements of Applied Bifurcation Theory. Applied Mathematical Sciences, vol. 112, 3rd edn. Springer Science+Business Media, New York (2004)
8. Lacitignola, D., Bozzini, B., Sgura,I.: Spatio-temporal organization in a morphochemical electrodeposition model: Hopf and Turing instabilities and their interplay. Eur. J. Appl. Math. (2014). https://doi.org/10.1017/S0956792514000370
9. Leiva, H.: Stability of a periodic solution for a system of parabolic equations. Appl. Anal. **60**, 277–300 (1996)
10. Lengyel, I., Epstein, I.R.: Global bifurcation and structure of Turing patterns in the 1-D Lengyel-Epstein model. Sciences **251**, 650–652 (1991)
11. Lengyel, I., Epstein, I.R.: A chemical approach to designing Turing patterns in reaction-diffusion system. Proc. Natl. Acad. Sci. USA **89**, 3977–3979 (1992)

12. Liu, P., Shi, J., Wang, Y., Feng, X.: Bifurcation analysis of reaction-diffusion Schnakenberg model. J. Math. Chem. **51**, 2001–2019 (2013). https://doi.org/10.1007/s10910-013-0196-x
13. Marsden, J.E., McCracken, M.: The Hopf Bifurcation and its Applications. Springer, New York (1976)
14. Meixner, M., De Wit, A., Bose, S., Schöll, E.: Generic spatiotemporal dynamics near codimension-two Turing-Hopf bifurcations. Phys. Rev. E **55**(6), 6690–6697 (1997)
15. Murray, J.D.: Mathematical Biology I: An Introduction. Interdisciplinary Applied Mathematics, vol. 17, 3rd edn. Springer, New York (2001)
16. Murray, J.D.: Mathematical Biology II: Spatial Models and Biomedical Applications. Interdisciplinary Applied Mathematics, vol. 18, 3rd edn. Springer, New York (2003)
17. Ricard, M.R.: On degenerate planar Hopf bifurcations. J. Phys. A. Math. Theor. **44**, 065202 (2011)
18. Ricard, M.R., Mischler, S.: Turing instabilities at Hopf bifurcation. J. Nonlinear Sci. **19**(5), 467–496 (2009). https://doi.org/10.1007/s00332-009-9041-6
19. Rudin, W.: Principles of Mathematical Analysis, 3rd edn. McGraw-Hill, New York (1976)
20. Sanders, J.A., Verhulst, F., Murdock, J.: Averaging Methods in Nonlinear Dynamical Systems. Applied Mathematical Sciences, vol. 59, 2nd edn. Springer, New York (2007)
21. Sarría-González, J., Ricard, M.R.: Twinkling patterns for the Lengyel-Epstein reaction-diffusion model (2018) (in preparation)
22. Settani, G., Sgura, I.: Devising efficient numerical methods for oscillating patterns in reaction-diffusion systems. J. Comput. Appl. Math. (2015). https://doi.org/10.1016/j.cam.2015.04.044
23. Sgura, I., Bozzini, B., Lacitignola, D.: Numerical approximation of Turing patterns in electrodeposition by ADI methods. J. Comput. Appl. Math **236**, 4132–4147 (2012)
24. Sgura, I., Bozzini, B., Lacitignola, D.: Numerical approximation of oscillating Turing patterns in a reaction-diffusion model for electrochemical material growth. In: AIP Conference Proceedings, vol. 1493, pp. 896–903. Melville, New York (2012). https://doi.org/10.1063/1.4765594
25. Strier, D.E., Ponce, D.S.: Turing patterns inside cells. PLoS ONE **2, (10)**, e1053 (2007). https://doi.org/10.1371/journal.pone.0001053
26. Turing, A.M.: The chemical basis for morphogenesis. Philos. Trans. R. Soc. Lond. B **237**, 37–72 (1952)
27. Wang, L., Zhao, H.: Hopf bifurcation and Turing instability of 2-D Lengyel-Epstein system with reaction-diffusion terms. Appl. Math. Comput. **21**, 9229–9244 (2013)
28. Wu, X.P., Eshete, M.: Bautin bifurcation for the Lengyel-Epstein system. J. Math. Chem. **52**, 2570–2580 (2014). https://doi.org/10.1007/s10910-014-0401-6
29. Yang, L., Berenstein, I., Epstein, I.R.: Segmented Waves from a Spatiotemporal Transverse Wave Instability. Phys. Rev. Lett. **95, 3**, 038303 (2005). https://doi.org/10.1103/PhysRevLett.95.038303

30. Yi, F., Wei, J., Shi, J.: Diffusion-driven instability and bifurcation in the Lengyel-Epstein system. Nonlinear Anal. Real World Appl. **9**, 1038–1051 (2008)
31. Zhou, J., Shi, J.: Pattern formation in a general glycolysis reaction-diffusion system. IMA J. Appl. Math. **13**, 1–36 (2015). https://doi.org/10.1093/imamat/hxv013

Correspondence Between Multiscale Frame Shrinkage and High-Order Nonlinear Diffusion

Haihui Wang[1(✉)], Qi Huang[1], and Bo Meng[2]

[1] School of Mathematics and System Science, Beihang University, No. 37, Xueyuan Road, Haidian District, Beijing 100191, China
whhmath@buaa.edu.cn, hqzx2003@126.com

[2] School of Energy and Power Enigineering, Beihang University, No. 37, Xueyuan Road, Haidian District, Beijing 100191, China
davidmeng0206@gmail.com

Abstract. Wavelet frame and nonlinear diffusion filters are two popular tools for signal denoising. The correspondence between Ron-Shen's framelet and high-order nonlinear diffusion has been proved at multilevel setting. However, for the general framelet, the correspondence is established only at one level. In this paper we extend the relationship between framelet shrinkage and high-order nonlinear diffusion in Jiang (Appl Numerical Math 51–66, 2012 [19]) from one level framelet shrinkage to the multilevel framelet shrinkage setting. Subsequently, we complete the correspondence between framelet shrinkage and high-order nonlinear diffusion. Furthermore, we propose a framelet-diffused denoising method for processing the dynamic pressure signals which are generated by a transonic axial compressor. Numerical results show that our algorithm has superior noise removal ability than traditional algorithms and presents the ability in analyzing the pressure signals from an axial transonic compressor.

Keywords: Wavelets · Nonlinear diffusion · Signal analysis

1 Introduction

In the past few decades, mathematical based approaches have been widely adopted in signal denoising, among which the partial differential equation (PDE) based approach [1] and tis generalizations, nonlinear diffusion [2,3], and wavelet frame based approach are some of successful examples. These approaches were developed through different paths and generally provided understandings from different angles of the same problem. Wavelet frame shrinkage and nonlinear diffusion filtering are two commonly used methods for denoising without sacrificing important features such as edges, see [4–12]. Its major purpose is to enhance the quality of a given signal that is corrupted in various ways during the process of

J. Delgado and M. Ruzhansky (Eds.): *Analysis and Partial Differential Equations: Perspectives from Developing Countries*, PROMS 275, pp. 159–171, 2018.
https://doi.org/10.1007/978-3-030-05657-5_10

imaging, acquisition and communication, and enable us to see crucial but subtle information that reside in the signal.

One of the most commonly nonlinear PDE based approach for the signal denoising is the nonlinear diffusion. See the referees [3] for the introduction of the 2nd-order nonlinear diffusion by Perona and Malik (PM) in 1990, see references [2,9,13] for a variety of 2nd-order nonlinear diffusions have been proposed. The theoretical properties of high order diffusion have been studied in [15]. What these models for signal denoising have in common is the seek of a good balance between the two seemingly contradictory objectives: smoothness at locations where noise or other artifacts have been removed, and preservation or enhancement of the peeks which are singularities.

Ever since the early 90's, a fundamental connection between wavelet based approach and a general variation model were established in [22]. It was shown in [22] that a special type of wavelet model using hear wavelet systems can be regarded as an approximation of the numerous image restoration models and algorithms based on PDEs and wavelets were proposed and studied in the literature .This correspondence can help to choose proper shrinkage functions from diffusivity, and vice versa, we refer to [16,17,20,21] for more discussions on the importance of this relationship. The equivalence of wavelet shrinkage and nonlinear diffusion under one-scale was shown in [16], the correspondence between wavelet shrinkage and nonlinear diffusion at multiscale, which permits larger time steps while preserving convergence was proven in [17].

Due to the efficiency of frame, there has been good progress in works on the frame shrinkage and nonlinear diffusion. In [18], Wang introduced a multiscale tight frame-inspired scheme for nonlinear diffusion, which constructed a one-to-one relation between the fourth-order nonlinear diffusion and Ron-Shen's tight frame (special, not arbitrary frame), which then successfully extended it to multiscale. For arbitrary frame, Jiang illustrated it in [19], but only at one scale. However, on the multiscale case, it has never been studied before. Motivated by Wang's [18] and Jiang's [19], this paper is to establish a fundamental connection between between arbitrary frame shrinkage and high-order nonlinear diffusion at multiscale, provide interpretations and analytical studies of such connection, and propose new algorithm which we call it framelet-based algorithm for signal denoising based on the new understandings. The connections automatically gives a wavelet frame approach a geometry explanation through nonlinear PDEs and , at the same time, it equips the PDE based approach with a time frequency analysis by the nature of the two approaches.

The rest of the paper is organized as follows. In Sect. 2 presents the frame shrinkage process at different scales. The introduce of high-order nonlinear diffusion equation and its approximation at different scales shows in Sect. 3. Sect. 4 shows out the correspondence between frame shrinkage and high-order nonlinear diffusion equation, two examples given out to help understand the theorem. The paper is concluded with a summary in Sect. 5.

2 Frame Shrinkage at Different scales

Set $u^{(k)} := (u_0^{(k)}, ..., u_{N-1}^{(k)})^T$ and let I be the identity matrix of order N. we use $u^k(z)$ to denote its z-transform:

$$u^{(k)}(z) = \sum_{i=0}^{N-1} u_i^{(k)} z^{-k}.$$

Let $\{p, q^1, \cdots, q^L\}$ and $\{\tilde{p}, \tilde{q}^1, \cdots, \tilde{q}^L\}$ be a pair of FIR frame filter banks. Then $q(z)$ can be written as $q(z) = \sum_{m=0}^{M-1} q_m z^{-m}$, where M denotes the length of the high-pass filter. Suppose q^l and $\tilde{q}^l$ have vanishing moment orders α_l and β_l, respectively, and satisfy the perfect reconstruction property, namely,

$$\tilde{p}(z^{-1})p(z) + \sum_{l=1}^{L} \tilde{q}^l(z^{-1})q^l(z) = 1. \tag{1}$$

Let j denotes the scale, and let $S_{lj}, l = 1, \cdots, L$, be the frame shrinkage functions. When $j = 1, L = 2$, Fig. 1 illustrates the shrinkage process.

The denoised signal $u^{(k+1)}$ after 1-step frame shrinkage is

$$u^{(k+1)}(z) = \tilde{p}(z^{-1})p(z)u^{(k)}(z) + \sum_{l=1}^{2} \tilde{q}^l(z^{-1})S_{l1}\left(q^l(z)u^{(k)}(z)\right). \tag{2}$$

Using perfect reconstruction property (1), we have

$$u^{(k+1)}(z) = u^{(k)}(z) - \sum_{l=1}^{2} \tilde{q}^l(z^{-1}) \left[q^l(z)u^{(k)}(z) - S_{l1}\left[q^l(z)u^{(k)}(z)\right]\right].$$

Without loss of generality, in this paper, we suppose $L = 2$. We introduce the circulant shift matrix V of order N, where N represents the length of the signal,

$$V = \begin{pmatrix} 0 & 1 & 0 & 0 & 0 \cdots & 0 & 0 & 0 \\ 0 & 0 & 1 & 0 & 0 \cdots & 0 & 0 & 0 \\ \cdots & \cdots & \cdots & \cdots & \cdots & \cdots & \cdots & \cdots \\ 0 & 0 & 0 & 0 & 0 \cdots & 0 & 1 & 0 \\ 0 & 0 & 0 & 0 & 0 \cdots & 0 & 0 & 1 \\ 1 & 0 & 0 & 0 & 0 \cdots & 0 & 0 & 0 \end{pmatrix} \tag{3}$$

$$V^2 = \begin{pmatrix} 0 & 0 & 1 & 0 & 0 \cdots & 0 & 0 & 0 \\ 0 & 0 & 0 & 1 & 0 \cdots & 0 & 0 & 0 \\ \cdots & \cdots & \cdots & \cdots & \cdots & \cdots & \cdots & \cdots \\ 0 & 0 & 0 & 0 & 0 \cdots & 0 & 0 & 1 \\ 1 & 0 & 0 & 0 & 0 \cdots & 0 & 0 & 0 \\ 0 & 1 & 0 & 0 & 0 \cdots & 0 & 0 & 0 \end{pmatrix} \tag{4}$$

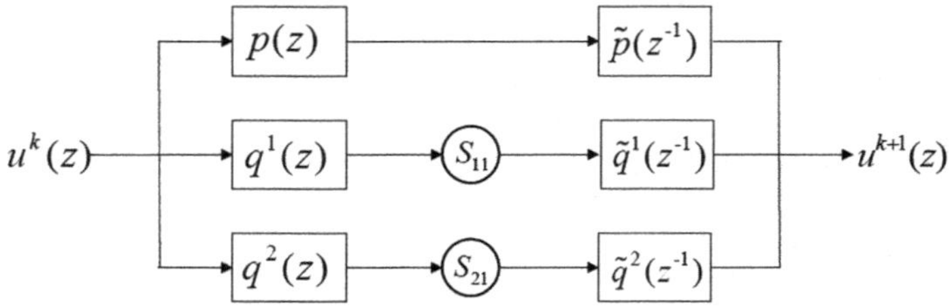

Fig. 1. Nonsampled three-channel filter bank with $p(z) = \frac{2+z+2z^2}{2}, q^1(z) = \frac{1-z^2}{\sqrt{2}}, q^2(z) = \frac{-1+2z-z^2}{2}$

$$I - 2V + V^2 = \begin{pmatrix} 1 & -2 & 1 & 0 & 0 \cdots & 0 & 0 & 0 \\ 0 & 1 & -2 & 1 & 0 \cdots & 0 & 0 & 0 \\ 0 & 0 & 1 & -2 & 1 \cdots & 0 & 0 & 0 \\ \cdots & \cdots & \cdots & \cdots & \cdots & \cdots & \cdots & \cdots \\ 0 & 0 & 0 & 0 & 0 \cdots & 1 & -2 & 1 \\ 1 & 0 & 0 & 0 & 0 \cdots & 0 & 1 & -2 \\ -2 & 1 & 0 & 0 & 0 \cdots & 0 & 0 & 1 \end{pmatrix}. \tag{5}$$

Note that $V^{-1} = V^T$, and replacing z with V, then

$$u^{k+1} = u^k - \sum_{l=1}^{2} \tilde{q}^l(V^T)[q^l(V)u^k - S_{l1}(q^l(V)u^k)] \tag{6}$$

where $q^l(V)u^k = \sum_{m=0}^{M-1} q^l_m \cdot u^k V^m$.

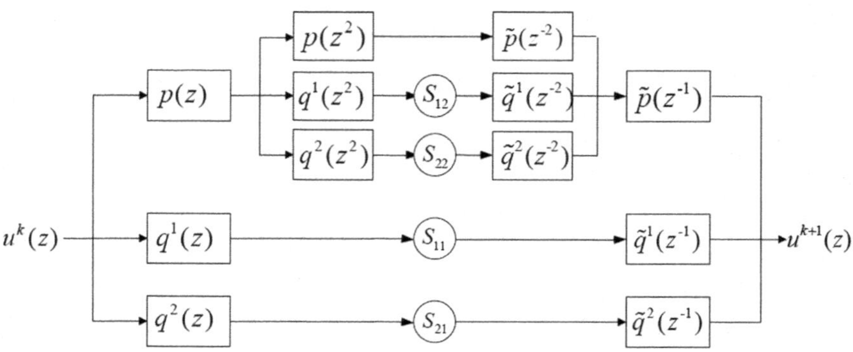

Fig. 2. Frame shrinkage at two scale

When $j = 2$, the frame shrinkage process at two scale is shown in Fig. 2. Let S_{l1} and $S_{l2}, l = 1, 2$, be the shrinkage functions at the first and the second scale, respectively. Then the denoised signal $u^{(k+1)}$ can be written as

$$u^{(k+1)}(z) = \tilde{p}(z^{-1})\tilde{p}(z^{-2})p(z^2)p(z)u^{(k)}(z) + \sum_{l=1}^{2} \tilde{q}^l(z^{-1})S_{l1}(q^l(z)u^{(k)}(z))$$
$$+ \sum_{l=1}^{2} \tilde{p}(z^{-1})\tilde{q}^l(z^{-2})S_{l2}(q^l(z^2)p(z)u^{(k)}(z)). \quad (7)$$

Since

$$\tilde{p}(z^{-1})\tilde{p}(z^{-2})p(z^2)p(z) + \sum_{l=1}^{2} \tilde{q}^l(z^{-1})q^l(z) + \sum_{l=1}^{2} \tilde{p}(z^{-1})\tilde{q}^l(z^{-2})q^l(z^2)p(z) = 1,$$

which indeed holds true for every filter bank with perfect reconstruction property, we can rewrite (7) as

$$u^{(k+1)}(z) = u^{(k)}(z) - \sum_{l=1}^{2} \tilde{p}(z^{-1})\tilde{q}^l(z^{-2})[q^l(z^2)p(z)u^{(k)}(z) - S_{l2}(q^l(z^2)p(z)u^{(k)}(z))]$$
$$- \sum_{l=1}^{2} \tilde{q}^l(z^{-1})[q^l(z)u^{(k)}(z) - S_{l1}(q^l(z)u^{(k)}(z))] \quad (8)$$

Identifying z with the shift matrix V, the convolution of $u^{(k)}$ with the filter H yields $H(V)u^{(k)}$, where $H(V) = \frac{I+2V+V^2}{2}$. Similarly, the convolution with the filters $G_1$and G_2can be written as multiiplication with $G_1(V) = \frac{I-V^2}{\sqrt{2}}, G_2(V) = \frac{-I+2V-V^2}{2}$ Replacing z with circulant matrix V of order N, where N is the length of signal, we obtain

$$u^{(k+1)} = u^{(k)} - \sum_{l=1}^{2} \tilde{p}(V^T)\tilde{q}^l((V^2)^T)[q^l(V^2)p(V)u^{(k)} - S_{l2}(q^j(V^2)p(V)u^{(k)}]$$
$$- \sum_{l=1}^{2} \tilde{q}^l(V^T)[q^l(V)u^{(k)} - S_{l1}(q^l(V)u^{(k)})]. \quad (9)$$

Generally, for $j = J$, by using the same steps as above, we have

$$u^{(k+1)} = u^{(k)} - \sum_{l=1}^{2} \tilde{q}^l(V^T)[q^l(V)u^{(k)} - S_{l1}(q^l(V)u^{(k)})]$$
$$- \sum_{l=1}^{2}\sum_{j=2}^{J}\prod_{m=2}^{j} \tilde{p}(V^{2^{m-2}})^T\tilde{q}^l(V^{2^{j-1}})^T[q^l(V^{2^{j-1}})\prod_{m=2}^{j} p(V^{2^{m-2}})u^{(k)}$$
$$-S_{lj}(q^l(V^{2^{j-1}})\prod_{m=2}^{j} p(V^{2^{m-2}})u^{(k)}]. \quad (10)$$

The formula above holds for any shrinkage functions $S_{l1}(l = 1, 2)$ and $S_{lj}(l = 1, 2; j = 1, \cdots, J)$.

3 High-Order Nonlinear Diffusion Equation

For any given signal f with noise, the goal of nonlinear diffusion filtering is to find the solution $u = u(x,t)$, subject to

$$u_t = \frac{\partial}{\partial x}(g(u_x)^2 u_x),$$

with f as the initial condition $u(x,0) = f(x)$, and with certain boundary condition. Solution $u(x,t)$ is the denoised signal, diffusivity g is a nonnegative decreasing function, more about diffusivity g can be found in [8]. Let u_x denote the first-order partial derivative of $u(x,t)$ with respect to x. A 1-D high-order diffusion equation corresponding to wavelet is an equation

$$u_t = (-1)^{n+1}\frac{\partial^n}{\partial x^n}\left(g\left((\frac{\partial^n u}{\partial x^n})^2\right)\frac{\partial^n u}{\partial x^n}\right),$$

for an integer $n \geq 2$, with f as initial condition $u(x,0) = f(x)$. More properties of high-order nonlinear diffusion can be found in [15]. Corresponding to framelet shrinkage, consider nonlinear diffusion equation

$$u_t = \sum_{l=1}^{L}(-1)^{\alpha_l+1}\frac{\partial^{\alpha_l}}{\partial x^{\alpha_l}}\left(g_l((\frac{\partial^{\beta_l} u}{\partial x^{\beta_l}})^2)\frac{\partial^{\beta_l} u}{\partial x^{\beta_l}}\right),$$

with initial condition $u(x,0) = f(x)$, where α_l and β_l are the vanishing moment orders of $q^{(l)}$ and $\tilde{q}^{(l)}$, respectively. Without loss of generality, in this paper, we only consider $L = 2$. So the high-order nonlinear diffusion equation u_t can be written as

$$u_t = (-1)^{\alpha_1+1}\frac{\partial^{\alpha_1}}{\partial x^{\alpha_1}}(g_1((\frac{\partial^{\beta_1} u}{\partial x^{\beta_1}})^2)\frac{\partial^{\beta_1} u}{\partial x^{\beta_1}}) + (-1)^{\alpha_2+1}\frac{\partial^{\alpha_2}}{\partial x^{\alpha_2}}(g_2((\frac{\partial^{\beta_2} u}{\partial x^{\beta_2}})^2)\frac{\partial^{\beta_2} u}{\partial x^{\beta_2}}) \quad (11)$$

We now discuss the approximation of derivatives at different scales.There we denote the vectors $u =: (u(ih))_{i=0}^N, u^{(\alpha)} =: (u^{(\alpha)}(ih))_{i=0}^N$ approximates $u_x =: (u_x(ih))_{i=0}^N$ and $u_x^{(\alpha)} =: (u_x^{(\alpha)}(ih))_{i=0}^N$, respectively.

Theorem 1. *For a function $u(x)$ smooth enough, q is a filter with vanishing moment of order α, the j scale decomposition of $u^{(\alpha)}$ is*

$$u^{(\alpha)} = \frac{1}{C_\alpha h^\alpha 2^{(\alpha+1)(j-1)}}\sum_{k\in Z} q_k u((i+2^{j-1})hk)\prod_{l=0}^{j-2}(I+V^{2^l}) + o(h^{\alpha+1}), \quad (12)$$

where $C_\alpha := \frac{1}{\alpha!}\sum_{k\in Z} k^\alpha q_k$.

Proof. Let $B_1 = B_1(i,l,j) = \sum_{k\in Z} q_k u((i+l+2^{j-1}k)h)$.Then

$$\frac{1}{C_\alpha h^\alpha 2^{(\alpha+1)(j-1)}}\sum_{k\in Z} q_k u(ih+2^{j-1}hk)\prod_{l=0}^{j-2}(I+V^{2^l})+o(h^{\alpha+1})$$

$$=\frac{1}{C_\alpha h^\alpha 2^{(\alpha+1)(j-1)}}[\sum_{l=0}^{2^{i-1}-1}(l+1)B_1(i,l,j)+\sum_{l=2^{i-1}}^{2^j-2}(2^j-l-1)B_1(i,l,j)]$$
$$\prod_{l=0}^{j-2}(I+V^{2^l})+o(h^{\alpha+1})$$

$$=\frac{1}{C_\alpha h^\alpha 2^{(\alpha+1)(j-1)}}[[\sum_{l=0}^{2^{i-1}-1}(l+1)[\sum_{k\in Z} q_k(u(ih)+(l+2^{j-1}k)h)u^{(1)}(ih)+\cdots$$
$$+\frac{[(l+2^{j-1})h]^\alpha}{\alpha}!]u^{(\alpha)}(ih)+o(h^{\alpha+1})]+[\sum_{l=2^{i-1}}^{2^j-2}(2^j-l-2)[\sum_{k\in Z} q_k(u(ih)$$
$$+(l+2^{j-1}k)h)u^{(1)}(ih)+\cdots+\frac{[(l+2^{j-1})h]^\alpha}{\alpha}!]u^{(\alpha)}(ih)+o(h^{\alpha+1})]]\prod_{l=0}^{j-2}(I+V^{2^l}).$$

As q_k has vanishing moment of order α, for $\forall p<\alpha$, $\sum_{k\in Z} k^p q_k = 0$, the first α terms of the sum above equal to zero. The coefficient of item $u^{(\alpha)}$ is
$\sum_{l=0}^{2^{i-1}-1}(l+1)\sum_{k\in Z} q_k\frac{[(l+2^{j-1})h]^\alpha}{\alpha!}+\sum_{l=2^{i-1}}^{2^j-2}(2^j-l-2)\sum_{k\in Z} q_k\frac{[(l+2^{j-1})h]^\alpha}{\alpha!}=C_\alpha h^\alpha 2^{\alpha(j-1)}$.

Moreover, $\prod_{l=0}^{j-2}(I+V^{2^l})u$ is the average term, its coefficient is 2^{j-1}, (8) holds.

We note that actually, (12) can be rewritten as

$$u^{(\alpha)}=\frac{1}{C_\alpha h^\alpha 2^{(\alpha+1)(j-1)}}q(V^{2^{j-1}})u\prod_{l=0}^{j-2}(I+V^{2^l})+o(h^{\alpha+1}).$$

Especially, when $j=1$, $u^{(\alpha)}(x)$ can be approximated as

$$u^{(\alpha)}(x)=\frac{1}{C_\alpha}\frac{1}{h^\alpha}\sum_{k\in Z} q_k u(x+hk)+o(h).$$

It is the same as Lemma 2 in [19].

Again, denote $u_i^{(0)}=u_i$,u_i^k denote the approximation to the value $u(x,t)$ at $(ih,k\tau)$ where h be the spatial step size, τ be the time step size. Set $M_\alpha^j = C_\alpha h^\alpha 2^{(\alpha+1)(j-1)}$. Then the discretized version of Eq. (12) at one scale is

$$\frac{u_i^{(k+1)}-u_i^{(k)}}{\tau}=\sum_{l=1}^{2}\frac{1}{M_{\alpha_l}^1}\tilde{q}^l(V^T)g_l((\frac{1}{M_{\beta_l}^1}q^l(V)u_i^{(k)})^2)(\frac{1}{M_{\beta_l}^1}q^l(V)u_i^{(k)}).$$

In particular, the above equation for $k = 0$ is

$$u_i^{(1)} = u_i - \tau \sum_{l=1}^{2} \frac{1}{M_{\alpha_l}^1} \tilde{q}^l(V^T) g_l((\frac{1}{M_{\beta_l}^1} q^l(V) u_i)^2)(\frac{1}{M_{\beta_l}^1} q^l(V) u_i). \tag{13}$$

For $j = 2$, the approximation of $u^{(\alpha)}(x)$ at the $(ih, k\tau)$ is

$$u^{(\alpha)} = \frac{1}{M_\alpha^2} q(V^2) u(I + V).$$

Using the above equation, (12) can be discretized as

$$\begin{aligned} u_i^{(k+1)} &= u_i^{(k)} - \tau[\lambda_1 \sum_{l=1}^{2} \frac{1}{M_{\alpha_l}^1} \tilde{q}^l(V^T) g_l((\frac{1}{M_{\beta_l}^1} q^l(V) u_i^{(k)})^2)(\frac{1}{M_{\beta_l}^1} q^l(V) u_i^{(k)}) \\ &+ \lambda_2 \sum_{l=1}^{2} \frac{1}{M_{\alpha_l}^2} \tilde{q}^l(V^2)^T g_l((\frac{1}{M_{\beta_l}^2} q^l(V^2) u_i^{(k)}(I+V))^2)(\frac{1}{M_{\beta_l}^2} q^l(V^2) u_i^{(k)}(I+V)(I+V^T))]. \end{aligned} \tag{14}$$

Especially, when $k = 0$, (10) can be rewritten as

$$\begin{aligned} u_i^{(1)} &= u_i - \sum_{l=1}^{2} \tilde{q}^l(V^T) g_l((\frac{1}{M_{\beta_l}^1} q^l(V) u_i)^2) \frac{\lambda_1 \tau}{M_{\alpha_l}^1 M_{\beta_l}^1} q^l(V) u_i \\ &- \sum_{l=1}^{2} \tilde{q}^l(V^2)^T g_l((\frac{1}{M_{\beta_l}^2} q^l(V^2) u_i(I+V))^2)(\frac{\lambda_2 \tau}{M_{\alpha_l}^2 M_{\beta_l}^2} q^l(V^2) u_i(I+V))(I+V^T), \end{aligned}$$

where $\lambda_1 + \lambda_2 = 1$.

By using the same method, when $j = J$, Eq. (10) for $k = 0$ can be discretized as

$$\begin{aligned} u_i^{(1)} &= u_i - \tau[\lambda_1 \sum_{l=1}^{2} \frac{1}{M_{\alpha_l}^1} \tilde{q}^l(V^T) g_l((\frac{1}{M_{\beta_l}^1} q^l(V) u_i)^2)(\frac{1}{M_{\beta_l}^1} q^l(V) u_i) \\ &+ \lambda_2 \sum_{l=1}^{2} \frac{1}{M_{\alpha_l}^2} \tilde{q}^l(V^2)^T g_l((\frac{1}{M_{\beta_l}^2} q^l(V^2) u_i(I+V))^2)(\frac{1}{M_{\beta_l}^2} q^l(V^2) u_i(I+V))(I+V^T) \\ &\cdots + \lambda_J \sum_{l=1}^{2} \frac{1}{M_{\alpha_l}^J} \tilde{q}^l(V^{2^{J-1}})^T g_l((\frac{1}{M_{\beta_l}^J} q^l(V^{2^{J-1}}) u_i \prod_{l=0}^{J-2}(I + V^{2^l}))^2) \\ &\cdot(\frac{1}{M_{\beta_l}^J} q^l(V^{2^{J-1}}) u_i \prod_{l=0}^{J-2}(I+V^{2^l})) \prod_{l=0}^{J-2}(I + (V^T)^{2^l})], \end{aligned} \tag{15}$$

where $\lambda_1 + \lambda_2 + \cdots + \lambda_J = 1$.

4 The Correspondence Between General Frame Shrinkage and High-Order Nonlinear Diffusion

In Sect. 3, we obtain a theorem which rewrite $u^{(k)}$ in a formula that is closely related to a discretized version of some high-order diffusion equation in J scale.In this section, we derive the correspondence between nonlinear diffusion equations and undecimated frame filter banks.

Theorem 2. *Let $u^{(k+1)}$ in (3) be the resulting signal after one step frame shrinkage with input $u^{(k)}, k \in Z$ with a pair of frame filter banks $\{p, q^1, q^2\}$, $\{\tilde{p}, \tilde{q}^1, \tilde{q}^2\}$, and shrinkage function S_{lj}. Let $u_k^{(1)}$ in (11) be the signal after 1-step diffusion, with $u_k, k \in Z$, as the initial input. For any $2 \le j \le J$, if*

$$\prod_{m=0}^{j-2} \tilde{p}(V^{2^m})^T S_{lj}(x \prod_{m=0}^{j-2} p(V^{2^m}))$$
$$= x[\prod_{m=0}^{j-2} \tilde{p}(V^{2^m})^T p(V^{2^m}) - \frac{\tau\lambda_j}{M_{\alpha_l}^j M_{\beta_l}^j} \prod_{m=0}^{j-2}(I + V^{2^m})^4 g_l((\frac{x}{M_{\beta_l}^j} \prod_{m=0}^{j-2}(I + V^{2^m})^2)^2)],$$

then $u^{(k+1)} = u_k^1$ for all k.

Proof. The frame shrinkage process for denoising at J scale is

$$u^{(k+1)} = u^{(k)} - \sum_{l=1}^{2} \tilde{q}^l(V^T)[q^l(V)u^{(k)} - S_{l1}(q^l(V)u^{(k)})]$$
$$- \sum_{l=1}^{2} \sum_{j=2}^{J} \prod_{m=2}^{j} \tilde{p}(V^{2^{m-2}})^T \tilde{q}((V^{2^{J-1}})^T)[q^l(V^{2^{J-1}}) \prod_{m=2}^{j} p(V^{2^{m-2}})u^{(k)}$$
$$-S_{lj}[q^l(V^{2^{j-1}})] \prod_{m=2}^{j} p(V^{2^{m-2}})u^{(k)}].$$

From the above formula, we get

$$u_i^{(k+1)} = u_i^{(k)} - \sum_{l=1}^{2} \tilde{q}^l(V^T)[q^l(V)u_i^{(k)} - S_{l1}(q^l(V)u_i^{(k)})]$$
$$- \sum_{l=1}^{2} \sum_{j=2}^{J} \prod_{m=2}^{j} \tilde{p}(V^{2^{m-2}})^T \tilde{q}((V^{2^{J-1}})^T)[q^l(V^{2^{J-1}}) \prod_{m=2}^{j} p(V^{2^{m-2}})u_i^{(k)}$$
$$-S_{lj}[q^l(V^{2^{j-1}})] \prod_{m=2}^{j} p(V^{2^{m-2}})u_i^{(k)}].$$

Especially, when $k = 0$,we obtain the following frame shrinkage process:

$$u_i^{(1)} = u_i - \sum_{l=1}^{2} \tilde{q}^l(V^T)[q^l(V)u_i - S_{l1}(q^l(V)u_i)]$$
$$- \sum_{l=1}^{2} \sum_{j=2}^{J} \prod_{m=2}^{j} \tilde{p}(V^{2^{m-2}})^T \tilde{q}((V^{2^{J-1}})^T)[q^l(V^{2^{J-1}}) \prod_{m=2}^{j} p(V^{2^{m-2}})u_i$$
$$-S_{lj}[q^l(V^{2^{j-1}})] \prod_{m=2}^{j} p(V^{2^{m-2}})u_i].$$

While the high-order nonlinear diffusion process is

$$u_i^{(1)} = u_i - \tau[\lambda_1 \sum_{l=1}^{2} \frac{1}{\tilde{M}_{\alpha_l}^1} \tilde{q}^l(V^T) g_l(\frac{1}{M_{\beta_l}^1} q^l(V)u_i)^2(\frac{1}{M_{\beta_l}^1} q^l(V))u_i)]$$
$$= +\lambda_2 \sum_{l=1}^{2} \frac{1}{\tilde{M}_{\alpha_l}^2} \tilde{q}^l((V^2)^T) g_l((\frac{1}{M_{\beta_l}^2} q^l(V^2)u_i(I+V)^2)(\frac{1}{M_{\beta_l}^1} q^l(V)u_i(I+V)(I+V^T))]$$
$$= \quad \cdots + \lambda_J \sum_{l=1}^{2} \frac{1}{\tilde{M}_{\alpha_l}^J} \tilde{q}^l((V^{2^{J-1}})^T) g_l((\frac{1}{M_{\beta_l}^J} q^l(V^{2^{J-1}})u_i \prod_{l=0}^{J-2}(I + V^{2^l}))^2)$$
$$((\frac{1}{M_{\beta_l}^J} q^l(V^{2^{J-1}})u_i \prod_{l=0}^{J-2}(I + V^{2^l})^2) \prod_{l=0}^{J-2}(I + (V^T)^{2^l})],$$

where $\lambda_1 + \lambda_2 + \cdots + \lambda_J = 1$. Observe the second term of the above two formulas, we get the following:

$$\sum_{l=1}^{2} \tilde{q}^l(V^T)[q^l(V)u_i - S_{l1}(q^l(V)u_i))] = \tau[\lambda_1 \sum_{l=1}^{2} \frac{1}{\tilde{M}_{\alpha_l}^1} \tilde{q}^l(V^T) g_l(\frac{1}{M_{\beta_l}^1} q^l(V)u_i)^2(\frac{1}{M_{\beta_l}^1} q^l(V))u_i)].$$

Let $q^l(V)u_i = x$, then

$$\sum_{l=1}^{2} \tilde{q}^l(V^T)[x - S_{l1}(x)] = \tau\lambda_1 \sum_{l=1}^{2} \frac{1}{\tilde{M}_{\alpha_l}^1} \tilde{q}^l(V^T) g_l((\frac{1}{M_{\beta_l}^1} x)^2) \frac{1}{M_{\beta_l}^1} x.$$

That is

$$x - S_{l1}(x) = \frac{\tau \lambda_1}{\tilde{M}^1_{\alpha_l} M^1_{\beta_1}} g_l((\frac{1}{M^1_{\beta_l}} x)^2) x.$$

So we get

$$S_{l1}(x) = x(1 - \frac{\tau \lambda_1}{\tilde{M}^1_{\alpha_l} M^1_{\beta_1}} g_l((\frac{1}{M^1_{\beta_l}} x)^2)).$$

Step by step ,we get the $j+1$ term as the following:

$$\begin{aligned} &\textstyle\prod_{m=0}^{j-2} \tilde{p}(V^{2^m})^T S_{lJ}(x \prod_{m=0}^{j-2} p(V^{2^m})) \\ = x[&\textstyle\prod_{m=0}^{j-2} \tilde{p}(V^{2^m})^T p(V^{2^m}) - \frac{\tau \lambda_J}{\tilde{M}^J_{\alpha_l} M_{\beta_l^J}} \prod_{m=0}^{j-2} (I + V^{2^m})^4 g_l((\frac{x}{M^J_{\beta_l}} \prod_{m=0}^{j-2} (I + V^{2^m})^2)^2)]. \end{aligned}$$

The proof is complete.

When $J = 1$, the corresponding shrinkage functions S_{l1} are

$$S_{l1}(x) = x \left(1 - \frac{\tau}{M^1_{\alpha_l} M^1_{\beta_l}} g_l(\frac{x^2}{(M^1_{\beta_l})^2})\right), 1 \leq l \leq 2,$$

which is the same conclusion as given in [19].

Example 1. Let (p, q^1, q^2) be Ron-Shen's tight frame filter bank defined as follows

$$p = (\frac{1}{2}, 1, \frac{1}{2}), q^1 = (\frac{\sqrt{2}}{2}, 0, -\frac{\sqrt{2}}{2}), q^2 = (-\frac{1}{2}, 1, -\frac{1}{2}).$$

Then S_{11} and S_{21} in Theorem 4.1 are

$$S_{11}(x) = x \left(1 - \frac{\tau}{2h^2} g_l(\frac{x^2}{2h^2})\right), 1 \leq l \leq 2,$$

$$S_{21}(x) = x \left(1 - \frac{4\tau}{h^4} g_l(\frac{4x^2}{h^4})\right), 1 \leq l \leq 2.$$

Example 2. Especially when the framelet is a wavelet, choosing Haar wavelet filter banks

$$p = (\frac{1}{\sqrt{2}}, \frac{1}{\sqrt{2}}), q = (\frac{1}{\sqrt{2}}, -\frac{1}{\sqrt{2}}), \tag{16}$$

the shrinkage function in Theorem 4.1 is reduced to

$$S_1(x) = x \left(1 - \frac{2\tau}{h^2} g_l(\frac{x^2}{h^2})\right), 1 \leq l \leq 2.$$

The two cases given above are close to [17,19] except the coefficient, which is caused by unsampling.

5 Application to Axial-flow Compressor

Figure 3 illustrates the signal acquired by sensor a1 when the compressor is near the stall state and the signals denoised by the soft, hard and our new frame-diffused denoising algorithms, based on the Ron-Shen framelet filter banks. As evident from the figure, the soft and hard threshold denoising algorithms remove insufficient noise to yield any information on the original signal. In contrast, frame-diffused filters most of the noise, revealing that spikes occur 23 revolutions before the compressor disturbance (see Fig. 4). Figure 5 a–d plot the comprehensive denoising results of signals acquired by a1-a4. From this figure, we find that disturbance begins at the location of a1 and develops along the circumferential direction. After 23 revolutions, rotations are stalled, concordant with the actual physical phenomenon.

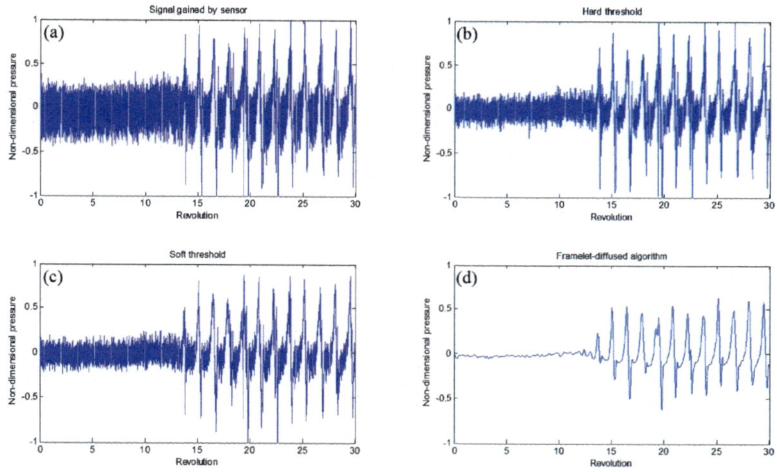

Fig. 3. (a) Signal acquired by sensor a1 at 65 percent rotation speed, near stall, uniform inlet. (b) Estimation with a wavelet hard thresholding. (c) Estimation with a wavelet soft thresholding. (d) Estimation with a frame-diffused thresholding

From above, we know our frame-diffused algorithm properly extracted and analyzed pressure signals from an axial transonic compressor. It captured the stall inception, and revealed the developmental progress of the compressor state from steady to unstable. As such, it presents as a reliable and robust post-processing tool for complex signal analysis.

6 Conclusion and Future Work

In this paper, we extend the correspondence between the special Ron-shen's frame shrinkage and high-order nonlinear diffusion to general frame at multi-scale. The connection between nonlinear diffusion and frame shrinkage opens

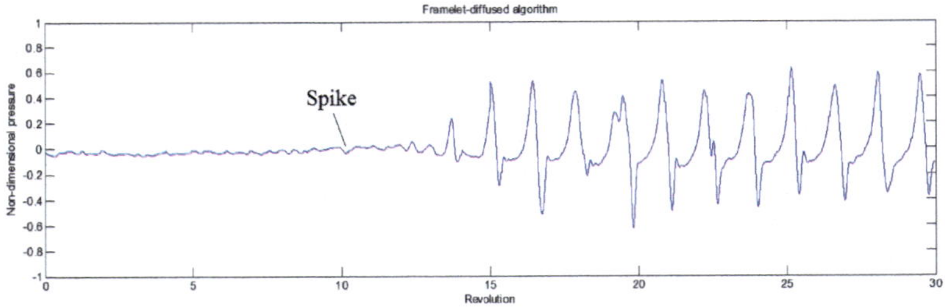

Fig. 4. Signal acquired by a1, denoised by frame-diffused algorithm

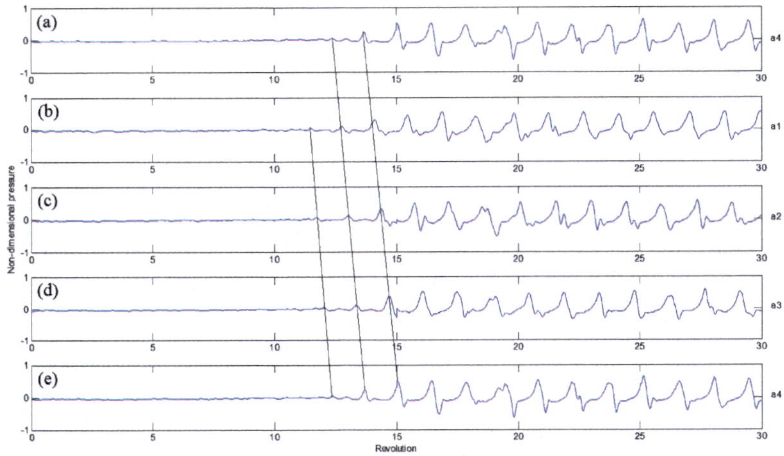

Fig. 5. (a) Signal acquired by a1 and denoised by frame-diffused algorithm. (b) Signal acquired by a2 and denoised by frame-diffused algorithm. (c) Signal acquired by a3 and denoised by frame-diffused algorithm. (d) Signal acquired by a4 and denoised by frame-diffused algorithm

the gate for fruitful exchange of ideas between the two worlds: we can construct shrinkage functions from diffusivity and vice versa. As the efficiency of the new diffusion-inspired shrinking approach, we can choose proper filter banks to construct suitable shrinkage function in signal denoising for the best capability.

Our further work involves the correspondence between special 2-D wavelet or framelet shrinkage and diffusion filtering, the design of diffusion-inspired shrinkage functions for image denoising, and the construction of dilation matrix which result in nonlinear diffusion equations with good performance in image noise removal.

Acknowledgement. The author would like to thank Professor Charles K. Chui for helpful discussions on the correspondence between wavelet shrinkage and diffusion filtering. Thanks to Dr. Qun Mo for some detailed and careful comments.

References

1. Rudin, L., Osher, S., Fatemi, E.: Nonlinear total variation based noise removal algorithms. Phys. D **60**, 259–268 (1992)
2. Catte, F., Lions, P., Morel, J., Coll, T.: Image selective smoothing and edge detection by nonlinear diffusion. SIAM J. Numer. Anal. **29**, 182–193 (1992)
3. Perona, P., Malik, J.: Scale-space and edge detection using anisotropic diffusion. IEEE Trans. Pattern Anal. Mach. Intell. **12**, 629–639 (1990)
4. Wei, G.W.: Generalized perona-malik equation for image restoration. IEEE Signal Process. Lett. **6**(7), 165–167 (2002)
5. Bates, P.W., Chen, Z., Sun, Y., et al. Geometric and potential driving formation and evolution of biomolecular surfaces. J. Math. Biol. **59**(2), 193–231 (2009)
6. Chambolle, A., DeVore, R.A., Lee, N., Lucier, B.L.: Nonlinear wavelet image processing: variationa problems, compression and noise removal through wavelet shrinkage. IEEE Trans. Image Process., 319–335 (1998)
7. Catte, F., L.Lions, P., Morel, J.M., Coll, T.: Image selective smoothing and edge detection by nonlinear diffusion. SIAM J. Numer. Anal., 845–866 (1992)
8. Donoho, D.: De-noising by soft thresholding. IEEE Trans. Inf. Theor., 613–627 (1995)
9. Mallat, S.: A Wavelet Tour of Signal Processing, 2nd edn. Academic Press, San Diego (1999)
10. Keeling, S.L., Stollberger, R.: Nonlinear anisotropic diffusion filtering for multi scale edge enhancement wide range edge. Inverse Probl., 175–190 (2002)
11. Coifman, R.R., Donoho, D.: Translation-invariant de-noising. In: Wavelets and Statistics. Springer Lecture Notes in Statistics, pp. 125–150 (1994)
12. Weickert, J.: Anisotropic diffusion in image processing. B.g. teubner Stuttgart, p. 272 (1998)
13. Didas, S., Denoising: Enhancement of digital imagesvariational methods, Integro differential Equations, and Wavelets, Ph.D. Dissertation, Saarland University (2008)
14. Cai, J., Chan, R., Shen, Z.: Simultaneous cartoon and texture inpainting. Inverse Proble. Imaging **4**, 379–395 (2010)
15. Didas, S., Weickert, J., Burgeth, B.: Properties of higher order nonlinear diffusion filtering. J. Math. Imaging Vis., 208–226 (2009)
16. Steidl, G., Weickert, J., Brox, T., Mrazek, P.: On the equivalence of soft wavelet shrinkage, total variation diffusion, total variation reguarization, and sides. SIJM J. Numer. Anal., 683–713 (2004)
17. Plonka, G., Steidl, G.: A multiscale wavelet-inspired scheme for nonlinear diffusion. Int. J. Wavelets, Multiresolution Inf. Process., 1–21 (2006)
18. Wang, H., Kong, X.: A multiscale tight frame inspired scheme for nonlinear diffusion. Int. J. Wavelets, Multiresolution Inf. Process., 1250041-1-1250041-22 (2012)
19. Jiang, Q.: Correspondence between frame shrinkage and high-order nonlinear diffusion. Appl. Numerical Math., 51–66 (2012)
20. Mrazek, P., Weickert, J., Steidl, G.: Diffusion-inspired shrinkage function and stability results for wavelet denoising. Int. J. Comput. Vis., 171–186 (2005)
21. Mrazek, P., Weickert, J., Steidl, G.: Correspondences between wavelet shrinkage and nonlinear diffusion. In: Proceedings of the Scale Space Methods in Computer Vision, International Conference, Scale-Space 2003, Isle of Skye, Uk, June 10–12, 2003, pp. 101–116 (2003)
22. Cai, J., Dong, B., Osher, S., Shen, Z.: Image restorations: total variation, wavelet frames and beyond. J. Am. Math. Soc. **25**, 1033–1089 (2012)

Pseudo-differential Operators Associated to General Type I Locally Compact Groups

Marius Măntoiu[1(✉)] and Maximiliano Sandoval[2]

[1] Name, Facultad de Ciencias, Departamento de Matematicas, Universidad de Chile, Las Palmeras 3645, Santiago, Chile
mantoiu@uchile.cl

[2] Facultad de Ciencias, Universidad de Chile, Las Palmeras 3645, Santiago, Chile
max.sandova@gmail.com

Abstract. In a recent paper by M. Măntoiu and M. Ruzhansky, a global pseudo-differential calculus has been developed for *unimodular* groups of type I. In the present article we generalize the main results to arbitrary locally compact groups of type I. Our methods involve the use of Plancherel's theorem for non-unimodular groups. We also make connections with a C^*-algebraic formalism, involving dynamical systems, and give explicit constructions for the group of affine transformations of the real line.

Keywords: Locally compact group · Lie group
Noncommutative Plancherel theorem · Modular function
Pseudo-differential operator

1 Introduction

Let G be a locally compact group with unitary dual $\widehat{\mathsf{G}}$, that is, the space of classes of unitary equivalence of (strongly continuous) unitary irreducible representations. It will be assumed that our groups are second countable and of type I. The formula (cf. [20], Eq. (1.1) for the unimodular case)

$$[\mathsf{Op}(A)u](x) = \int_{\mathsf{G}}\int_{\widehat{\mathsf{G}}} \mathrm{Tr}\left(A(x,\xi)D_\xi^{\frac{1}{2}}\pi_\xi(xy^{-1})^*\right)\Delta(y)^{-\frac{1}{2}}u(y)\,d\xi\,dy \qquad (1)$$

is the starting point for a global pseudo-differential calculus on G. It involves suitable operator-valued symbols A defined on $\mathsf{G}\times\widehat{\mathsf{G}}$, the modular function Δ of the group and the formal dimension operators D_ξ introduced by Duflo and Moore [8]. Formula (1) makes use of the Haar and Plancherel measures on G and $\widehat{\mathsf{G}}$ respectively. We also fixed a measurable field of representations $(\pi_\xi)_{\xi\in\widehat{\mathsf{G}}}$ such that π_ξ belongs to the class ξ and π_ξ acts on a Hilbert space $\mathcal{H}_\xi$.

J. Delgado and M. Ruzhansky (Eds.): *Analysis and Partial Differential Equations: Perspectives from Developing Countries*, PROMS 275, pp. 172–190, 2018.
https://doi.org/10.1007/978-3-030-05657-5_11

One of the advantages of using operator valued symbols is that one gets a global approach and a full symbol, free of localization choices, everything relying on harmonic analysis concepts attached to the group. Even for compact Lie groups there is no notion of full scalar-valued symbols for a pseudo-differential operator using local coordinates. For a more detailed discussion, for motivations and a full development of particular cases see [9, 20, 22].

In the present article we are not going to rely on properties such as compactness or nilpotency nor on smoothness or unimodularity, and most hypothesis will be on the measure theoretic side. The category of second countable type I locally compact groups has a nice integration theory and their unitary duals have an amenable integration theory. For an introduction to this topic refer to [10]. This framework allows a general form of Plancherel Theorem [8, 11, 27], which is all it is needed to develop the basic features of quantization even for a non-unimodular group.

The interest of our extension comes mainly from the fact there are many important examples of non-unimodular groups. The simplest one is perhaps the affine group consisting of all the affine transformations of the real line, which is actually the only non-unimodular group in dimension two. In dimension three there are many infinite families of non-isomorphic non-unimodular Lie groups. Many other examples arise in the study of parabolic subgroups of semisimple Lie groups, that are used to investigate irreducible representations using extensions of Mackey's machine.

Formula (1) is a generalization of the formula derived in [20, Eq. (1.1)] for unimodular groups, with a difference on the order of the factors that has to do with the choice of a convention for the Fourier transform (cf. Remark 7). Thus our quantization will cover right invariant operators whereas the one in [20] gives rise to left invariant operators.

Graded nilpotent Lie groups are treated systematically in [9] and in many other references. For a general treatment of pseudo-differential operators in a group setting, see [9, 22]. The compact Lie groups are also treated by using global operator-valued symbols in [5, 23–26]. The recent articles containing applications and developments are too many to be cited here. The idea of using the irreducible representations of a groups to define such calculus, seems to come from [29, Sect. 1.2], but it was not developed in this abstract setting. All the books and articles mentioned earlier contain historical background and references to the existing literature treating pseudo-differential operators and quantization in a group theoretic context. In many cases, specific properties of the group allow defining Hörmander-type classes of symbols, and this has far-reaching consequences.

Another approach to a quantization consist of using the formalism of C^*-algebras. Given a locally compact group G, there is an action by left (or right) translations on various C^*-algebras of functions on G. In such situations there are natural crossed products associated to them. Among the non-degenerate representations of these C^*-algebras stands the Schrödinger representation, acting on the Hilbert space $L^2(\mathsf{G})$. This formalism allows to take full advantage of the theory of C^*-algebras in the setting of pseudo-differential operators.

In the next two sections we introduce the notations and general harmonic analysis theory required for the quantization. In Sect. 4 we make a preliminary construction of the quantization Op. Section 5 includes a discussion on the difference and the connections between left and right quantizations, comming from the non-commutativity of the group, as well as the various τ-quantizations related to ordering issues. Section 6 exemplifies all these by showing how multiplication and convolution operators are covered by the calculus; some intricacies appear due to the presence of the modular function. The same can be said about the C^*-algebraic approach, which only gives a correction of the calculus Op by a factor defined by the modular function. In Sect. 8, for exponential groups, we put into evidence a new (but related) quantization applied to scalar symbols defined on the cotangent bundle of the group. The particular case of connected simply connected nilpotent groups has been trated in [20, Sect. 8]; see also [21] for refined results valid in the presence of flat coadjoint orbits. Other approaches in the nilpotent case can be found in [1,4,12,13,17,28]. In Sect. 9 we work out the case of the group of affine transformations of $\mathbb{R}$.

2 General Type I Locally Compact Groups

In this section we set up the general framework of the article.

We assume all Hilbert spaces $\mathcal{H}$ to be separable, using the convention that their scalar product $\langle\cdot,\cdot\rangle_{\mathcal{H}}$ is linear in the first variable and anti-linear in the second. By $\mathcal{H}^\dagger$ we denote the conjugate of $\mathcal{H}$, whose elements are the same, but the scalar product is defined as $\alpha\cdot u=\bar{\alpha}u$ and its inner product is conjugate to the one from $\mathcal{H}$, i.e. $\langle u,v\rangle_{\mathcal{H}^\dagger}=\langle v,u\rangle_{\mathcal{H}}$. If π is a strongly continuous unitary representation of a topological group G on a Hilbert space $\mathcal{H}_\pi$, its contragradient representation $\pi^\dagger$ acts on $\mathcal{H}_\pi^\dagger$ by $\pi^\dagger(x)f=\pi(x)f$; in general π and $\pi^\dagger$ are not equivalent.

By $\mathbb{B}(\mathcal{H})$ one denotes the C^*-algebra of all bounded linear operators on $\mathcal{H}$ and $\mathbb{K}(\mathcal{H})$ stands for the two sided *-ideal of compact operators on $\mathcal{H}$. We also make use of the Schatten-von Neumann classes $\mathbb{B}_p(\mathcal{H})$ for $p\geq 1$; these are Banach *-algebras with the norm $\| T\|_{\mathbb{B}_p}=\mathrm{Tr}\left((T^*T)^{p/2}\right)^{1/p}$. The most important cases are $\mathbb{B}_1(\mathcal{H})$, the space of trace-class operators, and, for $p=2$, the space of Hilbert-Schmidt operators, which endowed with the inner product $\langle T,S\rangle_{\mathbb{B}_2}=\mathrm{Tr}\,(TS^*)$ is unitarily isomorphic with the Hilbert tensor product $\mathcal{H}\otimes\mathcal{H}^\dagger$.

Let G be a (Hausdorff) locally compact group with unit e; we also assume that it is second countable and of type I. Recall that a second countable group is separable, σ-compact and completely metrizable; in particular as a measurable space it will be standard. Also recall that a group is of type I if every factor representation is quasi-equivalent to a an irreducible one. For these groups their C^*-enveloping algebra are postliminal.

Let us fix a left Haar measure μ on G, also denoted by $d\mu(x)=dx$. We get a right Haar measure μ^r defined by the formula $\mu^r(E)=\mu(E^{-1})$. Let $\Delta:\mathsf{G}\to(0,\infty)$ be the modular function of G satisfying $\mu(Ex)=\Delta(x)\mu(E)$ for measurable sets $E\subset\mathsf{G}$ and $x\in\mathsf{G}$; this implies in particular that $d\mu^r=\Delta^{-1}d\mu$.

The modular function is a continuous (smooth if G is a Lie group) homomorphism into the multiplicative group $\mathbb{R}_+$. We say that a group is *unimodular* if the modular function is constant. For the convenience of the reader we recall that the modular function plays the following role on integration by substitution of variables

$$\int_{\mathsf{G}} f(y)\, dy = \int_{\mathsf{G}} \Delta(x) f(yx)\, dy = \int_{\mathsf{G}} \Delta(y)^{-1} f(y^{-1})\, dy. \tag{2}$$

The spaces of p-integrable functions $L^p(\mathsf{G}) = L^p(\mathsf{G}, \mu)$ will always refer to the left Haar measure; these are separable Banach spaces for $p \in [1, \infty)$. By $C_c(\mathsf{G})$ we denote the space of complex continuous functions on G with compact support, which is dense in $L^p(\mathsf{G})$. One has a Banach *-algebra structure on $L^1(\mathsf{G})$, with convolution

$$(f * g)(x) = \int_{\mathsf{G}} f(y) g(y^{-1}x)\, dy = \int_{\mathsf{G}} \Delta(y)^{-1} f(xy^{-1}) g(y)\, dy\,,$$

and involution given by $f^*(x) = \Delta(x)^{-1}\overline{f(x^{-1})}$. In general, there is a p-dependent involution on $L^p(\mathsf{G})$ given by

$$f^*(x) = \Delta(x)^{-\frac{1}{p}}\overline{f(x^{-1})}. \tag{3}$$

But we reserve the notation f^* for $p = 2$.

Given a locally compact group G, its *unitary dual* $\widehat{\mathsf{G}}$ is the collection of all of its irreducible unitary representation modulo unitary equivalence. We endow $\widehat{\mathsf{G}}$ with the Mackey Borel structure [7]. It is known that being of type I is equivalent to $\widehat{\mathsf{G}}$ being countably separated and is also equivalent to being a standard Borel space.

Example 1. Some well-known examples [10] of type I groups are: (a) compact groups, (b) connected semisimple Lie groups, (c) Abelian groups, (d) exponentially solvable Lie groups, in particular connected simply connected nilpotent Lie groups, (e) real algebraic groups. It is known that a discrete group is of type I if and only if it possesses an Abelian normal subgroup of finite index.

For a representative $\pi_\xi \in \xi$ of an element of the unitary dual of G, we set $\mathcal{H}_\xi = \mathcal{H}_{\pi_\xi}$. The left and right regular representations on $L^2(\mathsf{G})$ are

$$\big[\lambda_y(f)\big](x) := f(y^{-1}x) \quad \text{and} \quad \big[\rho_y(f)\big](x) := \Delta(y)^{\frac{1}{2}} f(xy).$$

We say that a standard measure ν on $\widehat{\mathsf{G}}$ is a *Plancherel measure* if it yields a direct integral central decomposition of the regular representations into irreducible representations. We do not warry to put this in formal terms, since we are only going to use some of its properties indicated in Theorem 1.

Plancherel measures do exist for separable locally compact groups of type I and in fact they are all equivalent (cf. Theorem 1 below). From now on, for a Plancherel measure ν we adopt the notation $d\nu(\xi) = d\xi$.

There are various cases in which the Plancherel measure can be given explicitly. For Abelian groups, $\widehat{\mathsf{G}}$ is also an Abelian group (the Pontryagin dual) and the Plancherel measure coincides with one of its Haar measures. For connected, simply connected nilpotent Lie groups it corresponds to a measure on the space of coadjoint orbits arising from the Lebesgue measure on the dual $\mathfrak{g}^*$ of the Lie algebra. For compact groups the the Peter-Weyl theorem says that the irreducible representations form a discrete set and it describes the Plancherel measure.

For non-unimodular groups an important role is played by *the Duflo-Moore operators (also called formal dimension operators).* They are densely defined positive self-adjoint operator with dense image $D_\pi : \mathrm{Dom}(D_\pi) \to \mathcal{H}_\pi$ and satisfy almost everywhere *the semi-invariance condition*

$$\pi(x) D_\pi \pi(x)^* = \Delta(x)^{-1} D_\pi, \quad \forall x \in \mathsf{G}. \tag{4}$$

In [11, p. 97] an explicit construction of the operators D_π is made for square integrable representations. For unimodular groups, the operators D_π are just multiplication by some positive scalar d_π, which coincides with the dimension of $\mathcal{H}_\pi$ when the later is finite.

3 The Fourier and Plancherel Transforms

Now we recall some known results on the Fourier theory of non-unimodular groups of type I.

Suppose we have fixed a Plancherel measure ν in $\widehat{\mathsf{G}}$, a measurable field of representations $(\pi_\xi)_{\xi \in \widehat{\mathsf{G}}}$ and there is a family of densely defined self-adjoint positive operators $D_\xi : \mathcal{H}_\xi \to \mathcal{H}_\xi$ satisfying (4) for ν-almost all $\xi \in \widehat{\mathsf{G}}$. We define (in weak sense) the *operator-valued Fourier transform* of a function $f \in L^1(\mathsf{G})$ as

$$(\mathcal{F}f)(\xi) \equiv \pi_\xi(f) = \int_{\mathsf{G}} f(y) \pi_\xi(y)\, dy.$$

The Fourier transform is a non-degenerate *-representation of $L^1(\mathsf{G})$, but in the non-unimodular case it fails to intertwine the two-sided regular representation of G with $\int_{\widehat{\mathsf{G}}}^{\oplus} \xi \otimes \xi^\dagger\, d\xi$ and it also fails to be an L^2-unitary map. So one introduces the *Plancherel transform* of $f \in L^1(\mathsf{G}) \cap L^2(\mathsf{G})$ as the operator

$$(\mathcal{P}f)(\xi) = \pi_\xi(f) D_\xi^{1/2}.$$

In the following we denote by $\mathcal{P}f = \widehat{f}$ the Plancherel transform.

We are going to present a partial formulation of the Plancherel Theorem for non-unimodular groups; for a proof in the case where G is unimodular we refer to [7]. The non-unimodular Plancherel Theorem was developed by N. Tatsuuma in [27], and latter an extension of his theory, including some clarifications, has been obtained by Duflo and Moore [8].

Theorem 1. *Let* G *be a type I second countable locally compact group. There exists a* σ*-finite Plancherel measure* ν *on* $\widehat{\mathsf{G}}$*, a measurable field of irreducible* *representations* $(\pi_\xi)_{\xi\in\widehat{\mathsf{G}}}$ *with* $\pi_\xi \in \xi$, *a measurable field* $(D_\xi)_{\xi\in\widehat{\mathsf{G}}}$ *of densely defined self-adjoint positive operators on* $\mathcal{H}_\xi$ *with dense image, satisfying (4) for* ν*-almost every* $\xi \in \widehat{\mathsf{G}}$, *which have the following properties:*

1. *Let* $f \in L^1(\mathsf{G}) \cap L^2(\mathsf{G})$. *For* ν*-almost all* $\xi \in \widehat{\mathsf{G}}$, *the operator* $\widehat{f}(\xi)$ *extends to a Hilbert-Schmidt operator on* $\mathcal{H}_\xi$ *and*

$$\| f \|_2^2 = \int_{\widehat{\mathsf{G}}} \left\| \widehat{f}(\xi) \right\|_{\mathbb{B}_2}^2 d\xi.$$

The Plancherel transformation extends in a unique way to a unitary operator

$$\mathcal{P} : L^2(\mathsf{G}) \to \int_{\widehat{\mathsf{G}}}^{\oplus} \mathbb{B}_2(\mathcal{H}_\xi)\, d\xi.$$

2. *The Plancherel measure and the operator field satisfy the inversion formula*

$$f(x) = \int_{\widehat{\mathsf{G}}} \mathrm{Tr}\left(\widehat{f}(\xi) D_\xi^{\frac{1}{2}} \pi_\xi(x)^*\right) d\xi\,, \tag{5}$$

for all f *in the Fourier algebra of* G *(see below). The inversion formula converges absolutely in the sense that* $\widehat{f}(\xi)D_\xi^{\frac{1}{2}}$ *extends to a trace-class operator* ν*-a.e. and the integral of the trace-class norms is finite.*

3. *Suppose there is another Plancherel measure* ν' *on* $\widehat{\mathsf{G}}$ *and measurable fields* $(\pi_\xi{}', D_\xi{}')_{\xi\in\widehat{\mathsf{G}}}$ *that share the properties above. Then* ν *and* ν' *are equivalent measures, and there is a measurable field of unitary operators* $(U_\xi)_{\xi\in\widehat{\mathsf{G}}}$, *intertwining* π_ξ *and* $\pi_\xi{}'$, *such that for* ν*-almost all* $\xi \in \widehat{\mathsf{G}}$ *the Radon-Nikodym derivative of* ν' *with respect to* ν *satisfies*

$$\frac{d\nu'}{d\nu}(\xi)\, D_\xi{}' = U_\xi\, D_\xi\, U_\xi^*.$$

For simplicity we make use of the following notation

$$\mathscr{B}_2^{\oplus}(\widehat{\mathsf{G}}) = \int_{\widehat{\mathsf{G}}}^{\oplus} \mathbb{B}_2(\mathcal{H}_\xi)\, d\xi\,, \quad \mathscr{B}_1^{\oplus}(\widehat{\mathsf{G}}) = \int_{\widehat{\mathsf{G}}}^{\oplus} \mathbb{B}_1(\mathcal{H}_\xi) D_\xi^{-\frac{1}{2}}\, d\xi\,,$$

$$\mathscr{B}_2^{\oplus}(\mathsf{G}\times\widehat{\mathsf{G}}) = L^2(\mathsf{G}) \otimes \mathscr{B}_2^{\oplus}(\widehat{\mathsf{G}})\,, \quad \mathscr{B}_2^{\oplus}(\widehat{\mathsf{G}}\times\mathsf{G}) = \mathscr{B}_2^{\oplus}(\widehat{\mathsf{G}}) \otimes L^2(\mathsf{G}).$$

$\mathscr{B}_2^{\oplus}(\mathsf{G}\times\widehat{\mathsf{G}})$, one of the natural spaces of symbols, has the inner product

$$\langle A, B\rangle_{\mathscr{B}_2^{\oplus}} = \int_{\mathsf{G}} \int_{\widehat{\mathsf{G}}} \mathrm{Tr}\left[A(x,\xi)B(x,\xi)^*\right] d\xi\, dx.$$

Most of the results of this sections are presented in the work [11]. In order to shed some light on the trace-class hypothesis imposed to our symbols, we elaborate a little on the natural domain of the Plancherel transform such that formula (5) holds. We also give the natural domain for $\mathcal{P}^{-1}$.

Definition 1. The *Fourier algebra* $A(\mathsf{G})$ of a locally compact group G is

$$A(\mathsf{G}) = \{f * g^\flat \mid f, g \in L^2(\mathsf{G})\},$$

where $g^\flat(x) = \overline{g(x^{-1})}$. If we endow $A(G)$ with the norm

$$\| u \|_{A(\mathsf{G})} = \inf \left\{ \|f\|_2 \| g \|_2 \,\middle|\, u = f * g^\flat \right\},$$

it becomes a Banach *-algebra with convolution and $^\flat$ as the involution (it is the space of matrix coefficient functions of the left regular representation).

In [11, Th. 4.12] it is shown that the Plancherel transform induces an isomorphism between the Banach spaces $A(\mathsf{G})$ and $\mathscr{B}_1^{\oplus}(\widehat{\mathsf{G}})$ and an isomorphism

$$\mathcal{P} : A(\mathsf{G}) \cap L^2(\mathsf{G}) \to \mathscr{B}_1^{\oplus}(\widehat{\mathsf{G}}) \cap \mathscr{B}_2^{\oplus}(\widehat{\mathsf{G}}).$$

Next proposition (cf. [11] Theorem 4.15) shows that $A(\mathsf{G}) \cap L^2(\mathsf{G})$ is the natural domain of the Plancherel transform in such a way that the inversion formula holds. It also shows that in the Plancherel side, the natural domain for the inversion formula is $\mathscr{B}_1^{\oplus}(\widehat{\mathsf{G}}) \cap \mathscr{B}_2^{\oplus}(\widehat{\mathsf{G}})$.

Proposition 1. *Let $F \in \mathscr{B}_2^{\oplus}(\widehat{\mathsf{G}})$ and suppose that for ν-almost everywhere the operator $F(\xi)D_\xi^{\frac{1}{2}}$ extends to a trace-class operator. Suppose moreover that*

$$\int_{\widehat{\mathsf{G}}} \left\| F(\xi) D_\xi^{\frac{1}{2}} \right\|_{\mathbb{B}_1} d\xi < \infty.$$

If f is the inverse Plancherel transformation of F, then we have μ-almost everywhere

$$f(x) = \int_{\widehat{\mathsf{G}}} \operatorname{Tr}\left(F(\xi) D_\xi^{\frac{1}{2}} \pi_\xi(x)^* \right) d\xi. \tag{6}$$

4 The Basic Quantization

In this section we introduce a quantization leading to a pseudo-differential calculus for operator valued symbols defined on the whole group involving its irreducible representation theory. For this we fix a choice of a measurable field of representations $(\pi_\xi)_{\xi \in \widehat{\mathsf{G}}}$ and formal dimension operators $(D_\xi)_{\xi \in \widehat{\mathsf{G}}}$ such that Theorem 1 holds. Different choices of the measurable fields of representation or Duflo-Moore operators lead to isomorphic formulations.

For symbols $A \in L^2(\mathsf{G}) \otimes \left(\mathscr{B}_1^{\oplus}(\widehat{\mathsf{G}}) \cap \mathscr{B}_2^{\oplus}(\widehat{\mathsf{G}})\right)$, set $\mathsf{Op}(A) : L^2(\mathsf{G}) \to L^2(\mathsf{G})$ by

$$[\mathsf{Op}(A)u](x) = \int_{\mathsf{G}} \int_{\widehat{\mathsf{G}}} \operatorname{Tr}\left(A(x,\xi) D_\xi^{\frac{1}{2}} \pi_\xi(xy^{-1})^* \right) \Delta(y)^{-\frac{1}{2}} u(y)\, d\xi\, dy. \tag{7}$$

The operator $\mathsf{Op}(A)$ is called the *pseudo-differential operator* with symbol A. Let

$$\ker_A(x,y) = \Delta(y)^{-\frac{1}{2}} \int_{\widehat{\mathsf{G}}} \operatorname{Tr}\left(A(x,\xi) D_\xi^{\frac{1}{2}} \pi_\xi(xy^{-1})^* \right) d\xi.$$

Since A is in the domain of the inverse Plancherel transformation $\mathcal{P}_2$ in the second variable, the above integral converges absolutely and

$$\ker_A(x,y) = \left(\mathcal{P}_2^{-1}A\right)(x,xy^{-1})\Delta(y)^{-\frac{1}{2}}. \tag{8}$$

By Plancherel's theorem and the change of variables given by (2), we conclude that $\ker_A$ is a square integrable function on $\mathsf{G}\times\mathsf{G}$; hence $\mathsf{Op}(A)$ *is a Hilbert-Schmidt operator with kernel* $\ker_A$ *and Hilbert-Schmidt norm*

$$\| \mathsf{Op}(A) \|_{\mathbb{B}_2} = \| A \|_{\mathscr{B}_2^{\oplus}}.$$

So we may extend the definition of $\mathsf{Op}(A)$ for arbitrary symbols $A \in \mathscr{B}_2^{\oplus}(\mathsf{G}\times\widehat{\mathsf{G}})$, using the previous formula and the fact that $L^2(\mathsf{G})\otimes\left(\mathscr{B}_1^{\oplus}(\widehat{\mathsf{G}})\cap\mathscr{B}_2^{\oplus}(\mathsf{G})\right)$ is a dense subset of $\mathscr{B}_2^{\oplus}(\mathsf{G}\times\widehat{\mathsf{G}})$.

Remark 1. The factors $D_\xi^{\frac{1}{2}}$ and $\Delta(y)^{-\frac{1}{2}}$ disappear from (7) if G is unimodular. By setting $\mathsf{G}=\mathbb{R}^n$, under the identification of $\widehat{\mathsf{G}}$ with $\mathbb{R}^n$ given by $\xi(x) = e^{-2\pi i\langle\xi,x\rangle}$, we recover the Kohn-Nirenberg calculus.

Remark 2. Let us define by $\Lambda_{u,v}(w) = \langle w,u\rangle v\,,\ \forall\, w\in L^2(\mathsf{G})$, the rank-one operator associated to the pair (u,v). One shows easily that $\Lambda_{u,v}=\mathsf{Op}(\mathcal{V}_{u,v})$ for the Wigner transform of (u,v)

$$\mathcal{V}_{u,v}(x,\xi) = \int_{\mathsf{G}} \Delta(y^{-1}x)^{\frac{1}{2}}\overline{u(y^{-1}x)}v(x)\pi_\xi(y)D_\xi^{\frac{1}{2}}\,dy.$$

Remark 3. The fact that Op is an isomorphism allows us to define a product which we will call the *Moyal product*, and an involution on $\mathscr{B}(\mathsf{G}\times\widehat{\mathsf{G}})$ by

$$\mathsf{Op}(A\#B) = \mathsf{Op}(A)\mathsf{Op}(B) \quad\text{and}\quad \mathsf{Op}(A^{\#}) = \mathsf{Op}(A)^*.$$

Remark 4. We also note that the left regular representation of G induces a representation acting on $\mathscr{B}_2^{\oplus}(\mathsf{G}\times\widehat{\mathsf{G}})$. Let $A\in\mathscr{B}_2^{\oplus}(\mathsf{G}\times\widehat{\mathsf{G}})$ be a symbol and $y\in\mathsf{G}$, then $\lambda_y\mathsf{Op}(A)\lambda_y^*$ is a Hilbert-Schmidt operator, hence there is some other symbol $y.a\in\mathscr{B}_2^{\oplus}(\mathsf{G}\times\widehat{\mathsf{G}})$ such that

$$\lambda_y\mathsf{Op}(A)\lambda_y^* = \mathsf{Op}(y.A).$$

It is easy to see that this is an action of the group, composed of a left translation in the first variable and a unitary equivalence in the representation space:

$$(y.A)(x,\xi) = \pi_\xi(y)A(y^{-1}x,\xi)\pi_\xi(y)^*.$$

For compact Lie groups, in [22], and for graded nilpotent groups, in [9] (see in both cases references therein), Hörmander-type classes of functions have been developped for the global operator-valued pseudo-differential calculus. This leads to many interesting applications. Such a task is more difficult for larger classes of groups, and imposible in the generality of the present paper. We indicate, however, a way to extend the quantization procedure; it is still useful, but much less effective than the $S^m_{\rho,\delta}$-formalism, which is not available without strong extra structure.

Assume for simplicity that G is a type I Lie group. One sets $\mathcal{D}(\mathsf{G}) := C^\infty_c(\mathsf{G})$ with the usual inductive limit topology. The strong dual of $\mathcal{D}(\mathsf{G})$, denoted by $\mathcal{D}'(\mathsf{G})$, is composed of distributions. Let us define

$$\mathscr{D}(\widehat{\mathsf{G}}) := \mathscr{F}[\mathcal{D}(\mathsf{G})] \subset \mathscr{B}_2^\oplus(\widehat{\mathsf{G}}) \cap \mathscr{B}_1^\oplus(\widehat{\mathsf{G}})$$

with the locally convex topological structure transported from $\mathcal{D}(\mathsf{G})$ and then the projective tensor product

$$\mathscr{D}(\mathsf{G}\times\widehat{\mathsf{G}}) := \mathcal{D}(\mathsf{G})\,\overline{\otimes}\,\mathscr{D}(\widehat{\mathsf{G}}) \subset \mathscr{B}_2^\oplus(\mathsf{G}\times\widehat{\mathsf{G}}).$$

Also using its strong dual, one gets a Gelfand triple

$$\mathscr{D}\big(\mathsf{G}\times\widehat{\mathsf{G}}\big) \hookrightarrow \mathscr{B}_2^\oplus\big(\mathsf{G}\times\widehat{\mathsf{G}}\big) \hookrightarrow \mathscr{D}'\big(\mathsf{G}\times\widehat{\mathsf{G}}\big).$$

Then the pseudo-differential calculus $\mathsf{Op} : \mathscr{B}_2^\oplus(\mathsf{G}\times\widehat{\mathsf{G}}) \to \mathbb{B}^2\big[L^2(\mathsf{G})\big]$

- restricts to an isomorphism $\mathsf{Op} : \mathscr{D}(\mathsf{G}\times\widehat{\mathsf{G}}) \to \mathbb{B}\big(\mathcal{D}'(\mathsf{G});\mathcal{D}(\mathsf{G})\big)$,
- extends to an isomorphism $\mathsf{Op} : \mathscr{D}'(\mathsf{G}\times\widehat{\mathsf{G}}) \to \mathbb{B}\big(\mathcal{D}(\mathsf{G});\mathcal{D}'(\mathsf{G})\big)$.

We have denoted above by $\mathbb{B}(\mathcal{A};\mathcal{B})$ the space of all linear continuous mappings between the locally convex vector spaces $\mathcal{A}$ and $\mathcal{B}$. The proof is an easy adaptation of the proof of the corresponding result in [20, Sect. 5] and relies on the form (8) of the kernel and on Schwartz's Kernel Theorem. The partial inverse Plancherel transform $\mathcal{P}_2^{-1}$ is taken into account by the definition of $\mathscr{D}(\widehat{\mathsf{G}})$, while the change of variables $y \mapsto xy^{-1}$ and multiplication by $\Delta^{-1/2}$ are $\mathcal{D}$-isomorphisms.

Remark 5. If G is not a Lie group, one still can perform the same extension procedure using the Bruhat spaces $\mathcal{D}(\mathsf{G})$ and $\mathcal{D}'(\mathsf{G})$ introduced in [2]. This has been done in [20, Sect. 5] for unimodular groups, but one can adapt everything to the non-unimodular setting. We only sketched here the Lie case for space reasons and because, any way, this is the most important.

5 Other Quantizations

Having in mind the familiar Kohn-Nirenberg quantization for $\mathsf{G} = \mathbb{R}^n$, one notes that for non abelian groups there are two possible generalizations, connected to the non-commutativity in $\mathbb{B}(\mathcal{H})$: a left quantization $\mathsf{Op}_L \equiv \mathsf{Op}$, the one used so far, and a right quantization

$$[\mathsf{Op}_R(a)u]\,(x) = \int_{\mathsf{G}}\int_{\widehat{\mathsf{G}}} \mathrm{Tr}\left(a(x,\xi)D_\xi^{\frac{1}{2}}\pi_\xi(y^{-1}x)^*\right)\Delta(xy^{-1})^{\frac{1}{2}}u(y)\,d\xi\,dy.$$

Actually, this two quantizations are equivalent in the following sense:

Proposition 2. *Let A be a symbol, and consider the symbol defined by*

$$\tilde{A}(x,\xi) = \pi_\xi(x)^* A(x,\xi)\pi_\xi(x).$$

Then $\mathsf{Op}_L(A) = \mathsf{Op}_R(\tilde{A})$.

Proof. The result follows from the following computation

$$\begin{aligned}\operatorname{Tr}\left(\tilde{A}(x,\xi)D_\xi^{\frac{1}{2}}\pi_\xi(y^{-1}x)^*\right) &= \operatorname{Tr}\left(\pi_\xi(x)^* A(x,\xi)\pi_\xi(x)D_\xi^{\frac{1}{2}}\pi_\xi(y^{-1}x)^*\right)\\ &= \Delta(x)^{-\frac{1}{2}}\operatorname{Tr}\left(\pi_\xi(x)^* A(x,\xi)D_\xi^{\frac{1}{2}}\pi_\xi(x)\pi_\xi(y^{-1}x)^*\right)\\ &= \Delta(x)^{-\frac{1}{2}}\operatorname{Tr}\left(A(x,\xi)D_\xi^{\frac{1}{2}}\pi_\xi(xy^{-1})^*\right),\end{aligned}$$

based on the properties of the trace, of the representation π_ξ and on the semi-invariance formula (4).

Remark 6. The usual pseudo-differential calculus in $\mathbb{R}^n$ disposes of an extra parameter $\tau \in [0,1]$ connected to ordering issues in the quantization, arising from the non-commutativity of the operators (positions, momenta) that are behind its definition. This can also be implemented in our general situation (both for the left and for the right quantization), using a (any!) measurable function $\tau : \mathsf{G} \to \mathsf{G}$. In favorable cases there is even a symmetric quantization, having special properties, analog to the Weyl calculus ($\tau = 1/2$) for the particular group $\mathsf{G} = \mathbb{R}^n$. This has been explained in [20], it can be extended to our non-unimodular groups, but we are not going to indicate here the easy adaptations.

One encounters in the next section formulae showing that left (respectively right) convolution operators emerge naturally from the left (respectively right) quantization. The τ parameter deals (in the extreme cases $\tau(x) = \mathsf{e}$ and $\tau(x) = x$) with setting multiplication operators to the left or to the right of convolution operators.

Remark 7. Alternatively, we could also define the Plancherel transform as

$$[\dot{\mathcal{P}}(f)](\xi) = D_\xi^{\frac{1}{2}}\int_{\mathsf{G}} f(x)\pi_\xi(x)^*\,dx.$$

For unimodular groups, this was the choice in [20] and it lead to somehow different formulae than the present ones, so we indicate briefly the relation. By the semi-invariance relation and the involution (3) for $p = 2$, one has

$$\begin{aligned}[\dot{\mathcal{P}}(f)](\xi) &= \textstyle\int_{\mathsf{G}} f(x)\pi_\xi(x)^* D_\xi^{\frac{1}{2}}\Delta(x)^{-\frac{1}{2}}\,dx\\ &= \textstyle\int_{\mathsf{G}} f(x^{-1})\Delta(x)^{-\frac{1}{2}}\pi_\xi(x)D_\xi^{\frac{1}{2}}\,dx = [\mathcal{P}(\bar{f}^*)](\xi).\end{aligned}$$

So the two definitions differ by an automorphism $f \mapsto \bar{f}^*$ of $L^2(\mathsf{G})$. Another thing to have in mind is that the inversion formula (5) this time reads

$$\begin{aligned}f(x) = \bar{f}^*(x^{-1})\Delta(x)^{-\frac{1}{2}} &= \int_{\widehat{\mathsf{G}}}\operatorname{Tr}\left(\mathcal{P}(\bar{f}^*)D_\xi^{\frac{1}{2}}\pi_\xi(x^{-1})^*\right)\Delta(x)^{-\frac{1}{2}}\,d\xi\\ &= \int_{\widehat{\mathsf{G}}}\operatorname{Tr}\left(D_\xi^{\frac{1}{2}}\dot{\mathcal{P}}(f)\pi_\xi(x)\right)d\xi.\end{aligned}$$

6 Some Operators Arising from the Calculus

Important families of operators in $L^2(\mathsf{G})$ are formed of multiplication and convolution operators. We show now how to recover them using our pseudo-differential calculus; the non-unimodular case has some particular features, due to the presence of the modular function and of the formal dimension operators.

For two square integrable functions $f, g \in L^2(\mathsf{G})$ we define the operators

$$\big[\mathsf{Mult}_f(u)\big](x) = f(x)u(x)\,,$$
$$\big[\mathsf{Conv}_g^L(u)\big](x) = \int_{\mathsf{G}} g(y)u(y^{-1}x)\,dy.$$

Remark 8. In general Mult_f and Conv_g^L are not bounded in $L^2(\mathsf{G})$. In fact Mult_f is bounded if and only if f is essentially bounded and Conv_g^L is bounded if and only if $\operatorname{ess\,sup}_{\xi\in\widehat{\mathsf{G}}} \parallel [\mathcal{F}(g)](\xi) \parallel < \infty$. The second assertion follows from the easy formula (note that here $\mathcal{F}(g)$ is relevant, not $\widehat{g} = \mathcal{P}(g)$!)

$$\mathcal{P} \circ \mathsf{Conv}_g^L = \int_{\widehat{\mathsf{G}}}^{\oplus} [\mathcal{F}(g)](\xi) d\xi \circ \mathcal{P}.$$

But for unimodular groups the composition $\mathsf{Mult}_f\, \mathsf{Conv}_g^L$ does extend to a Hilbert-Schmidt operator. Assume G unimodular (so $\mathcal{F}$ and $\mathcal{P}$ coincide), let $f, g \in L^2(\mathsf{G})$ and define the symbol A by

$$A(x,\xi) = f(x)\,\widehat{g}(\xi).$$

Using Plancherel inversion formula one gets $\mathsf{Op}_L(A)u = f \cdot (g * u)$. For non-unimodular groups the picture changes dramatically. The operators $\mathsf{Mult}_f \circ \mathsf{Conv}_g$ are no longer Hilbert-Schmidt; in fact one has that

$$\Big\| \mathsf{Mult}_f\, \mathsf{Conv}_g^L \Big\|_{\mathbb{B}_2} = \Big\| \Delta^{-\frac{1}{2}} f \Big\|_2 \Big\| \Delta^{\frac{1}{2}} g \Big\|_2 .$$

In general $\mathsf{Mult}_f\, \mathsf{Conv}_g$ is not even a bounded operator if f and g are not chosen in a suitable manner.

One way to fix this is taking functions in appropriate dense subspaces. Choose $f, g \in L^2(\mathsf{G})$ such that the functions $\Delta^{-\frac{1}{2}} f$, $\Delta^{\frac{1}{2}} g$ are square integrable and set

$$A(x,\xi) = \big(\Delta^{-\frac{1}{2}} f\big)(x)\big(\widehat{\Delta^{\frac{1}{2}} g}\big)(\xi).$$

Then for $u \in L^2(G)$ we have $\mathfrak{O}p(A)u = f \cdot (g * u)$, which may be written

$$\mathsf{Mult}_f\, \mathsf{Conv}_g^L = \mathsf{Op}\big[\big(\Delta^{-\frac{1}{2}} f\big) \otimes \big(\widehat{\Delta^{\frac{1}{2}} g}\big)\big].$$

Indeed, by Plancherel inversion

$$\begin{aligned}[\mathsf{Op}(A)u](x) &= \textstyle\int_{\mathsf{G}} \Delta(x)^{-\frac{1}{2}} f(x) \Delta(xy^{-1})^{\frac{1}{2}} g(xy^{-1}) \Delta(y)^{-\frac{1}{2}} u(y)\,dy \\ &= f(x) \textstyle\int_{\mathsf{G}} \Delta(y)^{-1} g(xy^{-1}) u(y)\,dy = f(x)(g * u)(x).\end{aligned}$$

Another way to express the relation between symbols of the form $A = f \otimes \widehat{g}$ and operators of multiplication and convolution is given in the following formulas

$$\mathsf{Op}_L(f \otimes \widehat{g}) = \mathsf{Mult}_f\, \mathsf{Conv}_g^L\, \mathsf{Mult}_{\Delta^{1/2}} = \mathsf{Mult}_{\Delta^{1/2}f}\, \mathsf{Conv}^L_{\Delta^{-1/2}g}\,,$$
$$\mathsf{Op}_R\,(f \otimes \widehat{g}) = \mathsf{Mult}_f\, \mathsf{Conv}^R_{\Delta^{1/2}g} = \mathsf{Mult}_{\Delta^{1/2}f}\, \mathsf{Conv}_g^R\, \mathsf{Mult}_{\Delta^{-1/2}};$$

here Conv_g^R is the operator given by $\mathsf{Conv}_g^R(u) = u * g$.

There are other convolution operators that appear in the literature. In [6, Sect. 1.2] the author introduces (right) convolution operators by

$$\big[\check{\mathsf{Conv}}_g^R(u)\big](x) = \int_{\mathsf{G}} u(xy)\Delta(y)^{\frac{1}{2}} g(y)\,dy = (u * \bar{g}^*)(x).$$

These operators are then used to study the space of left invariant operators. We leave to the reader the task of finding the relevant connections with our left and right quantizations. Anyhow, when studying these operators in L^p-spaces, corrections of the powers of the factors Δ seem natural and useful.

7 The C^*-Algebraic Formalism

We introduce first some tools from the theory of crossed products of C^*-algebras.

Definition 2. A *C^*-dynamical system* is a triplet $(\mathcal{A}, \mathsf{G}, \alpha)$, where G is a locally compact group, $\mathcal{A}$ is a C^*-algebra and $\alpha : \mathsf{G} \to \mathrm{Aut}(\mathcal{A})$ is a strongly continuous representation of G.

To such a C^*-dynamical system we associate the space $L^1(\mathsf{G};\mathcal{A})$ of all Bochner-integrable functions $F : \mathsf{G} \to \mathcal{A}$; it is a Banach *-algebra with laws

$$(F \star G)(x) = \int_{\mathsf{G}} F(y)\, \alpha_y\big(G(y^{-1}x)\big)\, dy\,,$$
$$F^\star(x) = \Delta(x)^{-1}\alpha_x\big(F(x^{-1})^*\big)\,.$$

The Banach *-algebra $L^1(G;\mathcal{A})$ is naturally isomorphic to the projective tensor product $\mathcal{A} \otimes L^1(\mathsf{G})$. Consider the *universal norm* on $L^1(\mathsf{G};\mathcal{A})$ given by

$$\| F \|_{\mathcal{A}\rtimes\mathsf{G}} = \sup_\rho \| \rho(F) \|\,,$$

where the supremum is taken over all non-degenerate *-representations. The *crossed product* $\mathcal{A} \rtimes \mathsf{G}$ is the enveloping C^*-algebra of $L^1(\mathsf{G};\mathcal{A})$, that is, its completion under the norm $\|\cdot\|_{\mathcal{A}\rtimes\mathsf{G}}$.

Example 2. Let $\mathcal{A}$ be a C^*-algebra, $\mathsf{G} = \{\mathsf{e}\}$ the trivial group and α the trivial representation; then $\mathcal{A} \rtimes \mathsf{G}$ is naturally isomorphic to $\mathcal{A}$. The group C^*-algebra $C^*(\mathsf{G})$ is obtained taking $\mathcal{A} = \mathbb{C}$.

Definition 3. A *covariant representation* of a C^*-dynamical system $(\mathcal{A}, \mathsf{G}, \alpha)$ is composed of a unitary representation π of G and a non-degenerate *-representation ρ of $\mathcal{A}$, both acting on a Hilbert space $\mathcal{H}$, in such a way that they satisfy the relation

$$\pi(x)\rho(f)\pi(x)^* = \rho(\alpha_x f)\,, \qquad f \in \mathcal{A}\,, x \in \mathsf{G}.$$

We denote this data as the triplet $(\rho, \pi, \mathcal{H})$.

Example 3. The most interesting example is that attached to a continuous action of G by homeomorphisms of a locally compact space Ω; this induces an action on $\mathrm{C}_0(\Omega)$ given by $[\alpha_x(f)](\omega) = f(x^{-1}\cdot\omega)$. Then $(C_0(\Omega), \mathsf{G}, \alpha)$ is a C^*-dynamical system and it encapsulates all the information of the group action. A covariant representations is the same as a system of imprimitivity [15] Sect. 3.7. In fact there is a one-to-one correspondence between topological actions of a group G and C^*-dynamical systems $(\mathcal{A}, \mathsf{G}, \alpha)$ where the C^*-algebra $\mathcal{A}$ is Abelian; this can be easily seen from the fact that every abelian C^*-algebra is of the form $C_0(\Omega)$ for some locally compact space and there is a one to one correspondence between strongly continuous representations $\alpha : \mathsf{G} \to \mathrm{Aut}(C_0(\Omega))$ and continuous actions of G on Ω [30, Proposition 2.7].

Any covariant representation $(\rho, \pi, \mathcal{H})$ of a C^*-dynamical system induces a non-degenerate *-representation $\rho \rtimes \pi$ of the crossed product $\mathcal{A} \rtimes \mathsf{G}$ on $\mathcal{H}$, the unique extension of the representation of $L^1(G; \mathcal{A})$ given by

$$(\rho \rtimes \pi)(F) = \int_{\mathsf{G}} \rho\,(F(y))\,\pi(y)\,dy.$$

This process sets up a bijection between the covariant representation of a C^*-dynamical system and the non-degenerate *-representations of the crossed product associated to it [30, Proposition 2.40].

There is a natural covariant representation associated to any left-invariant C^*-algebra of functions defined on G. Let $\mathcal{A}$ be a left-invariant C^*-subalgebra of the space of bounded left uniformly continuous functions on G. For an $\mathcal{A}$-valued function F on G and elements $x, z \in \mathsf{G}$ we make the convenient identification $[F(x)](z) = F(z, x)$. The triplet $(\mathcal{A}, \mathsf{G}, \alpha)$ is a C^*-dynamical system when endowed with the action $\alpha : \mathsf{G} \to \mathrm{Aut}(\mathcal{A})$ given by $[\alpha_x(F)](z, y) = F(z, x^{-1}y)$. Then our convolution and involution laws are given by

$$(F \star G)(z, x) = \int_G F(z, y)\,G(y^{-1}z, y^{-1}x)\,dy,$$
$$F^\star(z, x) = \Delta(x)^{-1}\overline{F(x^{-1}z, x^{-1})}.$$

In $\mathcal{H} = L^2(\mathsf{G})$ we have a covariant representation of $(\mathcal{A}, \mathsf{G}, \alpha)$ given by

$$[\lambda_x(u)](y) = u(x^{-1}y), \quad [\mathsf{Mult}_f(u)](y) = f(y)u(y).$$

The *Schrödinger representation* is the integrated form $\mathsf{Sch} = \mathsf{Mult} \rtimes \lambda$ of $\mathcal{A} \rtimes \mathsf{G}$; more explicitly, for a function $F \in L^1(\mathsf{G}; \mathcal{A})$

$$[\mathsf{Sch}(F)u](x) = \int_{\mathsf{G}} F(x, y) u(y^{-1}x)\, dy = \int_{\mathsf{G}} F(x, xy^{-1}) \Delta(y)^{-1} u(y)\, dy. \tag{9}$$

One gets an integral operator $\mathsf{Sch}(F) = \mathsf{Int}(L_F)$ with kernel

$$L_F(x, y) = F(x, xy^{-1}) \Delta(y)^{-1}.$$

Using (2) and the left invariance of the Haar measure, one gets

$$\int_{\mathsf{G}}\int_{\mathsf{G}} |\, L_F(x, y)\,|^p\, dxdy = \int_{\mathsf{G}}\int_{\mathsf{G}} |\, F(x, y)\,|^p \Delta(x^{-1}y)^{p-1} dxdy.$$

Only for $p = 1$ the correspondence $F \mapsto L_F$ is an $L^p(\mathsf{G} \times \mathsf{G})$-isometry. This has consequences upon the relevance of the Schrödinger representation of the crossed product for our pseudo-differential calculus in the non-unimodular case. Extending the procedure in [20, Sect. 7], one would like to define

$$\mathfrak{Op}(A) = \mathsf{Sch}\left[\mathcal{P}_2^{-1}(A)\right],$$

composing the Schrödinger representation with the unitary inverse Plancherel transformation $\mathcal{P}_2^{-1} : \mathscr{B}_2^{\oplus}(\mathsf{G} \times \widehat{\mathsf{G}}) \to L^2(\mathsf{G} \times \mathsf{G})$. A direct computation leads to

$$\left[\mathfrak{Op}(A)u\right](x) = \int_{\mathsf{G}}\int_{\widehat{\mathsf{G}}} \mathrm{Tr}\left(A(x, \xi) D_\xi^{\frac{1}{2}} \pi_\xi(xy^{-1})^*\right) \Delta(y)^{-1} u(y)\, d\xi dy\,,$$

wich differs from (7) by a factor $\Delta(y)^{-1/2}$. Thus one has

$$\mathfrak{Op}(A) = \mathsf{Op}(A) \circ \mathsf{Mult}_{\Delta^{-1/2}}\,, \tag{10}$$

with a non-trivial correction if G is not unimodular, loosing in general the good square integrability properties of Op.

Remark 9. It is still legitimate to study the quantization $A \mapsto \mathfrak{Op}(A)$ even for groups that are not unimodular. Involving arbitrary left-invariant C^*-algebras $\mathcal{A}$ of functions defined on G, this allows a direct study of "pseudo-differential operators with coefficients of type $\mathcal{A}$". In addition, having the crossed product construction in the background, it allows extending the results concerning essential spectra and Fredholm properties of global pseudo-differential operators, that have been obtained in [16] only when $\Delta = 1$. The key role is played by the Gelfand spectrum $\Omega(\mathcal{A})$ of $\mathcal{A}$ which, under suitable assumptions, is a compactification of G on which G acts continuously. The relevant spectral information is contained in the orbit structure of the boundary $\Omega(\mathcal{A}) \setminus \mathsf{G}$.

Remark 10. Adding a 2-cocycle to the formalism and studying twisted operators is also possible, as done in [3] for unimodular groups. For $\mathsf{G} = \mathbb{R}^n$ this reproduces the gauge covariant magnetic Weyl calculus [14, 18, 19].

Remark 11. One can also extend [20, Subsect. 7.4], in which covariant families of pseudo-differential operators are introduced starting from suitable symbols associated to general Abelian C^*-dynamical systems, as in Example 3. This is a natural way to generate interesting classes of random Hamiltonians.

8 The Case of Exponential Lie Groups

Let G be an exponential group with Lie algebra $\mathfrak{g}$ and exponential map $\mathsf{exp} : \mathfrak{g} \to \mathsf{G}$ (a diffeomorphism, by definition) with inverse $\mathsf{log} : \mathsf{G} \to \mathfrak{g}$. Such a group is second countable and type I, so it fits in our setting.

Let us set $\sigma := \mathsf{log}(\mu)$ for the image through log of the (fixed) Haar measure on G. It is known how it is related to the Lebesgue measure on $\mathfrak{g}$: in terms of the adjoint action $\mathrm{ad} : \mathfrak{g} \to \mathrm{aut}(\mathfrak{g})$ one has $d\sigma(X) = \theta(X)dX$, where

$$\theta(X) = \Big| \det \frac{1 - e^{-\mathrm{ad}_X}}{\mathrm{ad}_X} \Big| .$$

Thus we have a unitary operator

$$U_\theta : L^2(\mathfrak{g}; \theta(X)dX) \to L^2(\mathfrak{g}; dX)\,, \quad U_\theta(v) := \theta^{1/2}\, v .$$

Consequently, one also gets the unitary operators

$$\mathsf{Exp} : L^2(\mathsf{G}) \to L^2(\mathfrak{g}; \theta(X)dX)\,, \quad \mathsf{Exp}(u) := u \circ \mathsf{exp}\,,$$

$$\mathsf{Exp}_\theta : L^2(\mathsf{G}) \to L^2(\mathfrak{g}; dX)\,, \quad \mathsf{Exp}_\theta(u) := (U_\theta \circ \mathsf{Exp})(u) = \theta^{1/2}(u \circ \mathsf{exp}).$$

Let $\mathfrak{g}^*$ be the dual of the Lie algebra. There is a unitary Fourier transformation $\mathcal{F}_{\mathfrak{g}\mathfrak{g}^*} : L^2(\mathfrak{g}; dX) \to L^2(\mathfrak{g}^*; d\mathcal{X})$ associated to the duality

$$\langle \cdot | \cdot \rangle : \mathfrak{g} \times \mathfrak{g}^* \to \mathbb{R}\,, \ \langle Y | \mathcal{X} \rangle := \mathcal{X}(Y).$$

It is defined by

$$\big[\mathcal{F}_{\mathfrak{g}\mathfrak{g}^*}(v)\big](\mathcal{X}) = \int_{\mathfrak{g}} e^{-i\langle X|\mathcal{X}\rangle} v(X) dX\,,$$

with inverse given (for a suitable normalization of $d\mathcal{X}$) by

$$\big[\mathcal{F}^{-1}_{\mathfrak{g}\mathfrak{g}^*}(w)\big](X) = \int_{\mathfrak{g}^*} e^{i\langle X|\mathcal{X}\rangle} w(\mathcal{X}) d\mathcal{X}.$$

Then the most important transformation

$$\mathcal{F}_{\mathsf{G}\mathfrak{g}^*} = \mathcal{F}_{\mathfrak{g}\mathfrak{g}^*} \circ \mathsf{Exp}_\theta : L^2(\mathsf{G}) \to L^2(\mathfrak{g}^*; d\mathcal{X})$$

is given explicitly by

$$\big[\mathcal{F}_{\mathsf{G}\mathfrak{g}^*}(u)\big](\mathcal{X}) = \int_{\mathfrak{g}} e^{-i\langle X|\mathcal{X}\rangle} u[\exp(X)]\, \theta(X)^{\frac{1}{2}} dX = \int_{\mathsf{G}} e^{-i\langle \mathsf{log}(x)|\mathcal{X}\rangle} u(x) \theta[\mathsf{log}(x)]^{-\frac{1}{2}} dx\,,$$

with inverse

$$\Big[\mathcal{F}^{-1}_{\mathsf{G}\mathfrak{g}^*}(w)\Big](x) \;=\; \theta[\mathsf{log}(x)]^{-1/2} \int_{\mathfrak{g}^*} e^{i\langle \mathsf{log}(x)|\mathcal{X}\rangle} w(\mathcal{X})\, d\mathcal{X}.$$

Thus, recalling the unitarity of the Plancherel transformation, the operator

$$\mathcal{L} := \mathcal{P} \circ \mathcal{F}^{-1}_{\mathsf{G},\mathfrak{g}^*} : L^2(\mathfrak{g}^*; d\mathcal{X}) \to \mathscr{B}_2^{\oplus}(\widehat{\mathsf{G}})$$

is also unitary. Just by using the definitions, one has

$$[\mathcal{L}(w)](\xi) = \int_{\mathsf{G}}\int_{\mathfrak{g}^*} e^{i\langle \log(x)|\mathcal{X}\rangle}\theta[\log(x)]^{-\frac{1}{2}} w(\mathcal{X})\pi_\xi(x) D_\xi^{\frac{1}{2}}\, dx d\mathcal{X},$$

with inverse given explicitly on $\mathscr{B}_1^{\oplus}(\widehat{\mathsf{G}}) \cap \mathscr{B}_2^{\oplus}(\widehat{\mathsf{G}})$ by

$$\left[\mathcal{L}^{-1}(v)\right](\mathcal{X}) = \int_{\mathsf{G}}\int_{\widehat{\mathsf{G}}} e^{-i\langle \log x|\mathcal{X}\rangle}\theta[\log(x)]^{-\frac{1}{2}} \mathrm{Tr}_\xi\left[v(\xi) D_\xi^{\frac{1}{2}}\pi_\xi(x)^*\right] dx d\xi.$$

We make a notational convention: If $\mathcal{T} : \mathcal{M} \to \mathcal{N}$ is a linear transformation between Hilbert spaces, we denote by

$$\mathbf{T} \equiv \mathsf{id} \otimes \mathcal{T} : L^2(\mathsf{G}) \otimes \mathcal{M} \to L^2(\mathsf{G}) \otimes \mathcal{N}$$

the obvious operator. Taking into account the unitary isomorphism $\mathcal{L}$, as well as the unitary pseudo-differential calculus $\mathsf{Op} : L^2(\mathsf{G}) \otimes \mathscr{B}_2^{\oplus}(\widehat{\mathsf{G}}) \to \mathbb{B}_2[L^2(\mathsf{G})]$, one defines

$$\mathsf{op} = \mathsf{Op} \circ \mathbf{L} = \mathsf{Op} \circ \mathbf{P} \circ \mathbf{F}_{\mathsf{G}\mathfrak{g}^*}^{-1} : L^2(\mathsf{G}) \otimes L^2(\mathfrak{g}^*; d\mathcal{X}) \to \mathbb{B}_2[L^2(\mathsf{G})],$$

seen as an attempt to quantize the cotangent bundle $T^*\mathsf{G} \cong \mathsf{G} \times \mathfrak{g}^*$. By construction, op is a unitary transformation. The simplest way to get an explicit form, is to recall formula (10):

$$\mathsf{op}(\mathsf{B}) = \mathsf{Op}[\mathbf{L}(\mathsf{B})] = \mathsf{Sch}\big(\mathbf{P}^{-1}[\mathbf{L}(\mathsf{B})]\big) \circ \mathsf{Mult}_{\Delta^{1/2}} = \mathsf{Sch}\big(\mathbf{F}_{\mathsf{G}\mathfrak{g}^*}^{-1}(\mathsf{B})\big) \circ \mathsf{Mult}_{\Delta^{1/2}},$$

which together with (9) leads to

$$[\mathsf{op}(\mathsf{B})u](x) = \int_{\mathsf{G}}\int_{\mathfrak{g}^*} e^{i\langle \log(xy^{-1})|\mathcal{X}\rangle}\theta[\log(xy^{-1})]^{-\frac{1}{2}} \mathsf{B}(x,\mathcal{X})\Delta(y)^{-\frac{1}{2}} dy d\mathcal{X}.$$

9 The Affine Group

In this section we indicate basic formulae for pseudo-differential calculi on the affine group of the real line. The theory of unitary representations of the Affine group has been worked out by Gelfand and Naïmark; see [10, Sect 6.7]. In this section

$$\mathsf{G} = \{(b,a) \in \mathbb{R}^2 \mid a > 0\} = \mathbb{R} \times \mathbb{R}_+,$$

denotes the affine group, with product law

$$(b,a) \cdot (b',a') = (ab' + b, aa').$$

The group G is a type I solvable Lie group, it is an obvious semi-direct product, and it is the connected component of the identity of a similar connected Lie group for which the restriction is only $a \neq 0$. The left Haar measure is $a^{-2}dadb$, and its right Haar measure is $a^{-1}dadb$, hence the modular function is given by $\Delta(b,a) = a^{-1}$.

Let $\mathfrak{g} \equiv \mathbb{R}^2$ be the Lie algebra of G, with bracket defined by

$$[(\beta,\alpha),(\beta',\alpha')] = (\alpha\beta' - \alpha'\beta, 0).$$

An easy computation gives

$$\theta(\beta,\alpha) = \frac{1-e^{-\alpha}}{\alpha}.$$

The exponential map

$$\mathsf{exp} : \mathfrak{g} \to \mathsf{G}\,, \quad \mathsf{exp}(\beta,\alpha) = \Big(\frac{\beta}{\alpha}(e^{\alpha}-1), e^{\alpha}\Big)$$

is a diffeomorphism. Its inverse is (a limit is necessary for $a=1$)

$$\mathsf{log}(b,a) = \Big(\frac{b}{a-1}\log(a), \log(a)\Big).$$

Thus we have all the elements needed to express the quantization. With the preliminary computation

$$\mathsf{log}\big[(b,a)(b_1,a_1)^{-1}\big] = \mathsf{log}\Big(b - \frac{ab_1}{a_1}, \frac{a}{a_1}\Big) = \Big(\frac{a_1 b - ab_1}{a-a_1}\log\big(\frac{a}{a_1}\big), \log\big(\frac{a}{a_1}\big)\Big),$$

one arrives at

$$\begin{aligned}[\mathsf{op}(\mathsf{B})u](b,a) = \int_{\mathbb{R}}\int_{\mathbb{R}_+}\int_{\mathbb{R}}\int_{\mathbb{R}} e^{i\log(\frac{a}{a_1})[\mathfrak{x}+\frac{a_1 b - ab_1}{a-a_1}\mathfrak{y}]}\mathsf{B}\big((b,a),(\mathfrak{y},\mathfrak{x})\big) \\ u(b_1,a_1)\left[\tfrac{a}{a-a_1}\log\big(\tfrac{a}{a_1}\big)\right]^{1/2} a_1^{-3/2}\,da_1\,db_1\,d\mathfrak{y}\,d\mathfrak{x}.\end{aligned} \tag{11}$$

One of the interesting properties of G is that its unitary dual consists only of two points with positive Plancherel measure equal to 1 each and a ν-null set of one-dimensional representations [10, Sect. 6.7], that we afford neglecting. Setting $\mathbb{R}_- = -\mathbb{R}_+ = (-\infty,0)$ and $\mathcal{H}_\pm = L^2(\mathbb{R}_\pm; ds)$, the two irreducible representations admit the realization

$$\pi_\pm : \mathsf{G} \to \mathbb{B}(\mathcal{H}_\pm)\,, \quad [\pi_\pm(b,a)\varphi](s) = a^{1/2}e^{2\pi i b s}\varphi(as).$$

Of course, they are the restrictions to $\mathcal{H}_\pm$ (respectively) of an obvious representation on $L^2(\mathbb{R})$. We denote the equivalence class of $\pi_\pm$ in $\widehat{\mathsf{G}}$ by $\xi_\pm$; these are square integrable irreducible representations, having positive Plancherel measure. The corresponding Duflo-Moore operators are

$$(D_{\xi_\pm}\varphi)(s) = |s|\,\varphi(s) = \pm s\varphi(s)\,, \quad \pm s > 0.$$

For a symbol $A \in \mathscr{B}_2^{\oplus}(\mathsf{G}\times\widehat{\mathsf{G}}) \cong L^2(\mathsf{G}) \oplus L^2(\mathsf{G})$, formula (1) reads

$$[\mathsf{Op}(A)u](b,a) = \sum_{\pm}\int_{\mathbb{R}}\int_{\mathbb{R}_+}\frac{1}{(a')^{3/2}}\mathrm{Tr}\left(A_\pm(b,a)D_\pm^{\frac{1}{2}}\pi_\pm\Big(b-\frac{a}{a'}b', \frac{a}{a'}\Big)^*\right)u(b',a')\,db'da'.$$

Since in this case $\widehat{\mathsf{G}}$ is measurably equivalent to a two-point space, our symbols can be seen as pairs of functions depending only on the group variable $(b, a) \in \mathsf{G}$. However, since these symbols have operator values in $\mathcal{H}_\pm = L^2(\mathbb{R}_\pm)$, they generate a non-commutative calculus; to get the commutative calculus of multiplication operators one has to restrict to scalar valued symbols only depending on the variable in G, that are very particular.

Remark 12. Let $Y = (0,1)$ and $Z = (1,0)$ be the generators of $\mathfrak{g}$; they satisfy the commutation relation $[Y, Z] = Z$. If $d\pi_+(X)\varphi = \frac{d}{dt}\pi_+[\mathsf{exp}(tX)]\varphi\,|_{t=0}$ gives the (densely defined) induced representation of $\mathfrak{g}$ on $L^2(\mathbb{R})$, then

$$[d\pi_+(Y)\varphi](s) = \frac{1}{2}\varphi(s) + s\varphi'(s)\,, \quad [d\pi_+(Z)\varphi](s) = 2\pi i s\varphi(s).$$

Clearly $[d\pi_+(Y), d\pi_+(Z)] = d\pi_+(Z)$. Note that

$$id\pi_+(Y) = \frac{i}{2}\left(s \cdot \frac{d}{ds} + \frac{d}{ds} \cdot s\right)$$

is (formally) the infinitesimal generator of dilations of $\mathbb{R}_+$. Similar statements hold for the sign $-$.

Acknowledgement. M. Sandoval has been supported by Beca de Magister Nacional 2016 Conicyt and partially supported by Núcleo Milenio de Física Matemática RC120002. M. Măntoiu is supported by the Fondecyt Project 1160359.

References

1. Bahouri, H., Fermanian-Kammerer, C., Gallagher, I.: Phase space analysis and pseudo-differential calculus on the Heisenberg group. Astérisque **342** (2012)
2. Bruhat, F.: Distributions sur un groupe localement compact et applications a l'étude des représentations des groupes p-adiques. Bull. Soc. Math. France **89**, 43–75 (1961)
3. Bustos, H., Măntoiu, M.: Twisted pseudo-differential operators on type I locally compact groups. Illinois J. Math. **60**(2), 365–390 (2016)
4. Christ, M., Geller, D., Glowacki, P., Polin, L.: Pseudo-differential operators on groups with dilations. Duke Math. J. **68**(1), 31–65 (1992)
5. Delgado, J., Ruzhansky, M.: L^p-nuclearity, traces, and Grothendieck-Lidskii formula on compact Lie groups. J. Math. Pures Appl. **102**(1), 153–172 (2014)
6. Derighetti, A.: Convolution Operators on Groups. Lecture Notes of the Unione Matematica Italiana, vol. 11, Springer, Heidelberg; UMI, Bologna (2011)
7. Dixmier, J.: Les C^*-algébres et leurs représentations, Cahiers Scientifiques, Fasc. XXIX, Gauthier-Villars Cie, Paris (1964)
8. Duflo, M., Moore, C.C.: On the regular representation of a nonunimodular locally compact group. J. Funct. Anal. **21**(2), 209–243 (1976)
9. Fischer, V., Ruzhansky, M.: Quantization on Nilpotent Lie Groups. Progress in Mathematics. Birkhäuser, Basel (2016)

10. Folland, G.B.: A Course in Abstract Harmonic Analysis, 2nd edn. Textbooks in Mathematics. CRC Press, Boca Raton, FL (2016)
11. Führ, H.: Abstract Harmonic Analysis of Continuous Wavelet Transforms. Lecture Notes in Mathematics, vol. 1863. Springer-Verlag, Berlin (2005)
12. Glowacki, P.: Invertibility of convolution operators on homogeneous groups. Rev. Mat. Iberoam. **28**(1), 141–156 (2012)
13. Glowacki, P.: The Melin calculus for general homogeneous groups. Ark. Mat. **45**(1), 31–48 (2007)
14. Iftimie, V., Măntoiu, M., Purice, R.: Magnetic pseudo-differential operators. Publ. RIMS. **43**, 585–623 (2007)
15. Mackey, G.W.: The Theory of Unitary Group Representations. University of Chicago Press, Chicago, London (1976)
16. Măntoiu, M.: Essential spectrum and Fredholm properties for operators on locally compact groups. J. Oper. Theory **77**(2), 481–501 (2017)
17. Melin, A.: Parametrix constructions for right invariant differential operators on nilpotent groups. Ann. Global Anal. Geom. **1**(1), 79–130 (1983)
18. Măntoiu, M., Purice, R.: The magnetic Weyl calculus. J. Math. Phys. **45**(4), 1394–1417 (2004)
19. Măntoiu, M., Purice, R., Richard, S.: Spectral and propagation results for magnetic Schrödinger operators; A C^*-algebraic framework. J. Funct. Anal. **250**, 42–67 (2007)
20. Măntoiu, M., Ruzhansky, M.: Pseudo-differential operators, Wigner transform and Weyl systems on type I locally compact groups. Doc, Math (2017)
21. Măntoiu, M., Ruzhansky, M.: Quantizations on nilpotent Lie groups and algebras having flat coadjoint orbits, Preprint ArXiV and submitted
22. Ruzhansky, M. and Turunen, V.: Pseudo-differential Operators and Symmetries Pseudo-differential Operators. Theory and Applications, vol. 2. Birkhäuser, Basel (2010)
23. Ruzhansky, M., Turunen, V.: Quantization of pseudo-differential operators on the torus. J. Fourier Anal. Appl. **16**, 943–982 (2010)
24. Ruzhansky, M., Turunen, V.: Global quantization of pseudo-differential operators on compact Lie groups, SU(2), 3-sphere, and homogeneous spaces. Int. Math. Res. Not. IMRN **11**, 2439–2496 (2013)
25. Ruzhansky, M., Turunen, V., Wirth, J.: Hörmander-class of pseudo-differential operators on compact Lie groups and global hypoellipticity. J. Fourier Anal. Appl. **20**, 476–499 (2014)
26. Ruzhansky, M., Wirth, J.: Global functional calculus for operators on compact Lie groups. J. Funct. Anal. **267**, 144–172 (2014)
27. Tatsuuma, N.: Plancherel formula for non-unimodular locally compact groups. J. Math. Kyoto Univ. **12**, 179–261 (1972)
28. Taylor, M.E.: Pseudo-differential Operators. Princeton Mathematical Series, vol. 34. Princeton University Press, Princeton, NJ (1981)
29. Taylor, M.E.: Noncommutative Microlocal Analysis I. American Mathematical Soc., vol. 52, no. 313 (1984)
30. Williams, D.P.: Crossed Products of C^*-Algebras, Mathematical Surveys and Monographs, vol. 34. American Mathematical Society, Providence, RI (2007)

Existence and Numerical Computation of Standing Wave Solutions for a System of Two Coupled Schrödinger Equations

Juan Carlos Muñoz Grajales[1(✉)] and Luisa Fernanda Vargas[2]

[1] Departamento de Matemáticas, Universidad del Valle, Calle 13 Nro 100-00, Cali, Colombia
jcarlmz@yahoo.com
[2] Departamento de Ciencias Naturales y Matemáticas, Universidad Javeriana, Calle 18 no. 118-250, Cali, Colombia
lfvargas@javerianacali.edu.co

Abstract. In this paper, we consider the existence of a type of stationary wave of a system of two coupled Schrödinger equations with variable coefficients, which can be employed to describe the interaction among propagating modes in nonlinear optics and Bose-Einstein condensates (BECs), for instance. To prove existence of these solutions, we use some existing fixed point theorems for completely continuous operators defined in a cone in a Banach space. Furthermore, some numerical approximations of stationary waves are computed by using a spectral collocation technique combined with a Newton's iteration.

Keywords: Schrödinger equations · Numerical computations · Standing waves

1 Introduction

In this paper, we will consider the system of two coupled Schrödinger equations with variable coefficients (henceforth called CNLS)

$$i\frac{\partial u}{\partial \xi} + K(x)\frac{\partial^2 u}{\partial x^2} + \sigma_1(x)u + a(x)|u|^2 u + b(x)|v|^2 u + e(x)v^2 u^* = 0, \tag{1}$$

$$i\frac{\partial v}{\partial \xi} + K(x)\frac{\partial^2 v}{\partial x^2} + \sigma_2(x)v + c(x)|v|^2 v + d(x)|u|^2 v + e(x)u^2 v^* = 0, \tag{2}$$

where $u(\xi, x), v(\xi, x)$ are the complex envelopes of the propagating beam of the two modes, x is the spatial coordinate, and ξ represents the retarded time, $K(x)$ is the group-velocity dispersion coefficient, $a(x), b(x), c(x), d(x), e(x)$ mean nonlinear interaction, and the functions $\sigma_1(x), \sigma_2(x)$ denote external potentials. Under some physical conditions, this system can be applied to model the interaction among propagating modes in some branches of nonlinear science, such as

J. Delgado and M. Ruzhansky (Eds.): *Analysis and Partial Differential Equations: Perspectives from Developing Countries*, PROMS 275, pp. 191–206, 2018.
https://doi.org/10.1007/978-3-030-05657-5_12

BECs and so on, [2–9,21], and in nonlinear optics for the case where all coefficients are constant (see for instance [20] and references therein). A Bose-Einstein condensate is a state of matter of a gas of bosons with extremely low density cooled to a temperature very close to absolute zero, i.e. very near $0K$ or $-273.15°$ C. This state was first predicted theoretically in 1924–25 by S. N. Bose and A. Einstein [31]. However, it was not until 1995 that a BCE was produced experimentally at the University of Colorado at Boulder NIST-JILA lab, by Cornell et al. [32]. BECs have proven to be useful in exploring a wide range of questions in fundamental physics, and since the initial discoveries by Cornell et al., the interest and experimental and theoretical activity on BCEs have increased and several applications of this phenomenon have been proposed : quantum information processing- concept of quantum computer, precision measurement by development of most sensitive detectors using BECs., and development of optical lattices which could be easily modifiable by varying the interplanar spacing, among others.

Coupled nonlinear Schrödinger equations (CNLS) with variable coefficients has been recently applied to study the effect of optical parameters on the modulational instability phenomenon in inhomogeneous fibers [30], to analyze the effect of the variable coefficients on the dynamics of the solitons of the system (i.e. intensities, velocities and accelerations) [28] and to describe the interaction among the modes in nonlinear optics and Bose-Einstein condensation and to study bright-dark and bright-bright solitons for the corresponding CNLS systems [29]. Mixed nonlinear terms in the forms v^2u^*, u^2v^* with a spatial and time dependent coefficient, appear for instance, in models for spinor BECs [11,13–19], and in [10] for a Bose-Einstein condensate model based on a system of coupled Gross-Pitaevskii (GP) equations [11,12], similar to system (1)–(2).

In the present paper, we are first interested in investigating the existence of standing wave solutions of system (1)–(2) in the form

$$u(\xi, x) = \widetilde{u}(x)e^{i\alpha\xi}, \quad v(\xi, x) = \widetilde{v}(x)e^{i\alpha\xi}, \tag{3}$$

where $\widetilde{u}$, $\widetilde{v}$ are real bounded functions such that $\lim\limits_{|x|\to\infty} \widetilde{u}(x) = \lim\limits_{|x|\to\infty} \widetilde{v}(x) = 0$, (localized solutions) and α is a constant. We assume that the inhomogeneous coefficients $a(x)$, $b(x)$, $c(x)$, $d(x)$, $e(x)$ are positive and compactly supported in $\mathbb{R}$ and $K(x) \geq k_*$, for some positive constant k_*. By adapting the techniques used in [22], the existence of solutions in the form (3) is obtained by applying the theory of fixed points of completely continuous operators defined in a cone in an appropriate Banach space [23–27]. The analysis of localized solutions and propagating waves in BECs with spatially inhomogeneous interactions has been a field of an enormous interest in Physics the last years [34–38], motivated by the discovery of new ways to control experimentally the interactions inside the BEC through optical manipulation of the Feshbach resonances [33].

In second place, we introduce a numerical solver for approximating standing wave solutions of system (1)–(2) by using a Newton's iteration together with a collocation spectral strategy analogous to that used by J.C. Muñoz [1] to discretize the corresponding travelling wave system. Furthermore, a numerical scheme for approximating the temporal evolution of solutions of system (1)–(2) is introduced. In that scheme the time stepping is performed by using a 4th- order Runge-Kutta strategy and a Fourier-spectral approximation for the spatial variable. To our best knowledge, this type of numerical study has not been considered in previous works on systems in the general form given by equations (1)–(2). We mention the work by Belmonte et al. [21] which used Lie group theory to compute some explicit solutions of a system of two coupled nonlinear Schrödinger equations with inhomogeneous coefficients analogous to system (1)–(2), but without the nonlinear mixed terms v^2u^*, u^2v^*.

The rest of this paper is organized as follows. In Sect. 2, we introduce the notation, definitions, functional spaces, and some fixed point theorems for completely continuous operators to be employed in the paper. In Sect. 3, we establish the existence of standing wave solutions of the CNLS system by reducing the problem to study the fixed points of an appropriate nonlinear operator. Finally, in Sect. 4, we introduce the Newton-type numerical solver employed to compute standing wave solutions of the CNLS system and the numerical scheme for approximating the temporal evolution of the system, which is used in the verification that the profiles computed correspond in fact to approximate solutions of the CNLS system.

2 Preliminaries

In this section, we review some existing results in the literature fixed points of completely continuous operators. These results will be used in this paper to establish the existence of standing wave solutions in the form (3) of system (1)–(2). Hereafter, the quantities $f_*, d_*, k_*, k_1*, k_2*$ will denote positive constants.

Definition 1. Let X be the Banach space and P be a nonempty closed subset of X. We say that P is a cone if

1. $\lambda x + \mu y \in P$, for all $x, y \in P$ and $\lambda,\ \mu \in [0, +\infty)$.
2. $x,\ -x \in P$ implies that $x = 0$.

Theorem 1. *([22]) Let X be a Banach space and $P \subset X$ be a cone in X. Suppose that Ω_1, Ω_2 are open subsets of X with $0 \in \Omega_1$, $\overline{\Omega_1} \subset \Omega_2$ and $T : P \cap (\overline{\Omega_2} \setminus \Omega_1) \to P$ be a completely continuous operator satisfying one of the following conditions:*

1. *$\|Tu\| \leq \|u\|$, if $u \in P \cap \partial\Omega_1$ and $\|Tu\| \geq \|u\|$, if $u \in P \cap \partial\Omega_2$.*
2. *$\|Tu\| \geq \|u\|$, if $u \in P \cap \partial\Omega_1$ and $\|Tu\| \leq \|u\|$, if $u \in P \cap \partial\Omega_2$.*

Then the operator T has at least one fixed point in $P \cap (\overline{\Omega_2} \setminus \Omega_1)$.

This result is due to Krasnosel'skii [39] and it has been already used for instance in [23,24,26] y [27].

Theorem 2. *([22]) Let $\Omega \subset BC(\mathbb{R})$. Suppose that the functions $u \in \Omega$ are an equicontinuous set in each compact interval of $\mathbb{R}$, and that for all $u \in \Omega$, we have $|u(x)| \leq \xi(x)$, for any $x \in \mathbb{R}$, where $\xi \in BC(\mathbb{R})$ satisfies $\lim_{|x| \to +\infty} \xi(x) = 0$. Then the set Ω is relatively compact.*

Lemma 1. *([25]) Let $f \in L^\infty(\mathbb{R})$, $f(x) \geq f_* > 0$. Let us consider the homogeneous problem*

$$- \phi'' + f(x)\phi = 0, \tag{4}$$

where $\lim_{x \to -\infty} \phi(x) = 0$, $\lim_{x \to \infty} \phi(x) = 0$.

The Green's function associated to this equation is given by

$$G_1(x,s) = \begin{cases} \phi_1(x)\phi_2(s), & -\infty < x \leq s < +\infty \\ \phi_1(s)\phi_2(x), & -\infty < s \leq x < +\infty, \end{cases} \tag{5}$$

where ϕ_1, ϕ_2 are solutions of (4) such that $\lim_{x \to -\infty} \phi_1(x) = 0$, $\lim_{x \to \infty} \phi_2(x) = 0$. Furthermore, ϕ_1, ϕ_2 can be chosen as positive increasing and positive decreasing functions, respectively.

The functions ϕ_1, ϕ_2 intersect in a unique point x_1. Therefore, we can define a function $p_1 \in BC(\mathbb{R})$ by

$$p_1(x) = \begin{cases} \frac{1}{\phi_2(x)}, & x \leq x_1 \\ \frac{1}{\phi_1(x)}, & x > x_1. \end{cases}$$

Similarly, for $h \in L^\infty(\mathbb{R})$, $h(x) \geq h_* > 0$, the Green's function associated to the equation

$$- \varphi'' + h(x)\varphi = 0, \tag{6}$$

and $\lim_{x \to -\infty} \varphi(x) = 0$, $\lim_{x \to \infty} \varphi(x) = 0$, is given by

$$G_2(x,s) = \begin{cases} \varphi_1(x)\varphi_2(s), & -\infty < x \leq s < +\infty \\ \varphi_1(s)\varphi_2(x), & -\infty < s \leq x < +\infty, \end{cases} \tag{7}$$

where the functions φ_1, φ_2 are solutions of (6) that can be chosen as positive increasing and positive decreasing functions, respectively, and satisfying that satisfy that $\lim_{x \to -\infty} \varphi_1(x) = 0$, $\lim_{x \to \infty} \varphi_2(x) = 0$. Further, let us define $p_2 \in BC(\mathbb{R})$ by

$$p_2(x) = \begin{cases} \frac{1}{\varphi_2(x)}, & x \leq x_2 \\ \frac{1}{\varphi_1(x)}, & x > x_2, \end{cases}$$

where x_2 is the unique point of intersection of the functions φ_1, φ_2.

Theorem 3. *(see [26]) The Green's functions defined by* (5) *and* (7) *have the following properties:*

1. $G_j(x,s) > 0$ *for all* $(x,s) \in \mathbb{R} \times \mathbb{R}$, $j = 1, 2$.
2. $G_j(x,s) \leq G_j(s,s)$ *for all* $(x,s) \in \mathbb{R} \times \mathbb{R}$, $j = 1, 2$.
3. *If* $P \subset \mathbb{R}$ *is a nonempty compact subset, let us define*

$$m_1(P) = \min\left(\phi_1(\inf P), \phi_2(\sup P)\right),$$
$$m_2(P) = \min\left(\varphi_1(\inf P), \varphi_2(\sup P)\right).$$

hen, $G_j(x,s) \geq m_j(P) p_j(s) G_j(s,s)$, $j = 1, 2$, *for all* $(x,s) \in P \times \mathbb{R}$.

3 Existence of Standing Wave Solutions

Our aim in this section is to establish the existence of standing wave solutions of system (1)–(2) in the form

$$u(\xi, x) = \widetilde{u}(x) e^{i\alpha\xi}, \quad v(\xi, x) = \widetilde{v}(x) e^{i\alpha\xi}, \tag{8}$$

where $\widetilde{u}$, $\widetilde{v}$ are real bounded functions such that $\lim\limits_{|x|\to\infty} \widetilde{u}(x) = \lim\limits_{|x|\to\infty} \widetilde{v}(x) = 0$, and α is a real constant by using the results on completely continuous operators outlined in the previous section. Let us define the coefficients

$$K_1(x) = \frac{\alpha - \sigma_1(x)}{K(x)},$$
$$K_2(x) = \frac{\alpha - \sigma_2(x)}{K(x)}.$$

We assume that $K(x) \geq k_* > 0$, $K_1(x) \geq k_1* > 0$, $K_2(x) \geq k_2* > 0$ are bounded functions, and that the variable coefficients $a(x)$, $b(x)$, $c(x)$, $d(x)$ and $e(x)$ are positive and compactly supported in $\mathbb{R}$. We see that the functions $\tilde{u}$ and $\tilde{v}$ must satisfy the system of ordinary differential equations

$$-\widetilde{u''}(x) + K_1(x)\widetilde{u}(x) = \left(\frac{a(x)}{K(x)}\widetilde{u}^2(x) + \frac{b(x)+e(x)}{K(x)}\widetilde{v}^2(x)\right)\widetilde{u}(x) \tag{9}$$

$$-\widetilde{v''}(x) + K_2(x)\widetilde{v}(x) = \left(\frac{c(x)}{K(x)}\widetilde{v}^2(x) + \frac{d(x)+e(x)}{K(x)}\widetilde{u}^2(x)\right)\widetilde{v}(x). \tag{10}$$

Hereafter, we will denote the space of continuous bounded functions by $BC(\mathbb{R})$. Let us denote by $\mathrm{Supp}(a)$ the support of a given function a. Henceforth, we will suppose that $M = \mathrm{Supp}(a) \cup \mathrm{Supp}(b) \cup \mathrm{Supp}(c) \cup \mathrm{Supp}(d) \cup \mathrm{Supp}(e)$ is a nonempty compact subset of $\mathbb{R}$. Let us consider the Banach space $X = BC(\mathbb{R}) \times$

$BC(\mathbb{R})$ with product norm $\|(u,v)\| = \max\{\|u\|_\infty, \|v\|_\infty\}$, for $(u,v) \in X$. It can be shown that the set P defined by

$$P = \big\{\mathbf{u} = (u,v) \in X : u(x), v(x) \geq 0, \text{ for all } x \in \mathbb{R}, \quad \min_{x\in M} u(x) \geq m_1 p_0^1 \|u\|, \\ \text{and } \min_{x\in M} v(x) \geq m_2 p_0^2 \|v\|\big\},$$

where $p_0^1 = \inf_M p_1(x)$, $p_0^2 = \inf_M p_2(x)$ and the constants $m_1 = m_1(M)$ and $m_2 = m_2(M)$ are defined by Eq. (8), is a cone in X. Observe that $p_0^1 > 0$ and $p_0^2 > 0$ due to the compactness of the set M.

Let $T : P \to X$ be an operator with components (T_1, T_2), where

$$\begin{aligned} T_1(\mathbf{u})(x) &= \int_{\mathbb{R}} G_1(x,s) \left(\frac{a(s)}{K(s)} u^2(s) + \frac{b(s)+e(s)}{K(s)} v^2(s) \right) u(s)ds \\ &= \int_M G_1(x,s) \left(\frac{a(s)}{K(s)} u^2(s) + \frac{b(s)+e(s)}{K(s)} v^2(s) \right) u(s)ds \end{aligned}$$

$$\begin{aligned} T_2(\mathbf{u})(\mathbf{x}) &= \int_{\mathbb{R}} G_2(x,s) \left(\frac{c(s)}{K(s)} v^2(s) + \frac{d(s)+e(s)}{K(s)} u^2(s) \right) v(s)ds \\ &= \int_M G_2(x,s) \left(\frac{c(s)}{K(s)} v^2(s) + \frac{d(s)+e(s)}{K(s)} u^2(s) \right) v(s)ds, \end{aligned}$$

with $\mathbf{u} = (\mathrm{u}, \mathrm{v})$, and G_1, G_2 the Green's functions defined in (5), (7) for equations (4), (6), with $f(x) = K_1(x)$ and $h(x) = K_2(x)$, respectively. We point out that the problem of finding a nontrivial solution of system (9)–(10) in the cone P has been reduced to that of finding a nontrivial fixed point of the operator $T : P \to X$ defined above. This is the advantage of introducing the Green's functions G_1, G_2.

Lemma 2. $T(P) \subset P$.

Proof. From Theorem 3 and the assumption that a, b, c, d y e are positive functions, it follows that $T_1(\mathbf{u})(x) \geq 0$ for any $x \in \mathbb{R}$. Let $x_m \in \mathbb{R}$ be the point where T_1 assumes the minimum. By using the properties of the Green's functions given in Theorem 3, we have that

$$\begin{aligned} \min_{x\in M} T_1\mathbf{u}(\mathbf{x}) = T_1 u(x_m) &= \textstyle\int_M G_1(x_m,s) \left(\frac{a(s)}{K(s)} u^2 + \frac{b(s)+e(s)}{K(s)} v^2(s) \right) u(s)\, ds \\ &\geq m_1 \textstyle\int_M p_1(s) G_1(s,s) \left(\frac{a(s)}{K(s)} u^2 + \frac{b(s)+e(s)}{K(s)} v^2(s) \right) u(s)\, ds \\ &\geq m_1 p_0^1 \textstyle\int_M G_1(x,s) \left(\frac{a(s)}{K(s)} u^2 + \frac{b(s)+e(s)}{K(s)} v^2(s) \right) u(s)\, ds \\ &= m_1 p_0^1 T_1(\mathbf{u})(x). \end{aligned} \tag{11}$$

From here, it follows that $\min_{x\in M} T_1\mathbf{u}(\mathbf{x}) \geq \mathbf{m_1 p_0^1} \|\mathbf{T_1 u}\|_\infty$. Analogously, it can be proved that $\min_{x\in M} T_2\mathbf{u}(\mathbf{x}) \geq \mathbf{m_2 p_0^2} \|\mathbf{T_2 u}\|_\infty$. Thus we have established that $T(P) \subset P$.

Lemma 3. *$T : P \to P$ is a completely continuous operator.*

Proof. Let $\Omega \subset P$ be a bounded subset and $C > 0$ a uniform bound for every function in Ω. Let us note that if $\mathbf{u} \in \mathbf{\Omega}$, then

$$|T_1\mathbf{u}(x)| \leq C^3 \int_{\mathbb{R}} G_1(x,s) \left(\frac{a(s)+b(s)+e(s)}{K(s)} \right) ds,$$

where the right side expression in the last inequality is a bounded function on $\mathbb{R}$ decaying to zero as $|x| \to \infty$. In analogous way, it can be shown that $T_2(\Omega)$ is also relatively compact and thus the desired result is obtained. Furthermore, it is easy to see that the sets of functions $T_1(\Omega), T_2(\Omega)$ are equicontinuous on each compact interval. Thus, from Theorem 2, $T_1(\Omega), T_2(\Omega)$ are relatively compact.

Theorem 4. *Let us assume that*

1. *The set $M = Supp(a) \cup Supp(b) \cup Supp(c) \cup Supp(d) \cup Supp(e)$ is a nonempty compact subset of $\mathbb{R}$.*
2. *$\int_M G_1(x,s) \left(\frac{a(s)+b(s)+e(s)}{K(s)} \right) ds < C$, and $\int_M G_2(x,s) \left(\frac{c(s)+d(s)+e(s)}{K(s)} \right) ds < C$, for all $x \in \mathbb{R}$, for some constant $C > 0$.*

Then, there exists a nontrivial solution $\mathbf{u} = (u,v) \in X$ of the system (9)–(10), and $R > 0, r > 0$, such that

$$\lim_{|x|\to\infty} u(x) = \lim_{|x|\to\infty} v(x) = 0, \tag{12}$$

$$\lim_{|x|\to\infty} u'(x) = \lim_{|x|\to\infty} v'(x) = 0, \tag{13}$$

and $r \leq \|\mathbf{u}\| \leq R$.

Proof. Let us consider the two open balls Ω_1 and Ω_2 de X centered at the origin and with radius r and R. Let $\mathbf{u} \in \mathbf{P} \cap \partial\mathbf{\Omega_1}$. Then

$$\begin{aligned}\|T\mathbf{u}\| &= \max_{x\in\mathbb{R}} \left\{ \int_M G_1(x,s) \left(\frac{a(s)}{K(s)} u^2(s) + \frac{b(s)+e(s)}{K(s)} v^2(s) \right) u(s) ds, \right.\\ &\qquad \left. \int_M G_2(x,s) \left(\frac{c(s)}{K(s)} v^2(s) + \frac{d(s)+e(s)}{K(s)} u^2(s) \right) v(s) ds \right\}\\ &\leq r^3 \max_{x\in\mathbb{R}} \left\{ \int_M G_1(x,s) \left(\frac{a(s)}{K(s)} + \frac{b(s)+e(s)}{K(s)} \right) ds, \right.\\ &\qquad \left. \int_M G_2(x,s) \left(\frac{c(s)}{K(s)} + \frac{d(s)}{K(s)} + \frac{e(s)}{K(s)} \right) ds \right\}\\ &< r = \|\mathbf{u}\|, \end{aligned} \tag{14}$$

for $r > 0$ small enough. In other words, $\|T\mathbf{u}\| \leq \|\mathbf{u}\|$. Now set $\mathbf{u} \in \mathrm{P} \cap \partial\Omega_2$. Without loss of generality, let us suppose that $\|u\|_\infty = R$. Therefore $m_1 p_0^1 R \leq u(x) \leq R$ for all $x \in M$. In consequence,

$$\|T\mathbf{u}\| = \max_{\mathbf{i=1,2,x\in\mathbb{R}}} (|\mathbf{T_i u(x)}|)_\mathbf{i} \geq \max_{x\in M} \int_M G_1(x,s) \left(\tfrac{a(s)}{K(s)} u^2(s) + \tfrac{b(s)+e(s)}{K(s)} v^2(s) \right) u(s) ds \quad (15)$$

$$\geq \tfrac{R^3}{k_*} \max_{x\in M} \int_M G_1(x,s) a(s) ds \quad (16)$$

$$\geq R = \|\mathbf{u}\|, \quad (17)$$

for $R > 0$ large enough. Notice that $\max_{x\in M} \int_M G_1(x,s)a(s)ds$ is finite since the set M is compact. From Theorem 1, T has a fixed point $\mathbf{u} \in P \cap (\overline{\Omega_2} - \Omega_1)$. Finally, notice that a fixed point $\mathbf{u} = (u, v)$ of the operator T is such that $u = T_1(\mathbf{u}), v = T_2(\mathbf{u})$. Thus, from the decay properties at infinity of the functions $\phi_1, \phi_2, \varphi_1, \varphi_2$ in the Green's functions G_1, G_2, we obtain the conditions (12)–(13).

4 Numerical results

To approximate standing wave solutions in the form (3) of system (1)–(2), we will use a spectral collocation discretization combined with a Newton's iteration method analogous to those employed by J.C. Muñoz in [1] to discretize the system of two coupled ordinary differential equations

$$-K(x)\widetilde{u''}(x) + (\alpha - \sigma_1(x))\widetilde{u}(x) = \left(a(x)\widetilde{u}^2(x) + (b(x) + e(x))\widetilde{v}^2(x)\right)\widetilde{u}(x) \quad (18)$$

$$-K(x)\widetilde{v''}(x) + (\alpha - \sigma_2(x))\widetilde{v}(x) = \left(c(x)\widetilde{v}^2(x) + (d(x) + e(x))\widetilde{u}^2(x)\right)\widetilde{v}(x), \quad (19)$$

where the functions $\widetilde{u}, \widetilde{v}$ decay to zero at infinity. On the other hand, for checking that the computed solutions correspond really to standing wave solutions of CNLS system, it is also necessary to formulate a numerical scheme for approximating the time evolution of a solution (u, v) to the initial value problem associated to this system. We adopt a Galerkin-spectral strategy to accomplish this goal. To do this, the computational domain $[0, L]$ of the variable x is discretized by N equidistant points with spacing $\Delta x = L/N$, and introduce approximations of the components u, v in the form of truncated Fourier series

$$u(\xi, x) \approx \sum_{k=-N/2+1}^{N/2} \hat{u}_k(\xi)\phi_k(x), \quad v(\xi, x) \approx \sum_{k=-N/2+1}^{N/2} \hat{v}_k(\xi)\phi_k(x), \quad (20)$$

where

$$\phi_k(x) = e^{iw_k x}, \quad w_k = \frac{2\pi k}{L}, \quad k = -N/2+1, ..., 0, ...N/2.$$

It is well known that the set $\{\phi_k\}$ is an orthonormal basis of $L^2(\mathbb{R})$ with respect to the inner product

$$\langle f, g \rangle = \frac{1}{L} \int_0^L f(x)\overline{g(x)} dx.$$

The time-dependent Fourier coefficients $\hat{u}_k(\xi), \hat{v}_k(\xi)$ in the sums given in (20) are computed using the Fast Fourier transform (FFT) algorithm with the expressions

$$\hat{u}_k(\xi) = \frac{1}{L}\int_0^L u(\xi, x)e^{-iw_k x}dx, \qquad \hat{v}_k(\xi) = \frac{1}{L}\int_0^L v(\xi, x)e^{-iw_k x}dx,$$

for $k = -N/2+1, ..., 0, ...N/2$. We point out that this periodic-type strategy can be used for approximating solutions decaying to zero at infinity, provided that the length L of the computational domain is taken large enough, so that the solutions do not reach their boundaries during the computer simulation. Replacing the expressions (20) in Eqs. (1)–(2) and projecting with respect to the basis $\{\phi_k\}$, we get for $k = -N/2+1, ..., 0, ...N/2$,

$$\frac{d\hat{u}_k}{d\xi} = iP_k\Big[K\frac{\partial^2 u}{\partial x^2} + \sigma_1 u + a|u|^2u + b|v|^2u + ev^2u^*\Big] := h_1(\hat{u}_k, \hat{v}_k),$$
$$\frac{d\hat{v}_k}{d\xi} = iP_k\Big[K\frac{\partial^2 v}{\partial x^2} + \sigma_2 v + c|v|^2v + d|u|^2v + eu^2v^*\Big] := h_2(\hat{u}_k, \hat{v}_k),$$

where $P_k[\cdot]$ denotes the operator

$$P_k[g] := \frac{1}{L}\int_0^L g(x)e^{-iw_k x}dx.$$

We point out that the derivatives $\frac{\partial^2 u}{\partial x^2}, \frac{\partial^2 v}{\partial x^2}$ are approximated by using the expressions

$$\frac{\partial^2 u}{\partial x^2} = \sum_{k=-N/2+1}^{N/2} -w_k^2\hat{u}_k e^{iw_k x}, \qquad \frac{\partial^2 v}{\partial x^2} = \sum_{k=-N/2+1}^{N/2} -w_k^2\hat{v}_k e^{iw_k x}.$$

The system of coupled ordinary differential equations (21) is solved for the unknown coefficients $\hat{u}_k, \hat{v}_k$ by using the classical fourth-order Runge-Kutta scheme (RK4)

$$\begin{aligned}
K_1u &:= h_1(\hat{u}_k^n, \hat{v}_k^n),\\
K_1v &:= h_2(\hat{u}_k^n, \hat{v}_k^n), \qquad (21)\\
K_2u &:= h_1\Big(\hat{u}_k^n + \tfrac{\Delta\xi}{2}K_1u, \hat{v}_k^n + \tfrac{\Delta\xi}{2}K_1v\Big),\\
K_2v &:= h_2\Big(\hat{u}_k^n + \tfrac{\Delta\xi}{2}K_1u, \hat{v}_k^n + \tfrac{\Delta\xi}{2}K_1v\Big),\\
K_3u &:= h_1\Big(\hat{u}_k^n + \tfrac{\Delta\xi}{2}K_2u, \hat{v}_k^n + \tfrac{\Delta\xi}{2}K_2v\Big),\\
K_3v &:= h_2\Big(\hat{u}_k^n + \tfrac{\Delta\xi}{2}K_2u, \hat{v}_k^n + \tfrac{\Delta\xi}{2}K_2v\Big),\\
K_4u &:= h_1\Big(\hat{u}_k^n + \Delta\xi K_3u, \hat{v}_k^n + \Delta\xi K_3v\Big),\\
K_4v &:= h_2\Big(\hat{u}_k^n + \Delta\xi K_3u, \hat{v}_k^n + \Delta\xi K_3v\Big),\\
\hat{u}_k^{n+1} &= \hat{u}_k^n + \tfrac{\Delta\xi}{6}(K_1u + 2K_2u + 2K_3u + K_4u),\\
\hat{v}_k^{n+1} &= \hat{v}_k^n + \tfrac{\Delta\xi}{6}(K_1v + 2K_2v + 2K_3v + K_4v). \qquad (22)
\end{aligned}$$

Here $\hat{u}_k^n, \hat{v}_k^n$ denote the approximations of the coefficients $\hat{u}_k, \hat{v}_k$ at the time $\xi_n = n\Delta\xi$. In the following numerical experiments, we can see that due to its simplicity and numerical stability, the RK4 method is appropriate for the treatment of a problem with variable coefficients, such as the CNLS system (1)–(2). The goal of the set of numerical experiments presented below for particular cases of synthetic model's coefficients, is to show that the numerical solvers introduced can be applied for simulating accurately the evolution of solutions of system (1)–(2) and for computing standing wave solutions in the form (3). With a scientific computing spirit, we expect that these computer simulations could help to initiate a complete numerical study of stationary waves in BECs, taking into account the effect of variations in the modelling parameters depending on the spatial variable. It is important to mention that the general system (1)–(2) can not be explicitly solved for arbitrary space-dependent model's coefficients $a(x), b(x), c(x), d(x), d(x), e(x)$, and thus a numerical strategy is a fundamental tool to explore and understand the physical phenomena involved.

Experiment 1: In first place, we will consider the case $a(x) = c(x) = \text{sech}(x-x_0)^p$, $b(x) = d(x) = 2\text{sech}(x-x_0)^p$, $e(x) = 3\text{sech}(x-x_0)^p$, $K(x) = 0.5\text{sech}(x-x_0)^p$, $\sigma_1(x) = 0.5e^{-0.5(x-x_0)^2}$, $\sigma_2(x) = -0.5e^{-0.5(x-x_0)^2}$, $\alpha = 1.5$, and $p = 3$, $x_0 = 35$. The starting values of Newton's iteration for approximating the solutions of the system of ODEs (18)–(19) are

$$\widetilde{u}(x) = \widetilde{v}(x) = \text{sech}(x-x_0)^p.$$

The resulting numerical approximations of $(\widetilde{u}, \widetilde{v})$ computed after 9 Newton's iterations are presented in Fig. 1. In this computation the spatial domain is $[0, 70]$, and the number of collocation points is 2^9. In order to make a verification that we have computed really a standing wave of system (1)–(2), we run the numerical solver (21) explained above, with time stepsize $\Delta\xi = 1e-3$, $N = 2^{10}$ FFT points, and spatial computational domain $[0, 70]$, using as initial data

$$u(0,x) = \widetilde{u}(x),$$
$$v(0,x) = \widetilde{v}(x).$$

The result of this numerical simulation at $\xi = 20$ is displayed in Fig. 2, superimposed with the expected standing wave solution given by

$$u(\xi,x) = e^{i\alpha\xi}\widetilde{u}(x),$$
$$v(\xi,x) = e^{i\alpha\xi}\widetilde{v}(x).$$

Observe a good accuracy with errors of magnitude $1e-3$ between the profiles of real and imaginary parts of the components u, v, corroborating that, in fact, we have an approximation of a standing wave solution to system (1)–(2).

Experiment 2: In second place, we run a numerical experiment using $a(x) = c(x) = \text{sech}(x-x_0)^p$, $b(x) = d(x) = 3\text{sech}(x-x_0)^p$, $e(x) = \text{sech}(x-x_0)^p$, $K(x) = 0.5\text{sech}(x-x_0)^{p_1}$, $\sigma_1(x) = e^{-0.5(x-x_0)^2}$, $\sigma_2(x) = e^{-0.5(x-x_0)^2}$, $\alpha = 1$,

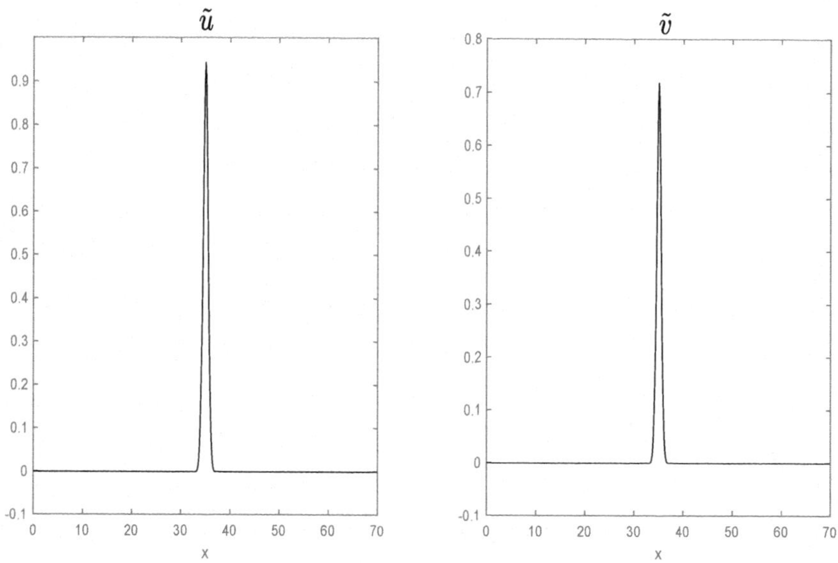

Fig. 1. Approximation of a solution $\widetilde{u}, \widetilde{v}$ to system (18)–(19).

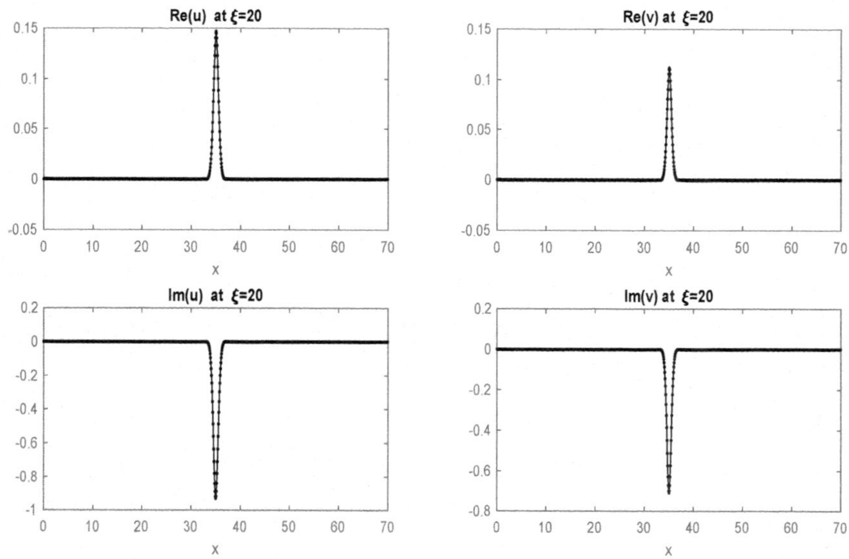

Fig. 2. Real and imaginary parts of the standing wave (u, v) of system (1)–(2) at $\xi = 20$. Solid line: numerical solution (u, v) of system (1)–(2) using the RK4 scheme (21). Pointed line: expected standing wave computed using (12.23).

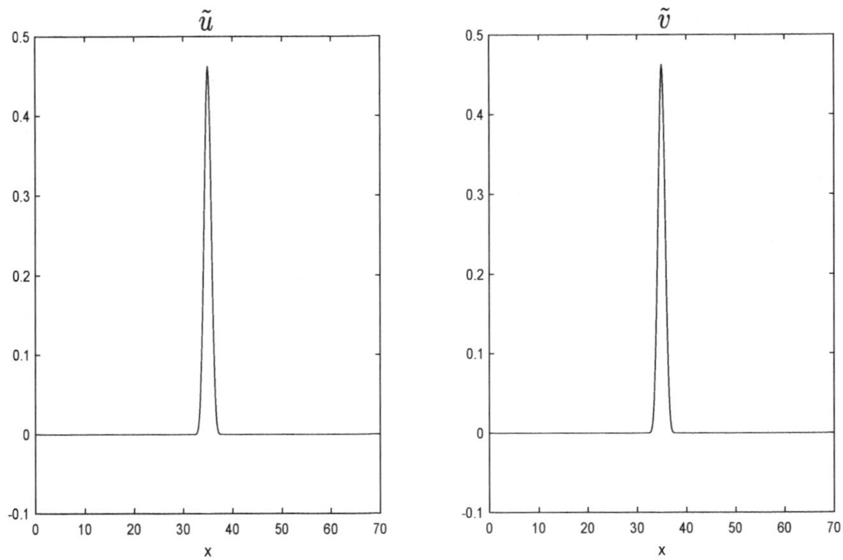

Fig. 3. Approximation of a solution $\widetilde{u}, \widetilde{v}$ to system (18)–(19).

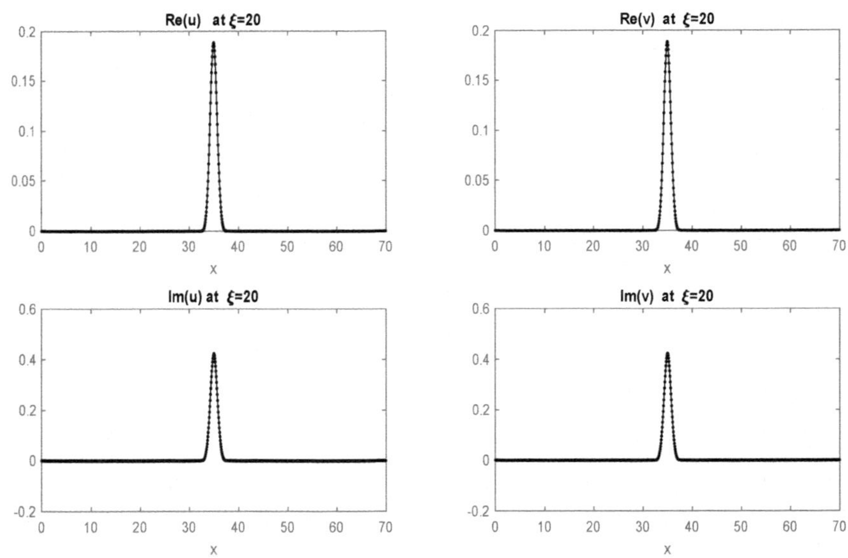

Fig. 4. Real and imaginary parts of the standing wave (u, v) of system (1)–(2) at $\xi = 20$. Solid line: numerical solution (u, v) of system (1)–(2) using the RK4 scheme (21). Pointed line: expected standing wave computed using (12.23).

and $p = 3, p_1 = 2$, $x_0 = 35$. Other parameters are left unchanged. The resulting profiles computed by Newton's iteration are displayed in Fig. 3. The verification that these profiles correspond in fact to an approximate standing wave solution (u, v) of the CNLS system is presented in Fig. 4.

Experiment 3: Finally, in Fig. 5 we display the profiles $\tilde{u}, \tilde{v}$ obtained by numerical simulation using as initial data for the Newton's iteration

$$\widetilde{u}(x) = \widetilde{v}(x) = -\text{sech}(x - 35)^2,$$

after 8 iterations. In this computer simulation we used the same parameters as in the experiment 2, except that $p = 1$ in the model's coefficients a, b, c, d, e.

We point out that in the numerical experiments presented, the model's coefficients $K(x), K_1(x), K_2(x)$, $\sigma_1(x), \sigma_2(x)$, $a(x), b(x), c(x), d(x), e(x)$ do not satisfy all the conditions required by the theoretical results developed in Section 3. This fact provides numerical evidence that the existence of standing wave solutions of the CNLS system is also possible when these technical assumptions are not satisfied.

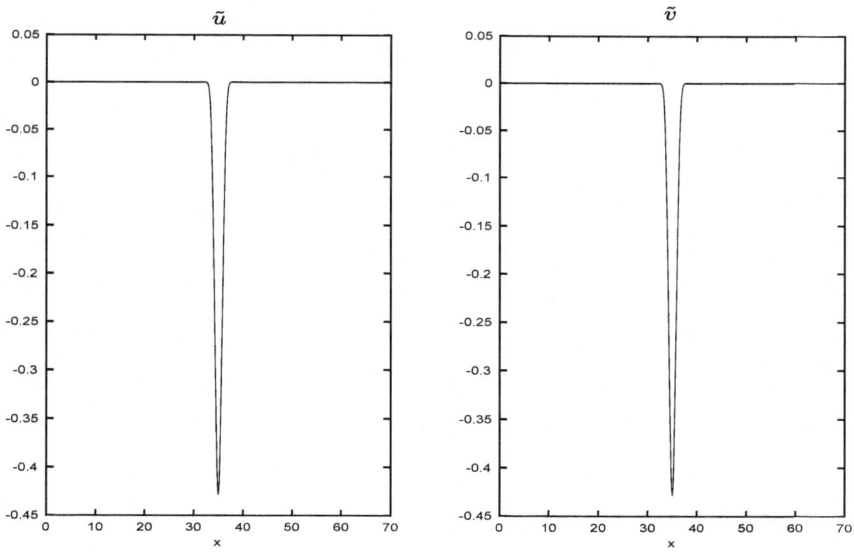

Fig. 5. Approximation of a solution $\widetilde{u}, \widetilde{v}$ to system (18)–(19).

5 Conclusions

In this paper, we established the existence of standing wave solutions of a system of two coupled Schödinger-type equations with variable coefficients (CNLS), which can be applied to model the interaction among propagating modes in a variety of physical applications, such as BECs and in nonlinear optics, for instance. To this end, we adapted the techniques used in [22] by applying the theory of fixed points of completely continuous operators defined in a cone in an appropriate Banach space. Furthermore, we introduced a numerical solver for approximating solutions of the travelling wave system (18)–(19) by using a Newton's iteration together with a collocation spectral strategy to discretize the corresponding travelling wave system. We also introduced a numerical scheme for computing the temporal evolution of solutions of the CNLS system, where the time stepping is performed by using a 4th-order Runge-Kutta strategy and a Fourier-spectral approximation for the spatial variable. This scheme was employed to verify that the profiles computed correspond in fact to approximate standing wave solutions of the CNLS system. The reliability of the numerical solvers introduced for approximation of stationary and time-dependent solutions was illustrated with some numerical simulations with synthetic localized model's coefficients.

Acknowledgments. This research was supported by Colciencias and Universidad del Valle, Calle 13 No. 100-00, Cali-Colombia, under the research project 1106-712-50006.

References

1. Muñoz Grajales, J.C.: Vector solitons of a coupled Schrödinger system with variable coefficients, Adv. Math. Phys. **2016**, Article ID 5787508, 19 (2016)
2. Kivshar, YuS, Agrawal, G.P.: Optical Solitons: From Fibers to Photonic Crystals. Academic Press, San Diego (2003)
3. Sulem, C., Sulem, P.: The Nonlinear Schrödinger Equation: Self-Focusing and Wave Collapse. Springer, Berlin (1999)
4. Malomed, B.A., Mihalache, D., Wise, F., Torner, L.: Spatio-temporal optical solitons. J. Opt. B Quantum Semiclassical Opt. **7**, R53 (2005)
5. Serkin, V.N., Hasegawa, A.: Novel soliton solutions of the nonlinear Schrödinger equation model. Phys. Rev. Lett. **85**, 4502–5 (2000)
6. Serkin, V.N., Hasegawa, A., Belyaeva, T.L.: Nonautonomous solitons in external potentials. Phys. Rev. Lett. **98**, 074102 (2007)
7. Salerno, M., Konotop, V.V., Bludov,Yu.V.: Long-living Bloch oscillations of matter waves in periodic potentials. Phys. Rev. Lett. **101**, 030405 (2008)
8. Ostrovskaya, E.A., Kivshar, Y.S., Lisak, M., Hall, B., Cattani, F.: Coupled-mode theory for Bose-Einstein condensates. Phys. Rev. A **61**, 031601(R) (2000)
9. Yang, Q., Zhang, J.F.: Bose-Einstein solitons in time-dependent linear potential. Opt. Commun. **258**, 35–42 (2006)
10. Radha, R., Vinayagam, P.S., Sudharsan, J.B., Malomed, B.A.: Persistent bright solitons in sign-indefinite coupled nonlinear Schrödinger equations with a time-dependent harmonic trap. Commun. Nonlinear Sci. Numer. Simul. **31**(1–3), 30–39 (2016)
11. Ieda, J., Miyakawa, T., Wadati, M.: Exact analysis of soliton dynamics in spinor Bose-Einstein condensates. Phys. Rev. Lett. **93**, 194102 (2004)
12. Feijoo, D., Paredes, A., Michinel, H.: Outcoupling vector solitons from a Bose-Einstein condensate with time-dependent interatomic forces. Phys. Rev. A **87**, 063619 (2013)

13. Yan, Z., Chow, K.W., Malomed, B.A.: Exact stationary wave patterns in three coupled nonlinear Schrödinger/Gross-Pitaevskii equations. Chaos, Solitons and Fractals **42**, 3013–3019 (2009)
14. Ho, T.L.: Spinor Bose condensates in optical traps. Phys. Rev. Lett. **81**, 742 (1998)
15. Law, C.K., Pu, H., Bigelow, N.P.: Quantum spins mixing in spinor Bose-Einstein condensates. Phys. Rev. Lett. **81**, 5257 (1998)
16. Ohmi, T., Machida, K.J.: Bose-Einstein condensation with internal degrees of freedom in alkali atom gases. J. Phys. Soc. Jpn. **67**, 1822 (1998)
17. Ieda, J., Miyakawa, T., Wadati, M.: Matter-wave solitons in an $F = 1$ spinor Bose-Einstein condensate. J. Phys. Soc. Jpn. **73**, 2996 (2004)
18. Wadati, M., Tsuchida, N.: Wave propagations in the $F = 1$ spinor Bose-Einstein condensates. J. Phys. Soc. Jpn. **75**, 014301 (2006)
19. Uchiyama, M., Ieda, J., Wadati, M.: Dark solitons in $F = 1$ spinor Bose-Einstein condensate. J. Phys. Soc. Jpn. **75**, 064002 (2006)
20. Agrawal, G.P.: Nonlinear Fiber Optics. Academic Press (2001)
21. Belmonte-Beitia, J., Pérez-García, V.M., Brazhnyi, V.: Solitary waves in coupled nonlinear Schrödinger equations with spatially inhomogeneous nonlinearities. Commun. Nonlinear Sci. Numer. Simul. **16**(1), 158–172 (2011)
22. Belmonte-Beitia, J., Pérez-García, V.M., Torres, P.J.: Solitary waves for linearly coupled nonlinear Schrödinger equations with inhomogeneous coefficients. J. Nonlinear Sci. **19**(4), 437–451 (2009)
23. Chu, J., O'Regan, D., Zhang, M.: Positive solutions and eigenvalue intervals for nonlinear systems. Proc. Indian Acad. Sci. (Math. Sci.) **117**(1), 85–94 (2003)
24. Jiang, D., Wei, J., Zhang, B.: Positive periodic solutions of functional differential equations and population models. Electron. J. Differ. Equ. **71**(1–13) (2002)
25. Stuart, C.A.: Guidance properties of nonlinear planar waveguides. Arch. Ration. Mech. Anal. **125**, 145–200 (1993)
26. Torres, P.J.: Guided waves in a multi-layered optical structure. Nonlinearity **19**, 2103–2113 (2006)
27. Zima, M.: On positive solutions of boundary value problems on the half-line. J. Math. Anal. Appl. **259**, 127–136 (2001)
28. Jing Su, J., Gao, Y.: Dark solitons for a (2+1)-dimension coupled nonlinear Schrödinger system with time-dependent coefficients in an optical fiber. Superlattices Microstruct. **104**, 498–508 (2017)
29. Han, L., Huang, Y., Liu, H.: Solitons in coupled nonlinear Schrödinger equations with variable coefficients. Commun. Nonlinear Sci. Numer. Simul. **19**(9), 3063–3073 (2014)
30. Wang, L., Zhang, L., Zhu, Y., Qi, F., Wang, P., Guo, R., Li, M.: Modulational instability, nonautonomous characteristics and semirational solutions for the coupled nonlinear Schrödinger equations in inhomogeneous fibers. Commun. Nonlinear Sci. Numer. Simul. **40**, 216–237 (2017)
31. Bose, S.N.: Plank's law and light quantum hypothesis. Z. Phys. **26**, 178–181 (1924)
32. Anderson, M.H., Ensher, J.R., Mattews, M.R., Weiman, C.E., Cornell, E.A.: Observation of Bose-Einstein condensation in a dilute atomic vapor. Science **269**, 198–201 (1995)
33. Theis, M., Thalhammer, G., Winkler, K., Hellwing, M., Ruff, G., Grimm, R., Hecker Denschlag, J.: Tuning the scattering length with an optically induced Feshbach resonance. Phys. Rev. Lett. **93**, 123001 (2004)
34. Rodas-Verde, M.I., Michinel, H., Pérez-García, V.M.: Controllable soliton emission from a Bose-Einstein condensate. Phys. Rev. Lett. **95**(15) 153903 (2005)

35. Teocharis, G., Schmelcher, P., Kevrekidis, P.G., Frantzeskakis, D.J.: Matter-wave solitons of collisionally inhomogeneous condensates. Phys. Rev. A **72**, 033614 (2005)
36. Primatorowa, M.T., Stoychev, K.T., Kamburova, R.S.: Interactions of solitons with extended nonlinear defects. Phys. Rev. E **72**, 036608 (2005)
37. Abdullaev, F.K., Garnier, J.: Propagation of matter-wave solitons in periodic and random nonlinear potentials. Phys. Rev. A **72**, 061605 (2005)
38. Garnier, J., Abdullaev, F.K.: Transmission of matter-wave solitons through nonlinear traps and barriers. Phys. Rev. A **74**, 013604 (2006)
39. Krasnosel'skii, M.A.: Positive Solutions of Operator Equations. P. Noordhoff, Groningen (1964)

Shannon Sampling and Weak Weyl's Law on Compact Riemannian Manifolds

Isaac Z. Pesenson(✉)

Department of Mathematics, Temple University, Philadelphia PA, 19122, USA
pesenson@temple.edu

Abstract. The well known Weyl's asymptotic formula gives an approximation to the number $\mathcal{N}_\omega$ of eigenvalues (counted with multiplicities) on an interval $[0, \omega]$ of an elliptic second-order differential self-adjoint non-negative operator on a compact Riemannian manifold $\mathbf{M}$. In this paper we approach this question from the point of view of Shannon-type sampling on compact Riemannian manifolds. Namely, we give a direct proof that $\mathcal{N}_\omega$ is comparable to cardinality of certain sampling sets for the subspace of ω-bandlimited functions on $\mathbf{M}$.

Keywords: Compact riemannian manifolds · Spectral geometry · Weyl's law

1 Introduction

1.1 Objectives

Spectral geometry concerned with questions which relate spectral properties of operators acting in function spaces on a Riemannian manifold and the geometry of the underlying manifold. One of the most famous results of such kind is the Weyl's asymptotic formula for the number of eigenvalues of an elliptic (pseudo-)differential operator on a compact Riemannian manifold. The goal of this paper is to demonstrate that in the case of a general compact Riemannian manifold the so-called weak Weyl's formula closely relates to cardinality of certain sampling sets for bandlimited functions. This fact was first noticed in [4].

1.2 Weyl's Asymptotic Formula on Compact Riemannian Manifolds

Let $\mathbf{M}$ be a compact connected Riemannian manifold without boundary and L is an elliptic second order differential operator which self-adjoint and non-negative in the corresponding $L_2(\mathbf{M})$ constructed with respect to Riemannian

J. Delgado and M. Ruzhansky (Eds.): *Analysis and Partial Differential Equations: Perspectives from Developing Countries*, PROMS 275, pp. 207–218, 2018.
https://doi.org/10.1007/978-3-030-05657-5_13

measure. One of the best known examples of such operators is the Laplace-Beltrami operator Δ. It is given in a local coordinate system by the formula [8]

$$\Delta f = \sum_{m,k} \frac{1}{\sqrt{det(g_{ij})}} \partial_m \left(\sqrt{det(g_{ij})} g^{mk} \partial_k f \right) \quad (1)$$

where g_{ij} are components of the metric tensor, $det(g_{ij})$ is the determinant of the matrix (g_{ij}), g^{mk} components of the matrix inverse to (g_{ij}). The operator is second-order differential self-adjoint and non-negative in the space $L_2(\mathbf{M})$ constructed with respect to Riemannian measure. Domains of the powers $L^{s/2}$, $s \in \mathbf{R}$, coincide with the Sobolev spaces $H^s(\mathbf{M})$, $s \in \mathbf{R}$. Since L is a second-order differential self-adjoint and non-negative definite operator on a compact connected Riemannian manifold it has a discrete spectrum $0 = \lambda_0 < \lambda_1 \leq \lambda_2, \ldots$ which goes to infinity without any accumulation points and there exists a complete family $\{u_j\}$ of orthonormal eigenfunctions which form a basis in $L_2(\mathbf{M})$ [7]. We will need the following definitions.

Definition 1. The space of ω-bandlimited functions $\mathbf{E}_\omega(L)$ is defined as the span of all eigenfunctions of L whose eigenvalues are not greater than ω. The dimension of the subspace $\mathbf{E}_\omega(L)$ will be denoted as $\mathcal{N}_\omega$.

One can easily verify that f belongs to $\mathbf{E}_\omega(L)$ if and only if the following Bernstein type inequality holds

$$\|L^k f\|_{L_2(M)} \leq \omega^k \|f\|_{L_2(M)}$$

for all natural k.

For a sufficiently small $\rho > 0$ we consider a finite set of points $M_\rho = \{x_j\}$, $x_j \in \mathbf{M}$, $\rho > 0$, such that

1. balls $B(x_j, \rho/2)$ are disjoint

$$B(x_j, \rho/2) \cap B(x_i, \rho/2) = \emptyset, \quad j \neq i,$$

2. balls $B(x_\nu, \rho)$ form a cover of $\mathbf{M}$.

It is shown in Lemma 1 below that for every compact Riemannian manifold there exists a constant $N_\mathbf{M}$ such that for every sufficiently small $\rho > 0$ one can construct a set $M_\rho = \{x_j\}$, $x_j \in \mathbf{M}$, $\rho > 0$, which in addition to properties (1)–(2) will also have the following property:

(3) multiplicity of the cover of $\mathbf{M}$ by balls $B(x_j, \rho)$ is not greater $N_\mathbf{M}$.

Definition 2. A set of points $M_\rho = \{x_j\}$, $x_j \in \mathbf{M}$, $\rho > 0$, which has properties (1)–(3) will be called a metric ρ-lattice.

According to the Weyl's asymptotic formula [7] one has for large ω

$$\mathcal{N}_\omega \sim A\,Vol(\mathbf{M})\omega^{d/2}, \quad (2)$$

where $d = dim\,\mathbf{M}$ and A is a constant which is independent on $\mathbf{M}$. To reveal meaning of the right-hand side of this formula let's rewrite it in the following form

$$\mathcal{N}_\omega \sim A\,Vol(\mathbf{M})\omega^{d/2} = A\frac{Vol(\mathbf{M})}{\left(\omega^{-1/2}\right)^d}. \tag{3}$$

Since in the case of a Riemannian manifold $\mathbf{M}$ of dimension n all the balls of the same radius ρ have essentially the same volume $\sim \rho^d$ the last fraction can be interpreted as a number of balls $B(x_\nu, \omega^{-1/2})$ whose centers $\{x_\nu\}$ form a lattice $M_{\omega^{-1/2}}$.

The main goal of our paper is to present a direct proof of the following Theorem 1 *(which we call the Weak Weyl's Law) without using the Weyl's asymptotic formula* (2).

Theorem 1. *(Weak Weyl's Law) In the case of a Riemannian manifold the number $\mathcal{N}_\omega$ of eigenvalues of L in $[0, \omega]$ counting with their multiplicities is equivalent to a number of points in a metric lattice $M_{\omega^{-1/2}}$. Namely, there are constants*

$$0 < a = a(\mathbf{M}) < 1, \quad 0 < \gamma = \gamma(\mathbf{M}) < 1,$$

such that for all sufficiently large ω the following double inequality holds

$$a \sup |M_{\omega^{-1/2}}| \leq \mathcal{N}_\omega \leq \inf |M_{\gamma\omega^{-1/2}}|, \tag{4}$$

where sup *is taken over all $\omega^{-1/2}$-lattices,* inf *is taken over all $\gamma\omega^{-1/2}$-lattices and $|M_s|$ denotes cardinality of a lattice.*

2 Covering Lemma

We consider a compact Riemannian manifold $\mathbf{M}, \dim \mathbf{M} = d$, with metric tensor g. It is known that the Laplace-Beltrami operator L which is defined in (1) is a self-adjoint positive definite operator in the corresponding space $L_2(\mathbf{M})$ constructed from g. Domains of the powers $L^{s/2}, s \in \mathbf{R}$, coincide with the Sobolev spaces $H^s(\mathbf{M}), s \in \mathbf{R}$. To choose norms on spaces $H^s(\mathbf{M})$, we consider a finite cover of $\mathbf{M}$ by balls $B(y_\nu, \sigma)$ where $y_\nu \in \mathbf{M}$ is the center of the ball and σ is its radius. For a partition of unity φ_ν subordinate to the family $\{B(y_\nu, \sigma)\}$ we introduce Sobolev space $H^s(\mathbf{M})$ as the completion of $C_0^\infty(\mathbf{M})$ with respect to the norm

$$\|f\|_{H^s(\mathbf{M})} = \left(\sum_\nu \|\varphi_\nu f\|^2_{H^s(B(y_\nu,\sigma))}\right)^{1/2}. \tag{5}$$

The regularity Theorem for the Laplace-Beltrami operator L states that the norm (1.1) is equivalent to the graph norm $\|f\| + \|L^{s/2}f\|$.

The volume of a ball $B(x, \rho)$ will be denoted by $|B(x, \rho)|$. Let us note that in the case of a compact Riemannian manifold of dimension d there exist constants

$a_1 = a_1(\mathbf{M}), a_2 = a_2(\mathbf{M})$ such that for a ball $B(x, \rho)$ of sufficiently small radius ρ and any center $x \in \mathbf{M}$ one has

$$a_1 \rho^d \leq |B(x, \rho)| \leq a_2 \rho^d, \tag{6}$$

where

$$|B(x, \rho)| = \int_{B(x,\rho)} dx, \;\; d = dim\, \mathbf{M}.$$

The inequality (6) implies the next inequality with the same a_1 and a_2:

$$\frac{a_1}{a_2}|B(x_2, \rho)| \leq |B(x_1, \rho)| \leq \frac{a_2}{a_1}|B(x_2, \rho)|, \;\; \rho < r, \tag{7}$$

where x_1, x_2 are any two points in $\mathbf{M}$ and r is the injectivity radius of the manifold. Since $\mathbf{M}$ is compact there exists a constant $c = c(\mathbf{M})$ such that for any $0 < \sigma < \lambda < r/2$ the following inequality holds true

$$|B(x, \lambda)| \leq (\lambda/\sigma)^d c|B(x, \sigma)|. \tag{8}$$

In what follows we will use the notation

$$N_{\mathbf{M}} = \frac{12^d c a_2}{a_1}.$$

The following Covering Lemma plays an important role for the paper.

Lemma 1. *If* $\mathbf{M}$ *is a compact manifold with the injectivity radius* $r > 0$ *then for any* $0 < \rho < r/6$ *there exists a finite set of points* $\{x_i\}$ *such that*

1) balls $B(x_i, \rho/4)$ *are disjoint,*
2) balls $B(x_i, \rho/2)$ *form a cover of* $\mathbf{M}$*,*
3) multiplicity of the cover by balls $B(x_i, \rho)$ *is not greater* $N_{\mathbf{M}}$*.*

Proof. Let us choose a family of disjoint balls $B(x_i, \rho/4)$ such that there is no ball $B(x, \rho/4), x \in \mathbf{M}$, which has empty intersections with all balls from our family. Then the family $B(x_i, \rho/2)$ is a cover of $\mathbf{M}$. Every ball from the family $\{B(x_i, \rho)\}$, that has non-empty intersection with a particular ball $\{B(x_j, \rho)\}$ is contained in the ball $\{B(x_j, 3\rho)\}$. Since any two balls from the family $B(x_i, \rho/4)$ are disjoint, it gives the following estimate for the index of multiplicity N of the cover $B(x_i, \rho)$:

$$N \leq \frac{\sup_{y \in \mathbf{M}} |B(y, 3\rho)|}{\inf_{x \in \mathbf{M}} |B(x, \rho/4)|}. \tag{9}$$

From here, according to (8) we obtain

$$N \leq \frac{\sup_{y \in \mathbf{M}} |B(y, 3\rho)|}{\inf_{x \in \mathbf{M}} |B(x, \rho/4)|} \leq 12^d c \frac{\sup_{y \in \mathbf{M}} |B(y, \rho/4)|}{\inf_{x \in \mathbf{M}} |B(x, \rho/4)|} \leq \frac{12^d c a_2}{a_1} = N_{\mathbf{M}}.$$

3 Sampling Sets for Bandlimited Functions and the Upper Estimate on the Number of Eigenvalues

3.1 Poincare-Type Inequality on Manifolds

One can prove the following Poincare type inequality (see [3], [5]). We sketch it's proof for completeness.

Theorem 2. *There exists a constant* $C = C(\mathbf{M}, k)$ *such that if* $\rho > 0$ *is sufficiently small then for all* ρ *lattices* $M_\rho = \{x_j\}$ *and all* $f \in H^k(\mathbf{M})$, $k > d/2$, $d = \dim \mathbf{M}$,

$$\|f\|_{L_2(\mathbf{M})} \leq C(\mathbf{M}, k) \left\{ \rho^{d/2} \left(\sum_{x_j \in M_\rho} |f(x_j)|^2 \right)^{1/2} + \rho^k \|L^{k/2} f\|_{L_2(M)} \right\}. \quad (10)$$

Proof. Let $M_\rho = \{x_i\}$ be a ρ-admissible set and $\{\varphi_\nu\}$ the partition of unity from (1.1). For any $f \in C^\infty(M)$, every fixed $B(x_i, \rho)$ and every $x \in B(x_i, \rho/2)$

$$(\varphi_\nu f)(x) = (\varphi_\nu f)(x_i) + \sum_{1 \leq |\alpha| \leq n-1} \frac{1}{\alpha!} \partial^\alpha (\varphi_\nu f)(x_i)(x - x_i)^\alpha +$$

$$\sum_{|\alpha|=n} \frac{1}{(n-1)!} \int_0^\tau t^{n-1} \partial^\alpha (\varphi_\nu f)(x_i + t\vartheta) \vartheta^\alpha dt, \quad (11)$$

where $x = (x_1, ..., x_d), x_i = (x_1^i, ..., x_d^i), \alpha = (\alpha_1, ..., \alpha_d), x - x_i = (x_1 - x_1^i)^{\alpha_1} ... (x_d - x_d^i)^{\alpha_d}, \tau = \|x - x_i\|, \vartheta = (x - x_i)/\tau$. By using the Sobolev embedding Theorem one can prove the following inequality

$$|\partial^\alpha (\varphi_\nu f)(x_i)| \leq C_{\mathbf{M},m} \sum_{|\mu| \leq m} \rho^{|\mu+\alpha|-d/2} \|\partial^{\mu+\alpha} (\varphi_\nu f)\|_{L_2(B(x_i,\rho))}, \quad (12)$$

where $\mu = (\mu_1, \mu_2, ..., \mu_d), m > d/2$. It allows the following estimation of the second term in (11).

$$\int_{B(x_i,\rho/2)} \left| \sum_{1 \leq |\alpha| \leq n-1} \frac{1}{\alpha!} \partial^\alpha (\varphi_\nu f)(x_i)(x - x_i)^\alpha \right|^2 dx \leq$$

$$C_{\mathbf{M},n} \sum_{|\gamma| \leq n+m-1} \rho^{2|\gamma|} \|\partial^\gamma (\varphi_\nu f)\|^2_{L_2(B(x_i,\rho))}.$$

Next, to estimate the third term in (11) we use the Schwartz inequality and the assumption $n > d/2$

$$\left| \int_0^\tau t^{n-1} \partial^\alpha (\varphi_\nu f)(x_i + t\vartheta) \vartheta^\alpha dt \right|^2 \leq$$

$$\left(\int_0^{\tau} t^{n-d/2-1/2}\left|t^{d/2-1/2}\partial^{\alpha}(\varphi_{\nu}f)(x_i+t\vartheta)\right|dt\right)^2 \leq$$

$$C_{\mathbf{M},n}\tau^{2n-d}\int_0^{\tau} t^{d-1}\left|\partial^{\alpha}(\varphi_{\nu}f)(x_i+t\vartheta)\right|^2 dt.$$

We integrate both sides of this inequality over the ball $B(x_i,\rho/2)$ using the spherical coordinate system (τ,ϑ).

$$\int_0^{\rho/2}\tau^{d-1}\int_{|\vartheta|=1}\left|\int_0^{\tau} t^{n-1}\partial^{\alpha}(\varphi_{\nu}f)(x_i+t\vartheta)\vartheta^{\alpha}dt\right|^2 d\vartheta d\tau \leq$$

$$C_{\mathbf{M},n}\int_0^{\rho/2} t^{d-1}\left(\int_{|\vartheta|=1}\int_0^{\rho/2}\tau^{2n-d}\left|\partial^{\alpha}(\varphi_{\nu}f)(x_i+t\vartheta)\right|^2\tau^{d-1}d\tau d\vartheta\right)dt \leq$$

$$C_{\mathbf{M},n}\rho^{2n}\|\partial^{\alpha}(\varphi_{\nu}f)\|^2_{L_2(B(x_i,\rho))},\quad \tau=\|x-x_i\|\leq\rho/2, |\alpha|=n.$$

Next, for $n>d/2$ and $k=n+m-1$,

$$\|\varphi_{\nu}f\|^2_{L_2(B(x_i,\rho/2))}\leq C_1(\mathbf{M},k)\left(\rho^d|f(x_i|^2+\sum_{j=1}^{k}\sum_{1\leq|\alpha|\leq j}\rho^{2|\alpha|}\|\partial^{\alpha}(\varphi_{\nu}f)\|^2_{L_2(B(x_i,\rho))}\right),$$

where $k>d-1$ since $n>d/2$ and $m>d/2$. Since balls $B(x_i,\rho/2)$ cover the manifold and the cover by $B(x_i,\rho)$ has a finite multiplicity $\leq N_{\mathbf{M}}$ the summation over all balls gives

$$\|f\|^2_{L_2(\mathbf{M})}\leq C_2(\mathbf{M},k)\left\{\rho^d\left(\sum_{i=1}^{\infty}|f(x_i)|^2\right)+\sum_{j=1}^{k}\rho^{2j}\|f\|^2_{H^j(\mathbf{M})}\right\}, k>d-1.$$

Using this inequality and the regularity theorem for Laplace-Beltrami operator we obtain

$$\|f\|_{L_2(\mathbf{M})}\leq C_3(\mathbf{M},k)\left\{\rho^{d/2}\left(\sum_{i=1}^{\infty}|f(x_i)|^2\right)^{1/2}+\sum_{j=1}^{k}\rho^j\left(\|f\|+\|L^{j/2}f\|\right)\right\}, k>d-1.$$

For the self-adjoint L for any $a>0,\rho>0,0\leq j\leq k$ we have the following interpolation inequality

$$\rho^j\|L^{j/2}f\|\leq a^{2k-j}\rho^{2k}\|L^k f\|+c_k a^{-j}\|f\|.$$

Because in the last inequality we are free to choose any $a>0$ we are coming to our main claim.

3.2 Sampling Sets for Bandlimited Functions and the Upper Estimate on the Number of Eigenvalues

Now, if a bandlimited function f belongs to $\mathbf{E}_\omega(L)$ the Bernstein inequality implies

$$\rho^k \|L^{k/2} f\|_{L_2(\mathbf{M})} \leq \left(\rho\omega^{1/2}\right)^k \|f\|_{L_2(\mathbf{M})}.$$

If $C(\mathbf{M}, k)$ is the same as in (10) and we pick a such ρ for which

$$\rho = \gamma\omega^{-1/2}, \ \ \gamma = \gamma(\mathbf{M}) = \frac{(C(\mathbf{M}, k))^{1/k}}{2} < 1,$$

we can move the second term on the right side in (10) to the left to obtain following Plancherel-Polya-type inequality which shows that in the spaces of bandlimited functions $\mathbf{E}_\omega(L)$ the regular $L_2(\mathbf{M})$ norm is controlled by a discrete one (in fact, they are equivalent).

Theorem 3. *There exists a $0 < \gamma = \gamma(\mathbf{M}) < 1$ and there exists a constant $C_1 = C_1(\mathbf{M})$ such that for any $\omega > 0$, every metric ρ-lattice $M_\rho = \{x_j\}$ with $\rho = \gamma\omega^{-1/2}$ the following inequality holds true*

$$\|f\|_{L_2(\mathbf{M})} \leq C_1 \rho^{d/2} \left(\sum_{x_j \in M_\rho} |f(x_j)|^2 \right)^{1/2} \tag{13}$$

for all $f \in E_\omega(L)$, $d = dim\,\mathbf{M}$.

Corollary 1. *There exists a $0 < \gamma = \gamma(\mathbf{M}) < 1$ such that for every $\omega > 0$ and every metric ρ-lattice $M_\rho = \{x_j\}$ with $\rho = \gamma\omega^{-1/2}$ the set $M_\rho = \{x_j\}$ is a sampling set for the space $\mathbf{E}_\omega(L)$.*

In other words, every function $f \in \mathbf{E}_\omega(L)$ is uniquely determined by its values $\{f(x_j)\}$ and can be reconstructed from this set of values in a stable way.

Since dimension $\mathcal{N}_\omega$ of the space $\mathbf{E}_\omega(L)$ cannot be bigger than cardinality of a sampling set for this space we obtain the following statement.

Corollary 2. *There exists a $0 < \gamma = \gamma(\mathbf{M}) < 1$ such that for any $\omega > 0$*

$$\mathcal{N}_\omega \leq \ \inf \left| M_{\gamma\omega^{-1/2}} \right|, \tag{14}$$

where $\left| M_{\gamma\omega^{-1/2}} \right|$ is the number of points in a lattice $M_{\gamma\omega^{-1/2}}$ and inf *is taken over all such lattices.*

4 The Lower Estimate

4.1 Kernels on Compact Riemannian Manifolds

Let $\sqrt{L}$ be the positive square root of a second order differential elliptic self-adjoint nonnegative operator L in $L_2(\mathbf{M})$. For any measurable bounded function

$F(\lambda),\ \lambda \in (-\infty, \infty)$ and any $t > 0$ one defines a bounded operator $F(t\sqrt{L})$ by the formula

$$F(t\sqrt{L})f(x) = \int_{\mathbf{M}} K_t^F(x,y)f(y)dy = \left\langle K_t^F(x,\cdot), f(\cdot)\right\rangle, \tag{15}$$

where $f \in L_2(\mathbf{M})$ and

$$K_t^F(x,y) = \sum_{l=0}^{\infty} F(t\sqrt{\lambda_l})u_l(x)\overline{u_l(y)} = K_t^F(y,x). \tag{16}$$

The function K_t^F is known as the kernel of the operator $F(t\sqrt{L})$.

We will need the following lemma.

Lemma 2. *If* $0 \le F_1 \le F_2$ *and both of them are bounded and have sufficiently fast decay at infinity then* $K_t^{F_1}(x,x) \le K_t^{F_2}(x,x)$ *for any* $x \in \mathbf{M}$ *and* $t > 0$.

Proof. Assume that $0 \le F_1 \le F_2$ and that both of them are bounded and have bounded supports. Clearly, $F_2 = F_1 + H$, where H is not negative. By (16) we have

$$K_t^{F_2}(x,x) = K_t^{F_1}(x,x) + K_t^{H}(x,x)$$

where each term is non-negative. The lemma is proven.

We are going to make use of the heat kernel

$$p_t(x,y) = \sum_{l=0}^{\infty} e^{-t\lambda_l}u_l(x)\overline{u_l(y)},$$

which is associated with the heat semigroup e^{-tL} generated by the self-adjoint operator L:

$$e^{-tL}f(x) = \int_{\mathbf{M}} p_t(x,y)f(y)dy.$$

Note, that in notations (15), (16)

$$p_t(x,y) = K_t^F(x,y),\ \ F(\lambda) = e^{-\lambda^2},\ \ e^{-tL} = F(t\sqrt{L}).$$

It is well known that in the case of a compact Riemannian manifold this kernel obeys the following short-time Gaussian estimates:

$$C_1 t^{-d/2} e^{-c_1 \frac{(dist(x,y))^2}{t}} \le p_t(x,y) \le C_2 t^{-d/2} e^{-c_2 \frac{(dist(x,y))^2}{t}} \tag{17}$$

where $0 < t < 1,\ d = dim\, \mathbf{M}$ and every constant depends on $\mathbf{M}$.

4.2 The Lower Estimate

We now sketch the proof of the opposite estimate by comparing $|M_{\omega^{-1/2}}|$ to the number of eigenvalues (counted with multiplicities) in the interval $[0, \omega]$.

Inequalities (7) in conjunction with (17) it gives for $0 < t < 1$

$$a_1 C_1 |B(x, t^{-1/2})| \le p_t(x, x) = \sum_{l=0}^{\infty} e^{-t\lambda_l^2} |u_l(x)|^2 \le a_2 C_2 |B(x, t^{-1/2})|. \quad (18)$$

In the following Lemma we follow some ideas of [1].

Lemma 3. *There exist constants $A_1 = A_1(\mathbf{M}) > 0$, $A_2 = A_2(\mathbf{M}) > 0$ such that for all sufficiently large $s > 0$*

$$\frac{A_1}{|B(x, s^{-1})|} \le \sum_{l,\, \lambda_l \le s} |u_l(x)|^2 \le \frac{A_2}{|B(x, s^{-1})|}. \quad (19)$$

Proof. First, we note that using the right-hand side of (18), Lemma 2 and the inequality

$$\chi_{[0,\, s]}(\lambda) \le e e^{-s^{-2}\lambda^2}$$

we obtain

$$\sum_{l,\, \lambda_l \le s} |u_l(x)|^2 \le e \sum_{l,\, \lambda_l \le s} e^{-s^{-2}\lambda_l^2} |u_l(x)|^2 \le$$

$$e \sum_{l \in \mathbf{N}} e^{-s^{-2}\lambda_l^2} |u_l(x)|^2 = e p_{s^{-2}}(x, x) \le \frac{A_2}{|B(x, s^{-1})|}, \quad A_2 = A_2(\mathbf{M}). \quad (20)$$

To prove the left-had side of (19) consider the inequality

$$e^{-t\lambda^2} = e^{-t\lambda^2} \chi_{[0,\, s]} + \sum_{j \ge 0} \chi_{[2^j s,\, 2^{j+1} s]}(\lambda) e^{-t\lambda^2} \le$$

$$\chi_{[0,\, s]} + \sum_{j \ge 0} \chi_{[0,\, 2^{j+1} s]}(\lambda) e^{-t 2^{2j} s^2},$$

which implies

$$p_t(x, x) \le K_1^{\chi_{[0,s]}}(x, x) + \sum_{j > 0} e^{-t 2^{2j} s^2} K_1^j(x, x), \quad (21)$$

where

$$K_1^{\chi_{[0,s]}}(x, x) = \sum_{l,\, \lambda_l \le s} |u_l(x)|^2,$$

$K_1^{\chi_{[0,s]}}(x, y)$ being the kernel of the operator $\chi_{[0,s]}\left(\sqrt{L}\right)$ and

$$K_1^j(x, x) = \sum_{l,\, \lambda_l \le 2^{j+1} s} |u_l(x)|^2,$$

$K_1^j(x,y)$ being the kernel of the operator $\chi_{[0,\,2^{j+1}s]}(\sqrt{L})$. In conjunction with (17) it gives

$$c_3|B(x,t^{-1/2})| \le p_t(x,x) \le K_1^{\chi_{[0,s]}}(x,x) + \sum_{j>0} e^{-t2^{2j}s^2} K_1^j(x,x) =$$

$$\sum_{l,\,\lambda_l\le s} |u_l(x)|^2 + \sum_{j>0} e^{-2^{2j}ts^2} \sum_{l,\,\lambda_l \le 2^{j+1}s} |u_l(x)|^2. \tag{22}$$

Note, that according to (8) if $\rho>1$ and ρs^{-1} is sufficiently small then

$$|B(x,\rho s^{-1}| \le c\rho^d |B(x,s^{-1})|, \;\; d = dim\,\mathbf{M}. \tag{23}$$

Next, given $s \ge 1$ and $m \in \mathbf{N}$ we pick t such that

$$s\sqrt{t} = 2^m. \tag{24}$$

The inequality (23) and the condition (24) imply

$$\frac{(c2^{md})^{-1}}{|B(x,s^{-1})|} \le \frac{1}{|B(x,2^m s^{-1})|} \le c_1|B(x,t^{-1/2})|, \;\; m\in\mathbf{N}, \tag{25}$$

and

$$\frac{1}{|B(x,2^{-m-1}s^{-1})|} \le \frac{c2^{(m+1)d}}{|B(x,s^{-1})|}, \;\; d = dim\,\mathbf{M}. \tag{26}$$

Thus according to (25), (20) and (22) we obtain that for a certain constant $c_2 = c_2(\mathbf{M})$

$$\frac{(c2^{md})^{-1}}{|B(x,s^{-1})|} \le c_1|B(x,t^{-1/2})| \le$$

$$c_2\left(\sum_{l,\,\lambda_l\le s} |u_l(x)|^2 + \sum_{j>0} e^{-2^{2j}ts^2} \sum_{l,\,\lambda_l\le 2^{j+1}s} |u_l(x)|^2\right) \le$$

$$c_2\left(\sum_{l,\,\lambda_l\le s} |u_l(x)|^2 + \sum_{j>0} \frac{e^{-2^{2j}ts^2}}{|B(x,2^{-j-1}s^{-1})|}\right). \tag{27}$$

Using (27), (24), and (26) we obtain that for a certain constant $a = a(\mathbf{M})$

$$\frac{(c2^{md})^{-1}}{|B(x,s^{-1})|} \le a\left(\sum_{l,\,\lambda_l\le s} |u_l(x)|^2 + \sum_{j>0} \frac{e^{-2^{2j}2^{2m}} 2^{(j+1)d}}{|B(x,s^{-1})|}\right) \le$$

$$c_2\left(\sum_{l,\,\lambda_l\le s} |u_l(x)|^2 + \frac{2^d}{|B(x,s^{-1})|} \sum_{j>0} e^{-2^{2j}2^{2m}} 2^{jd}\right).$$

Since

$$\frac{2^d}{|B(x,s^{-1})|}\sum_{j>0} e^{-2^{2j}2^{2m}}2^{jd} \leq \frac{2^d 2^{-md}}{|B(x,s^{-1})|}\sum_{j>0} e^{-2^{2j}2^{2m}}2^{(j+m)d} \leq$$

$$\frac{2^d 2^{-md}}{|B(x,s^{-1})|}\sum_{j>0} e^{-2^{2(j+m)}}2^{(j+m)d} \leq \frac{2^d 2^{-md}}{|B(x,s^{-1})|}\sum_{j>m} e^{-2^{2j}}2^{jd},$$

one has that there are positive constants c_3, c_4 such that for all sufficiently large s and $m \in \mathbf{N}$

$$\frac{2^{-md}}{|B(x,s^{-1})|}\left(c_3 - c_4 2^d \sum_{j\geq m} e^{-2^{2j}2^{jd}}\right) \leq \sum_{l,\,\lambda_l\leq s} |u_l(x)|^2,$$

where expression in parentheses is positive for sufficiently large $m \in \mathbf{N}$. It proves the left-had side of (19).

We apply this lemma when $t = \omega$ to obtain the following inequality for sufficiently large ω:

$$\frac{1}{|B(x,\omega^{-1/2})|} \leq c_5 p_\omega(x,x) \leq c_6 \sum_{l,\,\lambda_l\leq\omega} |u_l(x)|^2. \tag{28}$$

One has

$$|M_{\omega^{-1/2}}| = \sum_{x_j\in M_{\omega^{-1/2}}} \frac{|B(x_j,\omega^{-1/2})|}{|B(x_j,\omega^{-1/2})|} = \sum_{x_j\in M_{\omega^{-1/2}}} \frac{1}{|B(x_j,\omega^{-1/2})|}\int_{B(x_j,\omega^{-1/2})} dx,$$

and thanks to (7) we also have

$$\frac{1}{|B(x_j,\omega^{-1/2})|}\int_{B(x_j,\omega^{-1/2})} dx \leq \frac{a_2}{a_1}\int_{B(x_j,\omega^{-1/2})} \frac{dx}{|B(x,\omega^{-1/2})|}.$$

Now, the inequality (28) shows that for every sufficiently large $\omega > 0$ and every $\omega^{-1/2}$-lattice $M_{\omega^{-1/2}}$ the following inequalities hold true

$$|M_{\omega^{-1/2}}| \leq \frac{a_2}{a_1}\sum_{x_j\in M_{\omega^{-1/2}}}\int_{B(x_j,\omega^{-1/2})} \frac{dx}{|B(x,\omega^{-1/2})|} \leq$$

$$\frac{a_2}{a_1}\int_{\mathbf{M}} \frac{dx}{|B(x,\omega^{-1/2})|} \leq \frac{a_2}{a_1}c_6 \int_{\mathbf{M}}\left(\sum_{l,\,\lambda_l\leq\omega}|u_l(x)|^2\right)dx.$$

Since

$$\int_{\mathbf{M}}\left(\sum_{l,\,\lambda_l\leq\omega}|u_l(x)|^2\right)dx = \sum_{l,\,\lambda_l\leq\omega}\int_{\mathbf{M}}|u_l(x)|^2 dx = \mathcal{N}_\omega. \tag{29}$$

we receive the inequality

$$|M_{\omega^{-1/2}}| \leq c_7 \mathcal{N}_\omega,$$

for a certain $c_7 = c_7(\mathbf{M}) > 0$. Thus there exists an $a = a(\mathbf{M}) > 0$ such that

$$a \sup |M_{\omega^{-1/2}}| \leq \mathcal{N}_\omega, \tag{30}$$

where $|M_{\omega^{-1/2}}|$ is the number of points in a lattice $M_{\omega^{-1/2}}$ and sup is taken over all such lattices. The inequalities (14) and (30) show existence of constants

$$0 < a = a(\mathbf{M}) < 1, \quad 0 < \gamma = \gamma(\mathbf{M}) < 1,$$

such that for all sufficiently large ω the following double inequality holds

$$a \sup |M_{\omega^{-1/2}}| \leq \mathcal{N}_\omega \leq \inf |M_{\gamma\omega^{-1/2}}|, \tag{31}$$

where sup is taken over all $\omega^{-1/2}$-lattices and inf is taken over all $\gamma\omega^{-1/2}$-lattices and $|M_s|$ denotes cardinality of a lattice. Theorem 1 is proven.

References

1. Coulhon, T., Kerkyacharian, G., Petrushev, P.: Heat kernel generated frames in the setting of Dirichlet spaces. J. Fourier Anal. Appl. **18**(5), 995–1066 (2012)
2. Geller, D., Pesenson, I.: Band-limited localized Parseval frames and Besov spaces on compact homogeneous manifolds. J. Geom. Anal. **21**(2), 334–371 (2011)
3. Pesenson, I.: A sampling theorem on homogeneous manifolds. Trans. Amer. Math. Soc. **352**(9), 4257–4269 (2000)
4. Pesenson, I.: An approach to spectral problems on Riemannian manifolds. Pacific J. Math. **215**(1), 183–199 (2004)
5. Pesenson, I.: Poincare-type inequalities and reconstruction of Paley-Wiener functions on manifolds. J. Geom. Anal. **4**(1), 101–121 (2004)
6. Pesenson, I.Z.: Paley-Wiener approximations and multiscale approximations in Sobolev and Besov spaces on manifolds. J. Geom. Anal. **4**(1), 101–121 (2009)
7. Sogge, C.: Fourier Integrals in Classical Analysis. Cambridge University Press (1993)
8. Rosenberg, S.: The Laplacian on a Riemannian Manifold. Cambridge University Press (1997)

Well-posed Boundary Value Problems for New Classes of Singular Integral Equations in Cylindrical Domains

Nusrat Rajabov(✉)

Research Institute of the Tajik National University,
Rudaki Avenue 17, Dushanbe, Tajikistan
nusrat38@mail.ru

Abstract. In this work a class of three-dimensional complex integral equation in cylindrical domains is investigated in the case when the lateral surface may have singularity or super-singularity. For this type of integral equations condition for kernels are found under which the problem of finding solution is reduced to the problem of finding two splitting systems of integral equations which can be treated by existing methods. In this case the solution are obtained in an explicit form. In the case of more general kernels the, inversion formula is found in terms of the values on the surface of the cylinder. In model cases the solution of the integral equation is found in the form of absolutely and uniformly convergent generalised power series in powers of $(t-a)$ and the inversion formula is presented. It is used to investigate further Dirichlet-type boundary problems.

Keywords: Boundary value problems · Integral equations
Well-posedness problems

1 Introduction

Let Ω denote the cylindrical domain

$$\Omega = \{(z,t)\,;a<t<b,\ |z|<R\}.$$

The base of this cylinder will be denoted by

$$D = \{t=a,\ |z|<R\}$$

and the lateral surface will be denoted by

$$S = \{a<t<b,\quad |z|=R\},\ z=x+iy.$$

J. Delgado and M. Ruzhansky (Eds.): *Analysis and Partial Differential Equations: Perspectives from Developing Countries*, PROMS 275, pp. 219–241, 2018.
https://doi.org/10.1007/978-3-030-05657-5_14

We consider the integral equation in the domain Ω of the form

$$\varphi(t,z) + \int_a^t \frac{A(\tau)\varphi(\tau,z)}{\tau-a}d\tau - \frac{\lambda}{\pi}\iint_D \frac{\exp[i\theta]\varphi(t,s)\,ds}{(R-\rho)^{\beta}(s-z)} \tag{1}$$

$$-\frac{1}{\pi}\int_a^t \frac{d\tau}{\tau-a}\iint_D \frac{\exp[i\theta]B(\tau,s)\,ds}{(R-\rho)^{\beta}(s-z)} = f(t,z), \tag{2}$$

where $A(t)$, $B(t,z)$, $f(t,z)$ are given real-valued functions in the domains S and Ω, $\theta = \arg s$, $s = \xi + i\eta$, $ds = d\xi d\eta$, $\varphi(t,z)$ is an unknown function, $\rho^2 = \xi^2 + \eta^2$, $\lambda = const$, $\beta = const > 1$.

The solution of this equation is sought in the class of functions

$$\varphi(t,z) \in C(\overline{\Omega}),\ \varphi(a,z) = 0,\ \varphi\left(t, Re^{i,\theta}\right) = 0,\ \theta = \arg z,$$

and its asymptotic behaviour for $t \to a$ and R$\to r$ is given by the formulae

$$\varphi(t,z) = o\left[(t-a)^{\varepsilon}\right], \quad \varepsilon > 0 \quad \text{as} \quad t \to a, \tag{3}$$

and

$$\varphi(t,z) = 0\left[(R-r)^{\delta_1}\right], \quad \delta_1 > \beta - 1 \ \text{as} \ r \to R. \tag{4}$$

Assume that in the integral equation (1) we have

$$B(t,z) = \lambda A(t).$$

Then the problem of finding the solution of the integral equation (1) is equivalent to the problem of finding the solution of the following splitted system of integral equations:

$$T_{\lambda}^{\beta}(\psi) \equiv \psi(t,z) - \frac{\lambda}{\pi}\iint_D \frac{\exp[i\theta]\psi(t,s)\,ds}{(R-\rho)^{\beta}(s-z)} = f(t,s), \tag{5}$$

and

$$\Pi_A^1(\varphi) \equiv \psi(t,z) = \varphi(t,z) + \int_a^t \frac{A(\tau)}{\tau-a}\varphi(\tau,z)\,d\tau. \tag{6}$$

The theory of integral equations (5), (6) was developed in [1,2,4–10]. The research in this directions has a history, see [12–16].

For $A(a) < 0$, if the solution of the integral equation (6) exists, then it can be represented in the form (see [1])

$$\varphi(t,z) = \left(t-a)^{|A(a)|}\right)\exp\left[-W_A^1(t)\right]c_1(z) + \psi(t,z) -$$

$$\int_a^t \exp\left[W_A^1(\tau) - W_A^1(t)\right]\left(\frac{t-a}{\tau-a}\right)^{|A(a)|}\frac{A(\tau)\psi(s,z)\,ds}{\tau-a} \equiv \left(\Pi_{A,a}^{c_1}\right)^{-1}(\psi(t,z)), \tag{7}$$

where $W_A^1(t) = \int\limits_a^t \frac{A(\tau)-A(a)}{\tau-a} d\tau$ and where $c_1(z)$ is an arbitrary function in the domain D.

If $\lambda > 0$, $f(t,z) \in C(\overline{\Omega})$, if $f(t, Re^{i\theta}) = 0$ has the asymptotic behaviour

$$f(t,z) = o\left[\exp\left[-\lambda\omega_R^\alpha(r)\right](R-r)^{\gamma_1}\right], \quad \gamma_1 > \lambda - 1 \text{ as } r \to R, \tag{8}$$

and the solution of the integral equation (5) exists, then it can be represented in the following form (see [1])

$$\psi(t,z) = \exp\left[-2\lambda\omega_R^\beta(r)\right]\phi_1(t,z) + f(t,z) \tag{9}$$

$$+\frac{\lambda}{\pi}\iint\limits_D \exp\left[2\lambda\left(\omega_R^\beta(\rho) - \omega_R^\beta(r)\right) + i\theta\right]\frac{f(t,s)}{(R-\rho)^\alpha(t-s)} \equiv \left(T_\lambda^{\beta,\phi}\right)^{-1}(f), \tag{10}$$

where $\phi(t,z)$ is an arbitrary analytic function in z and continuous in t, and

$$\omega_R^\alpha(r) = \left[(\alpha-1)(R-r)^{\alpha-1}\right]^{-1}.$$

In order that the solution of the integral equation (1) exists and is a function $\varphi(t,z)$ which as $t \to a$ and $r \to R$ has the properties (3) and (4), it is necessary and sufficient that $f(t,z) \in C(\overline{\Omega})$, $f(a,z) = 0$, and that it has the asymptotic behaviour

$$f(t,z) = o\left[(t-a)^{\gamma_2}\right], \quad \gamma_2 > \varepsilon \text{ as } t \to a, \tag{11}$$

$f(t, Re^{i\theta}) = 0$, and that (8) holds.

Moreover we have $\phi(a,z) = 0$, it has the asymptotic behaviour (14) and $C_1(Re^{i\theta}) = 0$ is a function with the asymptotic behaviour (13). Thus, if the solution of the integral equation (1) exists, it can be represented in the form

$$\varphi(t,z) = \left(\Pi_{A,\alpha}^{c_1}\right)^{-1}\left(T_\lambda^{\beta,\phi}\right)^{-1}(f). \tag{12}$$

Theorem 1. *Consider the integral equation* (1)*. Assume that* $B(t,z) = A(t)\lambda$, *and that* $A(t)$ *in a neighbourhood of* $t = a$ *satisfies*

$$|A(t) - A(a)| \le H(t-a)^\varepsilon, \quad \varepsilon > 0 \quad \textit{as} \quad t \to a.$$

Assume that $f(t,z) \in C(\overline{\Omega})$, $f(a, Re^{i\theta}) = 0$, *and that it has the asymptotic behaviour* (8), (11)*. Then any solution of the integral equation* (1) *from the class* $C(\overline{\Omega})$ *can be represented in the form* (12)*, where* $C_1(z)$ *is an arbitrary continuous function in the domain* $\overline{D}$*, such that* $C_1(Re^{i\theta}) = 0$ *and*

$$C_1(z) = o\left[(R-r)^{\gamma_4}\right], \quad \gamma_4 > \beta - 1 \quad \textit{as} \quad r \to R. \tag{13}$$

Moreover, $\phi(t,z)$ *in the domain* $\overline{\Omega}$ *is analytic in* z *in* D *and continuous in* t *in* $\overline{\Omega}$*, and also* $\phi(a,z) = 0$ *has the asymptotic behaviour*

$$\phi(t,z) = o\left[(t-a)^{\gamma_3}\right], \quad \gamma_3 > 0 \quad \textit{as} \quad t \to a, \tag{14}$$

Theorem 2. *Suppose that the conditions of Theorem 1 hold. Then the integral equation* (12) *is invertible, that is, for arbitrary functions* $\phi(t,z)$, $c_1(z)$, *the unknown function* $\varphi(t,z)$ *can be found from equality*

$$\phi(t,z) = \exp\left[\lambda\omega_R^\alpha(r)\right]\left[\Pi_A^\alpha(\varphi) - f(t,z)\right] -$$

$$\frac{\lambda}{\pi}\iint\limits_D \frac{\exp\left[2\lambda\omega_R^\alpha(\rho) + i\theta\right] f(t,s)\,ds}{(R-\rho)^\alpha (t-z)}, \ (t,z)\in\Omega \tag{15}$$

$$C_1(z) = \left[(t-a)^{A(a)}\left(\varphi(t,z) - \Pi_A^1(\varphi)\right)\right]_{t=a} \tag{16}$$

Remark 1. Integral representation (12) and its inversion formula (15), (14) give a possibility to investigate different boundary problems for the integral equation (1).

1.1 Determination of Solution of Integral Equation (1), in the Case, When Coefficients are Disconnected Among Themselves

As before, denote by Ω the tube domain $\Omega = \{(x,t); a < t < b, \ |z| < R\}$. The base of this cylinder will be denoted by $D = \{t = a, \ |z| < R\}$, and the lateral surface will be denoted by $S = \{a < t < b, \ |z| = R\}$. We consider the following integral equation in Ω:

$$\varphi(t,z) + \mu\int\limits_a^t \frac{\varphi(\tau,z)}{\tau - a}d\tau - \frac{\lambda}{\pi}\iint\limits_D \frac{exp\left[i\theta\right]\varphi(t,s)\,ds}{(R-\rho)^\beta (s-z)} -$$

$$-\frac{\delta}{\pi}\int\limits_a^t \frac{d\tau}{\tau - a}\iint\limits_D \frac{exp\left[i\theta\right]\varphi(\tau,s)\,ds}{(R-\rho)^\beta (s-z)} = f(t,z), \tag{17}$$

where λ, μ, δ are given constants, $f(t,z)$ are given function, $\varphi(t,z)$ is an unknown function, $\theta = args$, $ds = d\xi d\eta$.

Assume that in the integral equation (17), the function $f(t,z)$ on the right hand side can be represented in the following uniformly convergent generalised power series form

$$f(t,z) = \sum_{k=1}^{\infty}(t-a)^{k+\gamma} f_k(z), \tag{18}$$

where $f_k(z)\,(k = 0,1,2,\dots)$ are known functions on D. The solution function $\varphi(t,z)$ to the integral equation (17) will be sought in the class of functions $\varphi(t,z)$ representable in the generalised power series form

$$\varphi(t,z) = \sum_{k=1}^{\infty}(t-a)^{k+\gamma}\varphi_k(z), \tag{19}$$

where $\varphi_k(z)\,(k = 0,1,2,\dots)$ are unknown functions on D.

Let us substitute these expressions $f(t,z)$ and $\varphi(t,z)$ using equalities (18) and (19) into integral equations (17), after calculating the corresponding integrals, we arrive at the following equality

$$\sum_{k=1}^{\infty}(t-a)^{k+\gamma}\left\{\left(1+\frac{\mu}{k+\gamma}\right)\varphi_k(z)-\left(\lambda+\frac{\delta}{k+\gamma}\right)\frac{1}{\pi}\iint\limits_D\frac{exp\,[i\theta]\,\varphi_k(s)\,ds}{(R-\rho)^{\beta}(s-z)}\right\}=$$

$$=\sum_{k=1}^{\infty}(t-a)^{k+\gamma}f_k(z).$$

From here we equate the coefficients in $(t-a)^{k+\gamma}$ at $k=0,1,2,\dots$ for determining unknown function $\varphi_k(z)\,(k=0,1,2,\dots)$, so that we obtaine the following infinitly split system of integral equations

$$\left(1+\frac{\mu}{k+\gamma}\right)\varphi_k(z)-\left(\lambda+\frac{\delta}{k+\gamma}\right)\frac{1}{\pi}\iint\limits_D\frac{exp\,[i\theta]\,\varphi_k(s)\,ds}{(R-\rho)^{\beta}(s-z)}=f_k(z),\quad k=0,1,2,\dots.$$

From here we get

$$\varphi_k(z)-\frac{\delta_k}{\pi}\iint\limits_D\frac{exp\,[i\theta]\,\varphi_k(s)\,ds}{(R-\rho)^{\beta}(s-z)}=\frac{(k+\gamma)}{k+\gamma+\mu}f_k(\mathbf{z}),\quad k=0,1,2,\dots,\tag{20}$$

where $\delta_k=\frac{\lambda(k+\gamma)+\delta}{k+\gamma+\mu}$.

If the following limits $\lim\limits_{n\to\infty}f_k(z)=f(z)$, $\lim\limits_{n\to\infty}\varphi_k(z)=\varphi(z)$ exist, then it follows from equality (20) that the limit function $\varphi(z)$ will be the solution of the following integral equation

$$\varphi(z)-\frac{\lambda}{\pi}\iint\limits_D\frac{exp\,[i\theta]\,\varphi(s)\,ds}{(R-\rho)^{\beta}(s-z)}=f(z).\tag{21}$$

If the solution of the integral equation (21) for $\lambda>0$ exists [1], then it is given by the formula

$$\varphi(z)=exp\left[-2\lambda\omega_R^{\beta}(r)\right]\varPhi(z)+f(z)+$$

$$+\frac{\lambda}{\pi}\iint\limits_D exp\left[2\lambda(\omega_R^{\beta}(\rho)-\omega_R^{\beta}(r))\right]\frac{exp\,[i\theta]\,f(s)\,ds}{(R-\rho)^{\beta}(s-z)},\tag{22}$$

where $\varPhi(z)$ is an arbitrary analytic function in the domain D.

For $\lambda<0$, if the solution of the integral equation (21) exists, then is unique and is given by the formula [1]:

$$\varphi(z)=f(z)+\frac{\lambda}{\pi}\iint\limits_D\frac{exp\left[2\lambda(\omega_R^{\beta}(\rho)-\omega_R^{\beta}(r))\right]exp\,[i\theta]\,f(s)\,ds}{(R-\rho)^{\beta}(s-z)}.\tag{23}$$

The solution of the type (22) exists if $f(z) \in C(\overline{D})$, $f(Re^{i\theta}) = 0$, with the asymptotic behaviour

$$f(z) = o\left[exp\left[-2\lambda\omega_R^{\beta}(r)\right](R-r)^{\omega_1}\right], \quad \omega_1 > \beta - 1 \quad at \quad r \to R. \qquad (24)$$

The solution of the type (23) exist, if $f(z) \in C(\overline{D})$, $f(Re^{i\theta}) = 0$ with asymptotic behavior

$$f(z) = o\left[(R-r)^{\omega_2}\right], \quad \omega_2 > \beta - 1 \quad at \quad r \to R. \qquad (25)$$

Analogously, if the solution of the system of integral equations (20) at $\delta_k > 0$ exists, then it is given by the formula

$$\varphi_k(z) = exp\left[-2\delta_k\omega_R^{\beta}(r)\right]\Phi_k(z) + \frac{(k+\gamma)}{k+\gamma+\mu}\{f_k(z) +$$

$$+\frac{\delta_k}{\pi}\iint\limits_D exp\left[2\delta_k(\omega_R^{\beta}(\rho) - \omega_R^{\beta}(r))\right]\frac{exp[i\theta]f_k(s)\,ds}{(R-\rho)^{\beta}(s-z)}\}, \quad k = 0,1,2,\ldots, \qquad (26)$$

where $\Phi_k(z)$ are arbitrary analytic functions in the domain D.

If the solution of the system of integral equations (20) at $\delta_k < 0$ exists, then the unique solution of this integral equation is given by the formula

$$\varphi_k(z) = \frac{(k+\gamma)}{k+\gamma+\mu}\{f_k(z) + \frac{2\delta_k}{\pi}\iint\limits_D exp\left[\delta_k(\omega_R^{\beta}(\rho) - \omega_R^{\beta}(r))\right]\frac{exp[i\theta]f_k(s)\,ds}{(R-\rho)^{\beta}(s-z)}\},$$

$$k = 0,1,2,... \qquad (27)$$

The solution of the type (26) exists if $f_k(z) \in C(\overline{D})$ and $f_k(Re^{i\theta}) = 0$, with asymptotic behaviour

$$f_j(z) = o\left[exp\left[-2\delta_j\omega_R^{\beta}(r)\right](R-r)^{\mu_j}\right], \mu_j > \beta_j \quad (j = o,1,2,3,\ldots) \quad at \quad r \to R. \qquad (28)$$

The solution of the type (27) exists if $f_j(z) \in C(\overline{D})$, $f_j(Re^{i\theta}) = 0$, with asymptotic behaviour

$$f_j(z) = \left[(R-r)^{\mu_j}\right], \quad \mu_j > 2\beta_j(j = o,1,2,3,\ldots) \quad at \quad r \to R. \qquad (29)$$

For $\delta_k > 0$ we obtain the values $\varphi_k(z)$ from formula (27), and substituting it in formula (19), we find the solution of the integral equation (1) in this case:

$$\varphi(t,z) = \sum_{k=0}^{\infty}(t-a)^{k+\gamma}\{exp\left[-2\delta_k\omega_R^{\beta}(r)\right]\Phi_k(z) + \frac{(k+\gamma)}{k+\gamma+\mu}\{f_k(z) +$$

$$\frac{\delta_k}{\pi}\iint\limits_D exp\left[2\delta_k(\omega_R^{\beta}(\rho) - \omega_R^{\beta}(r))\right]\frac{exp[i\theta]f_k(s)\,ds}{(R-\rho)^{\beta}(s-z)}, \qquad (30)$$

where $\Phi_k(z)(k = 0, 1, 2, 3, \dots)$ are arbitrary analytic functions of the complex variable z in the domain D.

A directly verification implies that if the series of type (18) converges absolutely and uniformly, and the series

$$\sum_{k=0}^{\infty} (t-a)^{k+\gamma} \{exp\left[-2\delta_k \omega_R^\beta(r))\right] \Phi_k(z)$$

converges absolutely and uniformly, than the series of type (30) also converges absolutely and uniformly.

Theorem 3. *Let in integral equation* (17) *the function* $f(t,z)$ *be expanded in a uniformly convergent generalised power series* (18)*. Let constants* λ*,* μ*,* δ *be such that* $\delta_k > 0(k = 0, 1, 2, 3, \dots)$*. Besides, let functions* $f_k(z) \in C(\overline{D})$*,* $f_k(Re^{i\theta}) = 0$ *have the asymptotic behaviour* (28)*. Then the integral equation* (17) *in the class of functions representable in the form* (18) *is always solvable, and is given by the formula* (30)*, where* $\Phi_k(z)(k = 0, 1, 2, 3, \dots)$ *are arbitrary analytic functions of complex variable* z *in the domain* D*.*

Moreover, let the solution of the integral equation (4) belong to class of infinitly differentiable functions in the variable t. Then from the integral representation (30) we have

$$\left[\varphi(t,z)(t-a)^{-\gamma}\right]_{t=a} = exp\left[-2\delta_0\omega_R^\beta(r)\right] \Phi_0(z) + K_0(f),$$

$$\left\{\frac{\partial}{\partial t}[\varphi(t,z)(t-a)^{-\gamma}]\right\}_{t=a} = exp\left[-2\delta_1\omega_R^\beta(r)\right] \Phi_1(z) + K_1(f),$$

$$\left\{\frac{\partial^2}{\partial t^2}[\varphi(t,z)(t-a)^{-\gamma}]\right\}_{t=a} = 2exp\left[-2\delta_2\omega_R^\beta(r)\right] \Phi_2(z) + K_2(f),$$

$$\left\{\frac{\partial^2}{\partial t^2}[\varphi(t,z)(t-a)^{-\gamma}]\right\}_{t=a} = 3!exp\left[-2\delta_3\omega_R^\beta(r)\right] \Phi_3(z) + K_3(f),$$

$$\dots\dots\dots\dots\dots\dots\dots,$$

$$\left\{\frac{\partial^n}{\partial t^n}[\varphi(t,z)(t-a)^{-\gamma}]\right\}_{t=a} = n!exp\left[-2\delta_n\omega_R^\beta(r)\right] \Phi_n(z) + K_n(f), \dots, \quad (31)$$

where

$$K_n(f) = \left(\frac{n+\gamma}{n+\gamma+\mu}\right) \times$$

$$\times \left\{ f_n(z) + \frac{2\delta_n}{\pi} \iint_D exp\left[2\delta_n(\omega_R^\beta(\rho) - \omega_R^\beta(r))\right] \frac{exp[i\theta] f_n(s) ds}{(R-\rho)^\beta (s-z)} \right\}, n = 0, 1, 2, \dots.$$

From this equalities we find

$$\Phi_0(z) = exp\left[2\delta_0\omega_R^\beta(r)\right] \left\{ \left[\varphi(t,z)(t-a)^{-\gamma}\right]_{t=a} - K_0(f) \right\},$$

$$\Phi_1(z) = exp\left[2\delta_1\omega_R^\beta(r)\right]\left\{\left\{\frac{\partial}{\partial t}\left[\varphi(t,z)(t-a)^{-\gamma}\right]\right\}_{t=a} - K_1(f)\right\},$$

$$\Phi_2(z) = \frac{1}{2!}exp\left[2\delta_2\omega_R^\beta(r)\right]\left\{\left\{\frac{\partial^2}{\partial t^2}\left[\varphi(t,z)(t-a)^{-\gamma}\right]\right\}_{t=a} - K_2(f)\right\},$$

$$\Phi_3(z) = \frac{1}{3!}exp\left[2\delta_3\omega_R^\beta(r)\right]\left\{\left\{\frac{\partial^3}{\partial t^3}\left[\varphi(t,z)(t-a)^{-\gamma}\right]\right\}_{t=a} - K_3(f)\right\},$$

$$\ldots\ldots\ldots\ldots\ldots\ldots\ldots\ldots,$$

$$\Phi_n(z) = \frac{1}{n!}exp\left[2\delta_n\omega_R^\beta(r)\right]\left\{\left\{\frac{\partial^n}{\partial t^n}\left[\varphi(t,z)(t-a)^{-\gamma}\right]\right\}_{t=a} - K_n(f)\right\},\ldots. \tag{32}$$

So, we prove the following statement.

Theorem 4. *Let conditions of Theorem 3 hold. Besides, let the solution of the integral equation* (17), *the function* $\varphi(t,z)$ *be infinity differentiable with respect to the variable* t. *Then the integral equation* (30) *is invertible, that is, the corresponding analytic functions* $\Phi_k(z)(k = 0,1,2,3,\ldots)$ *and their derivatives in the variable* t *can be found from formula* (32).

Theorem 5. *Let in the integral equation* (17), *the function* $f(t,z)$ *be expanded in a uniformly convergent generalised power series of type* (18). *Let constants* λ, μ, δ *be such that* $\delta_k < 0(k = 0,1,2,3,\ldots)$. *Let the function* $f_k(z) \in C(\overline{D})$, $f_k(Re^{i\theta}) = 0$, *have the asymptotic behaviour* (29). *Then the integral equation* (17) *in the class of functions representable in the form* (19) *has unique solution, which is given by the formula*

$$\varphi(t,z) = \sum_{k=0}^{\infty}(t-a)^{k+\gamma}\frac{(k+\gamma)}{(k+\gamma+\mu)}$$

$$\{f_k(z) + \frac{\delta_k}{\pi}\iint_D exp\left[2\delta_k(\omega_R^\beta(\rho) - \omega_R^\beta(r))\right]\frac{exp[i\theta]f_k(s)\,ds}{(R-\rho)^\beta(s-z)}\}.$$

Integral representation (30) and its inverse formula (30) give possibility for integral equation (17) in the case, when general solution contain arbitrary analytic functions, stand and investigate different boundary value problems.

2 Dirichlet Type Problem

Is required find the solution integral equation (17) represent able in form (19) by boundary conditions

$$Re\{exp\left[2\delta_0\omega_R^\beta(r))\right]\left[\varphi(t,z)(t-a)^{-\gamma}\right]_{t=a}\}_{r=R} = E_0(\tau), \tau = Re^{i\theta}$$

$$Re\left\{exp\left[2\delta_1\omega_R^\beta(r))\right]\frac{\partial}{\partial t}[\varphi(t,z)\,exp[\gamma\omega_a^\alpha(t)]]_{t=a}\right\}_{r=R} = E_1(\tau),$$

$$Re\left\{exp\left[2\delta_2\omega_R^{\beta}(r)\right)\right]\frac{\partial^2}{\partial t^2}[\varphi(t,z)\,exp\left[\gamma\omega_a^{\alpha}(t)\right]]_{t=a}\Big\}_{r=R}=E_2(\tau),$$

$$Re\left\{exp\left[2\delta_3\omega_R^{\beta}(r)\right)\right]\frac{\partial^3}{\partial t^3}[\varphi(t,z)\,exp\left[\gamma\omega_a^{\alpha}(t)\right]]_{t=a}\Big\}_{r=R}=E_3(\tau),$$

$$\cdots\cdots\cdots\cdots\cdots\cdots$$

$$Re\left\{exp\left[2\delta_n\omega_R^{\beta}(r)\right)\right]\frac{\partial^n}{\partial t^n}[\varphi(t,z)\,exp\left[\gamma\omega_a^{\alpha}(t)\right]]_{t=a}\Big\}_{r=R}=E_n(\tau), \tag{33}$$

$$\cdots\cdots\cdots\cdots\cdots\cdots$$

where $E_j(\tau)\,(j=0,1,2,3\ldots\ldots\ldots)$–are given functions point rib $S\cap D$.

Solution Dirichlet Type Problem. Let fulfillment all conditions of the Theorem 3. Then use integral representation (30), its inverse formula (32) and conditions (33), for obtained analytic functions $\Phi_j(z)(j=0,1,2,3,\ldots\ldots)$, we have the following conditions

$$(Re\Phi_0(z)_{r=R}=E_0(\tau)-\left[Re(exp(2\delta_0\omega_R^{\beta}(r))K_0(f))\right]_{r=R}\equiv F_0(\tau),$$

$$(Re\Phi_1(z))_{r=R}=E_1(\tau)-\left[Re(exp(2\delta_1\omega_R^{\beta}(r))K_1(f))\right]_{r=R}\equiv F_1(\tau),$$

$$(Re\Phi_2(z))_{r=R}=\frac{1}{2!}\left(E_2(\tau)-\left[Re(exp(2\delta_2\omega_R^{\beta}(r))K_2(f))\right]_{r=R}\right)\equiv F_2(\tau),\ (\Phi)$$

$$\cdots\cdots\cdots\cdots\cdots\cdots$$

$$(Re\Phi_n(z))_{r=R}=\frac{1}{n!}\left(E_n(\tau)-\left[Re(exp(2\delta_n\omega_R^{\beta}(r))K_n(f))\right]_{r=R}\right)\equiv F_2(\tau),$$

$$\cdots\cdots\cdots\cdots\cdots\cdots$$

By solving these problems, we found [11]

$$\Phi_0(\mathrm{z})=\frac{1}{2\pi}\int\limits_{\gamma}\frac{\tau+\mathrm{z}}{\tau(\tau-\mathrm{z})}\mathrm{F}_0(\tau)\,\mathrm{d}\tau+\mathrm{i}\mathrm{C}_0,$$

$$\Phi_1(\mathrm{z})=\frac{1}{2\pi}\int\limits_{\gamma}\frac{\tau+\mathrm{z}}{\tau(\tau-\mathrm{z})}\mathrm{F}_1(\tau)\,\mathrm{d}\tau+\mathrm{i}\mathrm{C}_1,$$

$$\Phi_2(\mathrm{z})=\frac{1}{2!2\pi}\int\limits_{\gamma}\frac{\tau+\mathrm{z}}{\tau(\tau-\mathrm{z})}\mathrm{F}_2(\tau)\,\mathrm{d}\tau+\mathrm{i}\mathrm{C}_2,$$

$$\Phi_3(\mathrm{z})=\frac{1}{3!2\pi}\int\limits_{\gamma}\frac{\tau+\mathrm{z}}{\tau(\tau-\mathrm{z})}\mathrm{F}_3(\tau)\,\mathrm{d}\tau+\mathrm{i}\mathrm{C}_3,$$

$$\cdots\cdots\cdots\cdots\cdots\cdots$$

$$\Phi_{\mathrm{n}}(\mathrm{z})=\frac{1}{\mathrm{n}!2\pi}\int\limits_{\gamma}\frac{\tau+\mathrm{z}}{\tau(\tau-\mathrm{z})}\mathrm{F}_{\mathrm{n}}(\tau)\,\mathrm{d}\tau+\mathrm{i}\mathrm{C}_{\mathrm{n}},\cdots\cdots\cdots\cdots \tag{34}$$

where $\mathrm{C_j}(j=0,1,2,\ldots)$- arbitrary constants, γ–circle $\gamma=\{\mathrm{t}=\mathrm{a},|\mathrm{z}|=\mathrm{R}\}$.

Substituting the finding values $\Phi_j(z)(j = 0, 1, 2, 3, \ldots)$– in formula (30), we be found the solution of the problem (33) in form

$$\varphi(t,z) = \sum_{k=0}^{\infty} \{(t-a)^{k+\gamma} \{exp\left[-2\delta_k \omega_R^\beta(r))\right] \; [\frac{1}{k!2\pi} \int_\gamma \frac{\tau + z}{\tau(\tau - z)} F_k(\tau)\,d\tau + iC_k]\} +$$

$$+ \{\frac{(k+\gamma)}{k+\gamma+\mu} \{f_k(z) + \frac{\delta_k}{\pi} \iint_D exp\left[\delta_k(\omega_R^\beta(\rho) - \omega_R^\beta(r))\right] \frac{exp[i\theta] f_k(s)\,ds}{(R-\rho)^\beta (s-z)}\}\}. \quad (35)$$

The function of the type (35) will be solution of the Dirichlet problem, if the series of the type

$$\sum_{k=0}^{\infty} (t-a)^{k+\gamma} \{exp\left[-2\delta_k \omega_R^\beta(r))\right] C_k \quad (36)$$

converge.

Series type (36) converge, if

$$\lim_{k\to\infty} \frac{|C_{k+1}|}{|C_k|} = L, \quad (37)$$

and

$$(b-a)\; L < 1. \quad (38)$$

So, we proof the following confirmation.

Theorem 6. *Let fulfillment all condition of the Theorem 3. In the conditions problem* (33) *all functions* $E_n(\tau)$ *bounded one constant. The constants* C_k*, such that exist limits of the type* (37) *and fulfillment conditions* (38)*. Then the problem* (33) *always solvable and its solution is given by formula* (35)*, where* C_k *(k=0,1,2,3,...)- arbitrary constants.*

Correct Formulation of Boundary Problem and its Investigation

From Theorem 6 it follows that the homogenous Dirichlet problems for integral equation (1) at $\delta_k < 0$ we have infinity number of linear independent solution types

$$(t-a)^{k+\gamma} \{exp\left[-2\delta_k \omega_R^\beta(r))\right], \quad k = 0, 1, 2, 3, \ldots$$

A question springs up for integral equation (1), if it is possible to state and investigate correct Dirichlet type boundary value problems. It is found that, besides condition (31), (33), additional conditions are given

$$\left\{exp\left[2\delta_0 \omega_R^\beta(r)\right]\left[(t-a)^{-\gamma}\varphi(t,z)\right]_{t=a}\right\}_{z=0} = A_0,$$

$$\left\{exp\left[2\delta_1 \omega_R^\beta(r)\right]\left[\frac{\partial}{\partial t}\left[(t-a)^{-\gamma}\varphi(t,z)\right]\right]_{t=a}\right\}_{z=0} = A_1$$

$$\left\{exp\left[2\delta_2 \omega_R^\beta(r)\right]\left[\frac{\partial^2}{\partial t^2}\left[(t-a)^{-\gamma}\varphi(t,z)\right]\right]_{t=a}\right\}_{z=0} = A_2, \quad (39)$$

$$\left\{exp\left[2\delta_3\omega_R^\beta(r)\right]\left[\frac{\partial^3}{\partial t^3}\left[(t-a)^{-\gamma}\varphi(t,z)\right]\right]_{t=a}\right\}_{z=0} = A_3,$$

$$\cdots\cdots\cdots\cdots\cdots\cdots\cdots\cdots\cdots\cdots\cdots\cdots\cdots$$

$$\left\{exp\left[2\delta_n\omega_R^\beta(r)\right]\left[\frac{\partial^3}{\partial t^3}\left[(t-a)^{-\gamma}\varphi(t,z)\right]\right]_{t=a}\right\}_{z=0} = A_n,$$

$$\cdots\cdots\cdots\cdots\cdots\cdots\cdots\cdots\cdots\cdots\cdots\cdots\cdots$$

where $A_j (j = 0, 1, 2, 3, \dots)$ - are given constants, then Dirichlet problem will be correct.

Thus, for correct Dirichlet type boundary value problems, it is necessary to give conditions in lower basis, in lateral surface and in principal axis, that is for integral equation (1), we can stand following correct Dirichlet type boundary value problems

Problem N_1. It is required to find the solution of integral equation (17) represented in form (19), by boundary conditions (33) in lower basis, in lateral surface, and condition (39) in principal axis, where $E_j(\theta) (j = 0, 1, 2, 3, 4, \dots, 0 \leqslant \theta \leqslant 2\pi)$ are given functions point $\varGamma = |z| = R, \quad 0 \le \theta \le 2\pi$, and $A_j (j = 0, 1, 2, 3, 4, \dots)$ –are given constants.

Solution Problem N$_1$. Let any conditions of the Theorem 3 be fulfilled. Then, using integral representation (30), boundary conditions (33) for determined analytics functions $\phi_j(z) (j = 0, 1, 2, 3, \dots)$, we come to solution of infinity number of Shvarts problems ($\varPhi$), solution of which is given by formula (34). By the further use of conditions (39), we found

$$\varPhi_0(0) = \frac{1}{2\pi}\int_\gamma \frac{1}{\tau}E_0(\tau)\,d\tau + iC_0 = A_0 - \{exp\left[2\delta_0\omega_R^\beta(r)\right]K_0(f)\}_{z=0},$$

$$\varPhi_1(0) = \frac{1}{2\pi}\int_\gamma \frac{1}{\tau}E_1(\tau)\,d\tau + iC_1 = A_1 - \{exp\left[2\delta_0\omega_R^\beta(r)\right]K_1(f)\}_{z=0},$$

$$\varPhi_2(0) = \frac{1}{2!2\pi}\int_\gamma \frac{1}{\tau}E_2(\tau)\,d\tau + iC_2 = A_2 - \{exp\left[2\delta_0\omega_R^\beta(r)\right]K_2(f)\}_{z=0},$$

$$\varPhi_3(0) = \frac{1}{3!2\pi}\int_\gamma \frac{1}{\tau}E_3(\tau)\,d\tau + iC_3 = A_3 - \{exp\left[2\delta_0\omega_R^\beta(r)\right]K_3(f)\}_{z=0},$$

$$\cdots\cdots\cdots\cdots\cdots\cdots\cdots\cdots\cdots\cdots\cdots\cdots\cdots$$

$$\varPhi_n(0) = \frac{1}{n!2\pi}\int_\gamma \frac{1}{\tau}E_n(\tau)\,d\tau + iC_n = A_n - \{exp\left[2\delta_n\omega_R^\beta(r)\right]K_n(f)\}_{z=0}, \dots.$$

From her we found

$$C_0 = -iA_0 - \frac{1}{2\pi i}\int_0^{2\pi} E_0\left(Re^{i\theta}\right)d\theta + i\{exp\left[2\delta_0\omega_R^\beta(r)\right]K_0(f)\}_{z=0},$$

$$C_1 = -iA_1 - \frac{1}{2\pi i}\int_0^{2\pi} E_1\left(Re^{i\theta}\right)d\theta + i\{exp\left[2\delta_1\omega_R^{\beta}(r)\right]K_1(f)\}_{z=0},$$

$$C_2 = -iA_2 - \frac{1}{2!2\pi i}\int_0^{2\pi} E_2\left(Re^{i\theta}\right)d\theta + i\{exp\left[2\delta_2\omega_R^{\beta}(r)\right]K_2(f)\}_{z=0},$$

$$C_3 = -iA_3 - \frac{1}{3!2\pi i}\int_0^{2\pi} E_3\left(Re^{i\theta}\right)d\theta + i\{exp\left[2\delta_3\omega_R^{\beta}(r)\right]K_3(f)\}_{z=0},$$

....

$$C_n = -iA_n - \frac{1}{n!2\pi i}\int_0^{2\pi} E_n\left(Re^{i\theta}\right)d\theta + i\{exp\left[2\delta_n\omega_R^{\beta}(r)\right]K_n(f)\}_{z=0}, \quad (40)$$

....

Substituting obtained value C_j $(j = 0, 1, 2, 3, \ldots\ldots.)$ to formula (35), we see that, if solution problem N_1 exists, then we can represent it in the following for

$$\varphi(t,z) =$$

$$\sum_{k=0}^{\infty}\{(t-a)^{k+\gamma}\{exp\left[-2\delta_k\omega_R^{\beta}(r)\right)\right][\frac{1}{k!2\pi}\int_{\gamma}\frac{\tau+z}{\tau(\tau-z)}F_k(\tau)\,d\tau + A_k + \frac{1}{k!2\pi}\int_0^{2\pi}F_k\left(Re^{i\theta}\right)d\theta]\}+$$

$$\frac{(k+\gamma)}{k+\gamma+\mu}\{f_k(z) + \frac{\delta_k}{\pi}\iint_D exp\left[2\delta_k(\omega_R^{\beta}(\rho) - \omega_R^{\beta}(r))\right]\frac{exp\,[i\theta]\,f_k(s)\,ds}{(R-\rho)(s-z)}\}\} \quad (41)$$

The function of the type (41) will be solution to problem N_1, if the following limit exists

$$\lim_{k\to\infty}\frac{|A_{k+1}|}{|A_k|} = L, \quad (42)$$

$$(b-a)\ L < 1 \quad (43)$$

and

$$|E_j(Re^{i\theta})| \le M(j = 0, 1, 2, 3, \ldots) \quad (44)$$

So we prove the following confirmation.

Theorem 7. *Let fulfillment all condition of the theorem 3.In the conditions problem* (N_1) *all functions* $E_n(\tau)$ *bounded one constant. The constants* A_k*, such that exist limits of the type* (42) *and fulfillment conditions* (43)*. Then the problem* (N_1) *have unique solution, which is given by formula* (41).

Clearing up the correct statement of boundary value problems in the case, when function is present in kernels among themselves

Let in integral equation (1) $\alpha = 1,\ \beta = 1,\ A(t) = \mu,\ B(t,z) = \delta$ and $\delta = \lambda\mu$. In this case integral equation (1) can be represented in form

$$\varphi(t,z)+\mu\int_a^t \frac{\varphi(\tau,z)}{\tau-a}d\,\tau-\frac{\lambda}{\pi}\iint_D \frac{\exp[i\theta]}{(R-\rho)(s-z)}\left[\varphi(t,s)+\mu\int_a^t \frac{\varphi(\tau,z)}{\tau-a}d\,\tau\right]ds = f(t,z). \tag{45}$$

Introducing

$$\psi(t,z) = \varphi(t,z)+\mu\int_a^t \frac{\varphi(\tau,z)}{\tau-a}d\,\tau \equiv T_\mu(\varphi), \tag{46}$$

we come to solution of the following two-dimensional complex integral equation

$$\psi(t,z)-\frac{\lambda}{\pi}\iint_D \frac{\exp[i\theta]\ \psi(t,s)}{(R-\rho)(s-z)}ds = f(t,z). \tag{47}$$

So in this case the problem found in solution of integral equation (1) is reduced to problem finding splitting system (46), (47).

According [1] if at $\lambda > 0$ and solution of the integral equation (47) exist, then we represent it in the following form

$$\psi(t,z) = (R-r)^{2\lambda}\phi(t,z)+f(t,z)+\frac{\lambda}{2\pi}\iint_D \left(\frac{R-r}{R-\rho}\right)^{2\lambda}\frac{\exp[i\theta]\ f(t,s)}{(R-\rho)(s-z)}ds \equiv (R-r)^{2\lambda}\phi(t,z)+(\Pi_\mu)^{-1}(f), \tag{48}$$

where $\phi(t,z)$-arbitrary function of two variables: analytical by variable z and continuously by variable t.

Solution form the (48) exists, if $f(t,z) \in C(\overline{\Omega})$, $f(t,Re^{i\theta}) = 0$ with asymptotic behavior

$$f(t,\ z) = o\left[(R-r)^{\delta_1}\right] \quad \delta_1 > 2\lambda \quad at \quad r \to R \tag{49}$$

If solution to integral equation (46) at $\mu < 0$ exists, it is given by formula [1]

$$\varphi(t,z) = (t-a)^{|\mu|}C_1(z)+\psi(t,z)-\mu\int_a^t \left(\frac{t-a}{\tau-a}\right)^{|\mu|}\frac{\psi(\tau,z)}{\tau-a}d\tau \equiv (t-a)^{|\mu|}C_1(z)+(T_\mu)^{-1}(\psi), \tag{50}$$

where $C_1(z)$ -arbitrary continuous function of the points of domain D.

The solution type (50) exist, if $\psi(t,z) \in C(\overline{\Omega})$, $\psi(a,z) = 0$ with asymptotic behavior

$$\psi(t,z) = o\left[(t-a)^{\delta_2}\right], \quad \delta_2 > |\mu| \quad at \quad t \to a. \tag{51}$$

Function $\psi(t,z)$ have asymptotic behavior (51), if $f(a,z) = 0$ and also

$$f(t,z) = o\left[(t-a)^{\delta_3}\right], \quad \delta_3 > |\mu| \quad at \quad t \to a \tag{52}$$

So, we proof the following confirmation.

Theorem 8. *Let in integral equation* (1) $\alpha = 1$, $\beta = 1$, $A(t) = \mu$, $B(t,z) = \delta$ *and* $\delta = \lambda\mu$, $\mu < 0$, $f(t,z) \in C(\overline{\Omega})$, $f(a,z) = 0$ *with asymptotic behavior*

$$f(t,z) = o[(t-a)^{\delta_3}], \quad \delta_3 > |\mu| \quad at \quad t \to a \tag{53}$$

and $f(t, Re^{i\theta}) = 0$ *with asymptotic behavior* (49). *Then any solution of integral equation* (1) *from class* $C(\overline{\Omega})$ *is repesentable in form*

$$\varphi(t,z) = (t-a)^{|\mu|}C_1(z) + (T_\mu)^{-1}\left[(R-r)^{\lambda}\phi(t,z) + (\Pi_\mu)^{-1}(f)\right] =$$

$$(t-a)^{|\mu|}C_1(z) + (R-r)^{2\lambda}(T_\mu)^{-1}[\phi(t,z)] + (T_\mu)^{-1}(\Pi_\mu)^{-1}(f), \tag{54}$$

where $C_1(z)$*–arbitrary continuous function of point* $\overline{D}$, *besides* $C_1\left(Re^{i\theta}\right) = 0$ *with asymptotic behavior*
$C_1(z) = o[(R-r)^{\varepsilon}]$, $\varepsilon > 0$ *at* $r \to R$,
$\phi(t,z)$*–arbitrary function of domain* $\overline{\Omega}$, *analytical by variable* z *and continuously by variable* t, *besides* $\phi(a,z) = 0$ *with asymptotic behavior*

$$\phi(t,z) = o\left[(t-a)^{\delta_4}\right], \quad \delta_4 > |\mu| \quad at \quad t \to a \tag{55}$$

and $\phi(t, Re^{i\theta}) = 0$ *with asymptotic behavior*

$$\phi(t,z) = o\left[(R-r)^{\delta_5}\right], \quad \delta_5 > 0, \quad at \quad r \to R. \tag{56}$$

Theorem 9. *Let any condition of the Theorem 8 fulfill. Then integral representation* (54) *convertible, that is corresponding arbitrary functions.* $C_1(z)$ *and* $\phi(t,z)$ *by valued unknown function* $\varphi(t,z)$ *is found by following formula*

$$C_1(z) = [(t-a)^{\mu}\varphi(t,z)]_{t=a} + [(t-a)^{\mu}f(t,z)]_{t=a} +$$

$$+\frac{2\lambda}{\pi}\iint\limits_{D}\left(\frac{R-r}{R-\rho}\right)^{\lambda}\frac{exp[i\theta]\,[(t-a)^{\mu}f(t,s)]_{t=a}}{(R-\rho)(s-z)}ds, \tag{57}$$

$$\phi(t,z) = (R-r)^{-2\lambda}\left[\varphi(t,z) - (t-a)^{|\mu|}C_1(z)\right] +$$

$$\mu\int\limits_a^t \frac{(R-r)^{-2\lambda}\left[\varphi(t,z) - (\tau-a)^{|\mu|}C_1(z)\right]}{(\tau-a)}d\tau - (R-r)^{-\lambda}f(t,z) + \tag{58}$$

$$\frac{\lambda}{\pi}\iint\limits_{D}(R-\rho)^{-\lambda-1}\frac{exp[i\theta]\,f(t,s)\,ds}{s-z}.$$

In the case, when $\mu > 0, \quad \lambda > 0$ from Theorem 8 follows.

Theorem 10. *Let in integral equation* (1) $\beta = 1, \quad A(t) = \mu, \quad B(t,z) = \delta$ *end* $\delta = \lambda\mu, \quad \mu > 0, \quad \lambda > 0, \quad f(t,z) \in C(\overline{\Omega}), \quad f(t, Re^{i\Theta}) = 0$ *with asymptotic behavior*

$$f(t,z) = o\left[(t-a)^{\varepsilon}\right], \quad \varepsilon > 0 \quad at \quad t \to a \tag{59}$$

end $f(a,z) = 0$ *with asymptotic behavior* (49)*. Then any solution of integral equation* (1) *from class* $C(\overline{\Omega})$ *is represent able in form*

$$\varphi(t,z) = (T_{\mu})^{-1}\left[(R-r)^{2\lambda}\phi(t,z) + (\Pi_{\mu})^{-1}(f)\right] =$$

$$= (R-r)^{2\lambda}(T_{\mu})^{-1}\left[\phi(t,z)\right] + (T_{\mu})^{-1}(\Pi_{\mu})^{-1}(f), \tag{60}$$

where $\Phi(t,z)$*-arbitrary function of domain* $\overline{\Omega}$*, analytical by variable* z *and continuously by variable* t*, besides* $\Phi(a,z) = 0$ *with asymptotic behavior*

$$f(t,z) = o\left[(t-a)^{\delta_3}\right], \quad \delta_3 > |\mu| \quad at \quad t \to a \tag{61}$$

and $f(t, Re^{i\Theta}) = 0$ *with asymptotic behavior* (56).

In the case, when $\mu < 0, \quad \lambda < 0$ from Theorem 8 follows.

Theorem 11. *Let in integral equation* (1) $\beta = 1, \quad A(t) = \mu, \quad B(t,z) = \delta$ *end* $\delta = \lambda\mu, \quad \mu < 0, \quad \lambda < 0, \quad f(t,z) \in C(\overline{\Omega}), \quad f(a,z) = 0$ *with asymptotic behavior*

$$f(t,z) = o\left[(t-a)^{\delta_3}\right], \quad \delta_3 > |\mu| \quad at \quad t \to a \tag{62}$$

end $f(t, Re^{i\Theta}) = 0$ *with asymptotic behavior*

$$f(t,z) = o\left[(R-r)^{\varepsilon}\right], \quad \varepsilon > 0 \quad at \quad r \to R. \tag{63}$$

Then any solution of integral equation (1) *from class* $C(\overline{\Omega})$ *is represent in form*

$$\varphi(t,z) = (t-a)^{|\mu|}C_1(z) + T_{\mu}^{-1}(\Pi_{\lambda})^{-1}(f), \tag{64}$$

where $C(z)$*-arbitrary continuous function of point* $\overline{D}$*, besides* $C_1(Re^{i\Theta}) = 0$ *with asymptotic behavior*

$$C_1(z) = o\left[(R-r)^{\varepsilon}\right], \quad \varepsilon > 0 \quad at \quad r \to R.$$

In the case, when $\mu > 0, \quad \lambda < 0$ *from Theorem 8 follows.*

Theorem 12. *Let integral equation* (1) $\beta = 1, \quad A(t) = \mu, \quad B(t,z) = \delta$ *end* $\delta = \lambda\mu, \quad \mu > 0, \quad \lambda < 0, \quad f(t,z) \in C(\overline{\Omega}),$ $f(a, z) = 0$ *with asymptotic behavior* (59) *end* $f(t, Re^{i\Theta}) = 0$ *with asymptotic behavior* (63)*. Then the integral equation* (1) *has a unique solution given by following formula*

$$\varphi(t,z) = T_{\mu}^{-1}(\Pi_{\lambda})^{-1}(f), \tag{65}$$

In the case, when $\mu > 0,\quad \lambda > 0$, and $\mu < 0,\quad \lambda < 0$, then general solution integral equation (1) also contain arbitrary function $\Phi(t,z)$ and $C_1(z)$. In this case we have the following confirmation.

Theorem 13. *Let any condition of the Theorem 10 fulfill. Then integral representation* (60) *convertible, that is corresponding arbitrary function. by valued unknown function is found by following formula*

$$\phi\,(t,z) = (R-r)^{-2\lambda}\varphi\,(t,z) +$$

$$+\,\mu\int_a^t \frac{(R-r)^{-2\lambda}\varphi\,(t,z)}{(\tau-a)}d\tau - (R-r)^{-2\lambda}f\,(t,z) + \tag{66}$$

$$+\frac{\lambda}{\pi}\iint_D (R-\rho)^{-2\lambda-1}\frac{exp\,[i\theta]\,f\,(t,s)\,ds}{s-z}.$$

Theorem 14. *Let any condition of the Theorem 11 fulfill. Then integral representation* (64) *invertible, that is corresponding arbitrary function* $C_1(z)$ *by valued unknown function* $\varphi(t,z)$ *is found by following formula*

$$C_1(z) = [(t-a)^{\mu}\varphi\,(t,z)]_{t=a} + [(t-a)^{\mu}f\,(t,z)]_{t=a} +$$

$$+\,\frac{\lambda}{\pi}\iint_D\left(\frac{R-r}{R-\rho}\right)^{2\lambda}\frac{exp\,[i\theta]\,[(t-a)^{\mu}f\,(t,s])\,ds}{(R-\rho)(s-z)}. \tag{67}$$

Integral representation (54) and its inverse formula (57), (58) give possibility for integral equation (1) statement and investigation of correct Dirichlet type boundary value problems.

Problem $\mathbf{N_2}$. *It is required to find solution integral equation* (17) *from class* $C\left(\overline{\Omega}\right)$ *at* $\delta = \mu\lambda, \lambda > 0, \mu < 0$ *by boundary conditions*

$$\begin{cases} [(t-a)^{\mu}\varphi\,(t,z)]_{t=a} = E_1\,(z)\,, \quad z \in D \\ Re\Big[(R-r)^{-2\lambda}\varphi\,(t,z)\Big]_{r=R} = E_2\,(t,\theta) \quad a \le t \le b \quad 0 \le \theta \le 2\pi \end{cases} \tag{68}$$

and following condition

$$\Big[(R-r)^{-2\lambda}\varphi\,(t,z)\Big]_{z=0} = E_3\,(t)\,,\ \ a \le t \le b, \tag{69}$$

in cylinder axis, where $E_1\,(z)$ *–are given function points of domain* D, $E_2\,(t,\theta)$*– are given function lateral surface* S, $E_3\,(t)$*–are given function points of cylinder axis* $\Gamma = \{a \le t \le b,\ \ z = 0\}$.

Solution Problem $\mathbf{N_2}$. Use inverse formula (57) and first condition from (59), we have

$$C_1(z) = E_1(z) + [(t-a)^{\mu} f(t,z)]_{t=a} +$$
$$\frac{\lambda}{\pi} \iint\limits_D \left(\frac{R-r}{R-\rho}\right)^{2\lambda} \frac{exp\,[i\theta]\,[(t-a)^{\mu} f(t,s)]_{t=a}}{(R-\rho)(s-z)} ds. \tag{70}$$

From formula (58) we have

$$\{\mathrm{Re}\,[\phi(t,z)]\}_{r=R} = \left\{\mathrm{Re}[(R-r)^{-2\lambda} \varphi(t,z)]\right\}_{rl=R} +$$
$$+\mu \int\limits_a^t \frac{\left\{\mathrm{Re}[(R-r)^{-2\lambda}\varphi(\tau,z)]\right\}_{r=R}}{\tau-a} d\,\tau - \left\{\mathrm{Re}[(R-r)^{-\lambda} f(t,z)]\right\}_{r=R} +$$
$$+\frac{\lambda}{\pi} \iint\limits_D (R-\rho)^{-2\lambda-1} \left\{Re\left[\frac{exp\,[i\theta]\, f(t,s)\, ds}{s-z}\right]\right\}_{r=R}. \tag{71}$$

Using second condition (68) we have

$$\{\mathrm{Re}\,[\phi(t,z)]\}_{r=R} = E_2(t,\theta) + \mu \int\limits_a^t \frac{E_2(\tau,\theta)}{\tau-a} d\,\tau -$$
$$\left\{\mathrm{Re}[(R-r)^{-\lambda} f(t,z)]\right\}_{r=R} + \frac{\lambda}{\pi} \iint\limits_D (R-\rho)^{-2\lambda-1} \left\{Re\left[\frac{exp\,[i\theta]\, f(t,s)\, ds}{s-z}\right]\right\}_{r=R} \equiv E_4(t,\theta). \tag{72}$$

So, problem of finding solution to *problem* $\mathbf{N_2}$ is reduced to following problem: It is required to find function two variables $\phi(t,z)$–analytical by variable z, continuous by variable t over boundary value problem

$$\{\mathrm{Re}\,[\phi(t,z)]\}_{r=R} = E_4(t,\theta), \quad a \le t \le b, \quad 0 \le \theta \le 2\pi, \tag{73}$$

where $E_3(t,\theta)$–are given function points of lateral surface

$$S = \{a < t < b,\ |z| = R, 0 \le \theta \le 2\pi\}.$$

Problem (73), is Schwarz problem with parameters t.

Solution to this problem according [11] is given by formula

$$\phi(t,z) = \frac{1}{2\pi} \int\limits_{\gamma} \frac{\tau + z}{\tau(\tau - z)} \mathrm{E}_4(t,\theta)\, d\tau + i\phi_1(t), \tag{74}$$

where $\phi_1(t)$– arbitrary function point $t(a < t < b)$.

From formula (58), we have ($C_1(z) = 0$)

$$\phi(t,0) = \left[(R-r)^{-2\lambda}\varphi(t,z)\right]_{z=0} + \mu\int\limits_a^t \frac{\left[(R-r)^{-2\lambda}\varphi(\tau,z)\right]_{z=0}}{\tau-a}d\,\tau -$$

$$\left[(R-r)^{-2\lambda}f(t,z)\right]_{z=0} + \frac{\lambda}{\pi}\iint\limits_D (R-\rho)^{-2\lambda-1}\frac{exp\,[i\theta]\,f(t,s)\,ds}{s}. \qquad (75)$$

In this equality by using condition (65) and equality (75), we have

$$\frac{1}{2\pi}\int\limits_\gamma \frac{1}{\tau}\mathrm{E}_4(\mathrm{t},\theta)\,\mathrm{d}\tau + \mathrm{i}\phi_1(t) = E_3(t) + \mu\int\limits_a^t \frac{E_3(\tau)}{\tau-a}d\,\tau +$$

$$\left[(R-r)^{-2\lambda}f(t,z)\right]_{z=0} + \frac{\lambda}{\pi}\iint\limits_D (R-\rho)^{-2\lambda-1}\frac{exp\,[i\theta]\,f(t,s)\,ds}{s}.$$

From there

$$i\phi_1(t) = E_3(t) + \mu\int\limits_a^t \frac{E_3(\tau)}{\tau-a}d\,\tau +$$

$$\left[(R-r)^{-2\lambda}f(t,z)\right]_{z=0} + \frac{\lambda}{\pi}\iint\limits_D (R-\rho)^{-2\lambda-1}\frac{exp\,[i\theta]\,f(t,s)\,ds}{s} + \frac{1}{2\pi\mathrm{i}}\int\limits_0^{2\pi} \mathrm{E}_4(\mathrm{t},\theta)\mathrm{d}\theta$$

Substituting this value $i\phi_1(t)$ into formula (74), we receive

$$\phi(t,z) = -\frac{1}{2\pi\mathrm{i}}\int\limits_0^{2\pi} \frac{\mathrm{Re}^{i\theta}+\mathrm{z}}{\mathrm{Re}^{i\theta}-\mathrm{z}}\mathrm{E}_4(\mathrm{t},\theta)\,\mathrm{d}\theta + E_3(t) + \mu\int\limits_a^t \frac{E_3(\tau)}{\tau-a}d\,\tau +$$

$$\left[(R-r)^{-2\lambda}f(t,z)\right]_{z=0} + \frac{\lambda}{\pi}\iint\limits_D (R-\rho)^{-2\lambda-1}\frac{exp\,[i\theta]\,\mathbf{f}(t,s)\,ds}{s} + \frac{1}{2\pi\mathrm{i}}\int\limits_0^{2\pi} \mathrm{E}_4(\mathrm{t},\theta)\mathrm{d}\theta$$

Or

$$\phi(t,z) = -\frac{2\mathrm{z}}{2\pi\mathrm{i}}\int\limits_0^{2\pi} \frac{\mathrm{E}_4(\mathrm{t},\theta)}{\mathrm{Re}^{i\theta}-\mathrm{z}}\mathrm{d}\theta + E_3(t) + \mu\int\limits_a^t \frac{E_3(\tau)}{\tau-a}d\,\tau +$$

$$\left[(R-r)^{-2\lambda}f(t,z)\right]_{z=0} + \frac{\lambda}{\pi}\iint\limits_D (R-\rho)^{-2\lambda-1}\frac{exp\,[i\theta]\,f(t,s)\,ds}{s}. \qquad (76)$$

Substituting value $C_1(z)$ and $\phi(t,z)$ from formula (70) and (76) in formula (54), we find solution to problem N_2 in the following form

$$\varphi(t,z) =$$

$$= (t-a)^{|\mu|}\{E_1(z) + [(t-a)^{\mu} f(t,z)]_{t=a} + \frac{\lambda}{\pi}\iint\limits_D \left(\frac{R-r}{R-\rho}\right)^{2\lambda} \frac{exp\,[i\theta]\,[(t-a)^{\mu} f(t,s)]_{t=a}}{(R-\rho)(s-z)} ds\} +$$

$$+(R-r)^{2\lambda}\int\limits_a^t \left(\frac{t-a}{\tau-a}\right)^{|\mu|}$$

$$\{-\frac{2z}{2\pi i}\int\limits_0^{2\pi}\frac{E_4(\tau,\theta)}{Re^{i\theta}-z}d\theta + E_3(\tau) + \mu\int\limits_a^{\tau}\frac{E_3(\sigma)}{\sigma-a}d\,\sigma + \left[(R-r)^{-2\lambda} f(\tau,z)\right]_{z=0}\} +$$

$$+\{\frac{\lambda}{\pi}\iint\limits_D (R-\rho)^{-2\lambda-1}\frac{exp\,[i\theta]\,f(\tau,s)\,ds}{s}\}\frac{d\tau}{\tau-a}d\tau + (T_{\mu})^{-1}(\Pi_{\lambda})^{-1}(f)\,. \tag{77}$$

The solution of the type (77) exists, if exists the following limit

$$\lim_{t\to a}\,[(t-a)^{\mu} f(t,z)] = F(z)\,,\ F(z)\in C\left(\overline{D}\right), F\left(Re^{i\theta}\right) = 0, \tag{78}$$

with asymptotic behaviors:

$$F(z) = o\left[(R-r)^{\delta_1}\right], \quad \delta_1 > \lambda \quad at \quad R\to r, \tag{79}$$

$$E_1(z)\in C\left(\overline{D}\right), \quad E_1\left(Re^{i\theta}\right) = 0$$

$$E_1(z) = o\left[(R-r)^{\delta_1}\right], \quad \delta_1 > \lambda \quad at \quad R\to r, \tag{80}$$

$$E_3(t)\in C\left(\overline{\Gamma}\right), \quad E_3(a) = 0$$

$$E_3(t) = o\left[(t-a)^{\delta_5}\right], \quad \delta_5 > |\mu| \quad at \quad t\to a. \tag{81}$$

Besides, for convergence of last integrals on the right part of equality (77) it is necessary to have $f(t,z)\in C\left(\overline{\Omega}\right)$, $f\left(a, Re^{i\theta}\right) = 0$ with asymptotic behavior

$$f(t,z) = o\left[(t-a)^{\delta_6}(R-r)^{\delta_7}\right], \delta_6 > |\mu|, \quad \delta_7 > \lambda \quad at \quad (t,z)\to\left(a, Re^{i\theta}\right). \tag{82}$$

So we proved the following theorem

Theorem 15. *Let any condition of the Theorem 8 fulfill. Besides, let function* $f(t,z), F(z),\ E_1(z),\ E_3(t)$ *satisfy conditions* (78)–(82). *Then, problem* $\mathbf{N_2}$ *have unique solution, which is given by formula* (77).

Analogously form stand and investigation boundary valued problems in the case when $\mu > 0$, $\lambda > 0$ and $\mu < 0$, $\lambda < 0$.

Problem N_3. *It is required to find solution integral equation* (17) *from class* $C(\overline{\Omega})$ *at* $\delta = \mu\lambda, \lambda < 0, \mu < 0$ *by boundary conditions*

$$[(t-a)^{\mu}\varphi(t,z)]_{t=a} = E_4(z), \quad z \in D \tag{83}$$

where $E_4(z)$ are given function points of domain D.

Solution Problem N_3. Use inverse formula (59) and condition (83), we have

$$C_1(z) = E_4(z) + [(t-a)^{\mu} f(t,z)]_{t=a} + \frac{\lambda}{\pi}\iint_D \left(\frac{R-r}{R-\rho}\right)^{2\lambda} \frac{exp[i\theta]\,[(t-a)^{\mu} f(t,s)]_{t=a}}{(R-\rho)(s-z)} ds, \tag{84}$$

Substituting this value $C_1(z)$ into formula (64), we receive

$$\varphi(t,z) =$$

$$= (t-a)^{|\mu|}\left\{E_n(z) + [(t-a)^{\mu} f(t,z)]_{t=a} + \frac{\lambda}{\pi}\iint_D \left(\frac{R-r}{R-\rho}\right)^{2\lambda} \frac{exp[i\theta]\,[(t-a)^{\mu} f(t,s)]_{t=a}}{(R-\rho)(s-z)} ds\right\}$$
$$+(T_{\mu})^{-1}(\Pi_{\lambda})^{-1}(f) \tag{85}$$

So we proved the following theorem

Theorem 16. *Let any condition of the Theorem 11 fulfil. Besides, let functions* $E_4(z)$. *Satisfy condition:* $E_4(z) \in C(\overline{D})$, $E_4(Re^{i\Theta}) = 0$ *with asymptotic behavior*

$$E_4(z) = o\left[(R-r)^{\varepsilon}\right], \quad \varepsilon > 0 \quad at \quad R \to r.$$

Then problem N_3 *have unique solution, which is given by formula* (85).

Problem N_4. *It is required to find solution integral equation* (17) *from class* $C(\overline{\Omega})$ *at* $\delta = \mu\lambda, \lambda > 0, \mu > 0$ *by boundary conditions*

$$\left[Re(R-r)^{-\lambda}\varphi(t,z)\right]_{r=R} = E_5(t,\theta), \quad a < t < b, \quad 0 \leqslant \theta \leqslant \pi \tag{86}$$

in lateral surface, and following condition

$$\left[(R-r)^{-2\lambda}\varphi(t,z)\right]_{r=R} = F_3(t), \quad a \leqslant t \leqslant b \tag{87}$$

in cylinder axis, where $E_5(t,\theta)$ *are given function lateral surface* S, $F_3(t)$ *are given function points of cylinder axis* $\Gamma = \{a \leqslant t \leqslant b, z = 0\}$.

Solution Problem N_4. Use inverse formula (66) and condition (86), we have

$$\{Re[\Phi(t,z)]\}_{r=R} = E_5(t,\theta) + \mu \int_a^t \frac{E_5(t,\theta)}{\tau - a} d\tau - \{Re[(R-r)^{-2\lambda}]f(t,z)\}_{r=R} +$$

$$+ \frac{\lambda}{\pi} \iint_D (R-\rho)^{-2\lambda-1} \left\{ Re \left[\frac{exp[i\theta] f(\tau,s)ds}{s-z} \right] \right\}_{r=R} \equiv E_6(t,\theta). \qquad (88)$$

Problem (88) is Schwarz problem with parameters t.

Solution to this problem according to [11] is given by formula

$$\Phi(t,z) = \frac{1}{2\pi} \int_\gamma \frac{\tau + z}{\tau(\tau - z)} E_6(t,\theta) d\tau + i\Phi_2(t) \qquad (89)$$

where $\Phi_2(t)$ arbitrary function point $t (a < t < b)$.

Use formulas (89), (66) condition (87) we have

$$\Phi(t,0) \equiv F_3(t) + \mu \int_a^t \frac{F_2(t)}{\tau - a} d\tau -$$

$$- \left[(R-r)^{-2\lambda} f(t,z) + \frac{\lambda}{\pi} \iint_D (R-\rho)^{-\lambda-1} \frac{exp[i\theta] f(t,s)ds}{s-z} \right]_{z=0} = \frac{\mu}{2\pi} \int_\gamma \frac{E_6(t,\theta)}{\tau} d\tau + i\Phi_2(t)$$

From here we find

$$i\Phi_2(t) = F_3(t) + \mu \int_a^t \frac{F_3(t,)}{\tau - a} d\tau -$$

$$- \left[(R-r)^{-2\lambda} f(t,z) + \frac{\lambda}{\pi} \iint_D (R-\rho)^{-2\lambda-1} \frac{exp[i\theta] f(t,s)ds}{s-z} \right]_{z=0} - \frac{1}{2\pi} \int_\gamma \frac{E_6(t,\theta)}{\tau} d\tau$$

Substituting this value $i\Phi_2(t)$ into formula (89), we receive

$$\Phi(t,z) = \frac{1}{2\pi} \int_\gamma \frac{\tau + z}{\tau(\tau - z)} E_6(t,\theta) d\tau + F_3(t) + \mu \int_a^t \frac{F_3(t)}{\tau - a} d\tau -$$

$$- \left[(R-r)^{-2\lambda} f(t,z) + \frac{\lambda}{\pi} \iint_D (R-\rho)^{-2\lambda-1} \frac{exp[i\theta] f(t,s)ds}{s-z} \right]_{z=0} - \frac{1}{2\pi} \int_\gamma \frac{E_6(t,\theta)}{\tau} d\tau. \qquad (90)$$

Substituting this value $\Phi(t,z)$ from formula (90) to (60) we have

$$\varphi(t,z) = (R-r)^{2\lambda} (T_\mu)^{-1} \left\{ \frac{1}{2\pi} \int_\gamma \frac{\tau + z}{\tau(\tau - z)} E_6(t,\theta) d\tau + F_3(t) + \mu \int_a^t \frac{F_3(t)}{\tau - a} d\tau \right\} -$$

$$- (R-r)^{2\lambda}(T_\mu)^{-1}\left\{\left[(R-r)^{-2\lambda}f(t,z)+\frac{\lambda}{\pi}\iint\limits_D (R-\rho)^{-2\lambda-1}\frac{exp[i\theta]f(t,s)ds}{s-z}\right]_{z=0}\right\}-$$
$$-(R-r)^{2\lambda}(T_\mu)^{-1}\left\{\frac{1}{2\pi}\int\limits_\gamma \frac{E_6(t,\theta)}{\tau}d\tau\right\}+(T_\mu)^{-1}(\Pi_T)^{-1}(f) \tag{91}$$

The solution of the type (91) exists, if exists the following limit

$$\lim_{t\to a}\left[(t-a)^\mu f(t,z)\right] = F(z),\ F(z)\in C\left(\overline{D}\right), F\left(Re^{i\theta}\right)=0, \tag{92}$$

with asymptotic behavior

$$F(z)=o\left[(R-r)^{\delta_2}\right], \delta_2>\lambda, \quad at \quad R\to r, \tag{93}$$

$E_6(t,\theta)\in C(\overline{S})$, $F_3(t)\in C(\overline{\Gamma})$, $F_3(a)=0$ with asymptotic behavior

$$F_3(t)=o\left[(t-a)^\varepsilon\right], \quad \varepsilon>0, \quad at \quad t\to a. \tag{94}$$

So we proved the following theorem

Theorem 17. *Let any condition of the Theorem 10 fulfil. Besides, let functions* $f(t,s)$, $F(z)$, $E_6(t,\theta)$, $F_3(t)$ *satisfy conditions* (92)–(94). *Then problem* N_4 *have unique solution, which is given by formula* (91).

Acknowledgement. The author is grateful to Professor Michael Ruzhansky for discussions and remarks about the paper.

References

1. Rajabov, N.: Volterra Type Integral Equations with Boundary and Interior Fixed Singularity and Super-singularity Kernels and their Application. LAP LAMBERT Academic Publishing, Germany (2011)
2. Rajabov, N., Rajabova, L.: Introduction to Theory Many-Dimensional Volterra Type Integral Equation with Singularity and Super-singularity Kernels and its Application. LAP LAMBERT Academic Publishing, Germany (2012)
3. Rajabov, N.: An explicit solution to a class of a second kind complex integral equation with singular and super-singular kernel. In: Functional–Analytic and Complex Methods, their Interactions, and Applications to Partial Differential Equations, pp. 313–329. World Scientific (2001)
4. Rajabov, N.: An explicit solution to a class of a second kind linear systems of complex integral equations with singular and super-singular kernel(Abstract). In: Functional–Analytic and Complex Methods, their Interactions, and Applications to Partial Differential Equations, pp. 329–332. World Scientific (2001)
5. Rajabov, N.: To theory one class of two dimensional complex integral equation for which all boundary is singular line. Report of the Adyghe (Circassia) Int. Acad. Sci. **16**(1), 16–23 (2014)

6. Rajabov, N.: To theory one class of Volterra type singular integral equation by cylindrical domain. Report of the Adyghe(Circassia) Int. Acad. Sci. **17**(4), 77–83 (2015)
7. Rajabov, N.: Generalized two-dimensional complex integral equation with boundary singular kernels. In: Books “Modern Problems Mathematical Analysis and Function Theory” (Materials International Scientific Conference Dedicate to 60th on the Occasion of his 60th Anniversary Academician AS Republic of Tajikistan Shobozov M. Sh.), Dushanbe, 29–30 June 2012, Dushanbe-2012, pp. 129–131 (2012)
8. Rajabov, N.: To theory one class singular integral equation by cylindrical domain. In: Materials International Scientific Conference on Mathematical Analysis, Differential Equation and Number Theory, Dushanbe-2015, pp. 140–143 (2015)
9. Rajabov, N.: Riemann-Hilbert boundary value problems for one class of two-dimensional complex integral equation with boundary singular lines. In: Muchsinov, A., Nazimov A.B., Bayzoev, S., Ocimova, D., Tuchliev, K. (eds.) Materials of the International Scientific Conference “Modern Problems in Mathematics and its Teaching” dedicated to Constitution Republic of Tajikistan on the his 20th Anniversary and Learned Mathematicians on the occasion of his 60th anniversary Transactions learned Chujand University in the name of B. Gafurov, no 2, Part I, Chujand 2014, pp. 227–229 (2014)
10. Rajabov, N.: Ney methods for Volterra type integral equation with boundary singular point. Ney trends in analysis and interdisciplinary applications. In: Selected Contributions of the 10th ISAAC Congress, Macau 2015, Birkhauser. Springer International Publishing AG, pp. 121–127 (2017)
11. Gachov, F.D.: Boundary Value Problems. Dover Publications Inc., New York (1990)
12. Tricomi, F.G.: Integral Equations. Dover Publications Inc., New York (1985)
13. Sabitov, K.: Functional, differential and integral equations
14. Mikhlin, S.G.: Multidimensional Singular Integrals and Integral Equations. Translated from the Russian by Whyte, W.J.A.. Translation edited by Sneddon, I.N. Pergamon Press, Oxford, New York, Paris (1965)
15. Muskhelishvili, N.I.: Singular Integral Equations. Boundary Problems of Function Theory and their Application to Mathematical Physics. Dover Publications Inc., New York (1992)
16. Vekua, I.N.: Generalized Analytic Functions. Pergamon Press, London, Paris, Frankfurt; Addison-Wesley Publishing Co., Inc., Reading, Mass (1962)

Weighted Stepanov-Like Pseudo Almost Automorphic Solutions of Class r for Some Partial Differential Equations

Hamidou Toure(✉) and Issa Zabsonre

Unité de Recherche et de Formation en Sciences Exactes et Appliquées, Département de Mathématiques, Université de Ouaga 1 Pr Joseph Ki-Zerbo, Ouagadougou 03 B.P.7021, Burkina Faso
toureh98@yahoo.fr, zabsonreissa@yahoo.fr

Abstract. The aim of this work is to study weighted Stepanov-like pseudo almost automorphic functions using the measure theory. We present a new concept of weighted ergodic functions which is more general than the classical one. Then we establish many interesting results on the functional space of such functions. We also study the existence and uniqueness of (μ, ν) -Weighted Stepanov-like pseudo almost automorphic solutions of class r for some neutral partial functional differential equations in a Banach space when the delay is distributed using the spectral decomposition of the phase space developed by Adimy and co-authors. Here we assume that the undelayed part is not necessarily densely defined and satisfies the well-known Hille-Yosida condition, the delayed part are assumed to be pseudo almost automorphic with respect to the first argument and Lipschitz continuous with respect to the second argument.

Keywords: Automorphic solutions · Pseudo periodic functions
Lipschitz condition

1 Introduction

In this work, we study the existence and uniqueness of Stepanov-like pseudo almost automorphic solutions of class r for the following neutral partial functional differential equation

$$u'(t) = Au_t + L(u_t) + f(t) \text{ for } t \in \mathbb{R}, \tag{1}$$

where A is a linear operator on a Banach space X satisfying the Hille-Yosida condition,

J. Delgado and M. Ruzhansky (Eds.): *Analysis and Partial Differential Equations: Perspectives from Developing Countries*, PROMS 275, pp. 242–269, 2018.
https://doi.org/10.1007/978-3-030-05657-5_15

$C = C([-r, 0]; X)$ denotes the space of continuous functions from $[-r, 0]$ to X endowed with the uniform norm topology. For every $t \geq 0$, u_t denotes the history function of C defined by

$$u_t(\theta) = u(t + \theta) \text{ for } \quad -r \leq \theta \leq 0.$$

L is a bounded linear operator from C into X and $f : \mathbb{R} \to X$ is a continuous function.

Almost automorphic functions are more general than almost periodic functions and they were introduced by Bochner [4,5], for more details about this topics we refer to the recents books [10,12] where the author gave an important overview about the theory of almost automorphic functions and their applications to differential equations. In [11] the authors introduced and studied a new class of Stepanov-like almost automorphic functions with values in a Banach space. Almost automorphic solutions in the context of differential equations has been studied by many authors. In [15], the authors presented a new approach dealing with weighted pseudo almost periodic functions and their applications in evolution equations and partial functional differential equations by using the measure theory.

The aim of this work is to generalise [15] by proving the existence of Stepanov-like pseudo almost automorphic solutions of Eq. 1 when the delay is distributed on $[-r, 0]$. Our approach is based on the variation of constants formula and the spectral decomposition of the phase space developed in [2].

This work is organised as follow, in Sect. 2 we recall some preliminary results on variation of constants formula and spectral decomposition. In Sects. 3 and 4, we recall some preliminary results on Stepanov like pseudo almost automorphic functions that will be used in this work. In Sect. 5, we prove some properties of S^p-pseudo almost periodic function of class r. In Sect. 6, we discuss the main result of this paper. Using the strict contraction principle we show the existence and uniqueness of Stepanov-like pseudo almost automorphic periodic solution of class r for Eq. 1. Finally, for illustration, we propose to study the existence and uniqueness of S^p-pseudo almost automorphic solution for some model arising in the population dynamics.

2 Variation of Constants Formula and Spectral Decomposition

To Eq. 1, we associate the following initial value problem

$$\begin{cases} \dfrac{d}{dt}u(t) = Au(t) + L(u_t) + f(t) \text{ for } t \geq 0 \\ \\ u_0 = \varphi \in C = C([-r, 0]; X), \end{cases} \tag{2}$$

where $f : \mathbb{R}^+ \to X$ is continuous. Let us introduce the part A_0 of the operator A in $\overline{D(A)}$ which is defined by

$$\begin{cases} D(A_0) = \{x \in D(A) : \ Ax \in \overline{D(A)}\} \\ A_0 x = Ax \text{ ;for } \ x \in D(A_0) \end{cases} \tag{3}$$

We make the following assumption:
($\mathbf{H_0}$) A satisfies the Hille-Yosida condition.

Lemma 1. ([1]) *A_0 generates a strongly continuous semigroup $(T_0(t))_{t\geq 0}$ on $\overline{D(A)}$.*

The phase space C_0 of Eq. 2 is defined by

$$C_0 = \{\varphi \in C : \ \varphi(0) \in \overline{D(A)}\}. \tag{4}$$

For each $t \geq 0$, we define the linear operator $\mathcal{U}(t)$ on C_0 by

$$\mathcal{U}(t) = v_t(., \varphi) \tag{5}$$

where $v(., \varphi)$ is the solution of the following homogeneous equation

$$\begin{cases} \dfrac{d}{dt}v(t) = Av(t) + L(v_t) \text{ for } t \geq 0 \\ v_0 = \varphi \in C. \end{cases} \tag{6}$$

Theorem 1. ([2]) *Let $\mathcal{A}_{\mathcal{U}}$ defined on C_0 by*

$$\begin{cases} D(\mathcal{A}_{\mathcal{U}}) = \left\{\varphi \in C^1([-r, 0]; X); \ \varphi(0) \in D(A), \ \varphi(0)' \in \overline{D(A)} \ \textit{and} \ \varphi(0)' = A\varphi(0) + L(\varphi)\right\} \\ \mathcal{A}_{\mathcal{U}}\varphi = \varphi' \ \textit{for} \ \varphi \in D(\mathcal{A}_{\mathcal{U}}). \end{cases}$$

Then $\mathcal{A}_{\mathcal{U}}$ is the infinitesimal generator of the semigroup $(\mathcal{U}(t))_{t\geq 0}$ on C_0.

Let $\langle X_0 \rangle$ be the space defined by

$$\langle X_0 \rangle = \{X_0 c : \ c \in X\} \tag{7}$$

where the function $X_0 c$ is defined by

$$(X_0 c)(\theta) = \begin{cases} 0 & \text{if} \quad \theta \in [-r, 0[, \\ c & \text{if} \quad \theta = 0. \end{cases}$$

The space $C_0 \oplus \langle X_0 \rangle$ equipped with the norm $|\phi + X_0 c| = |\phi|_C + |c|$ for $(\phi, c) \in C_0 \times X$ is a Banach space and consider the extension $\mathcal{A}_{\mathcal{U}}$ defined on $C_0 \oplus \langle X_0 \rangle$ by

$$\begin{cases} D(\widetilde{\mathcal{A}_{\mathcal{U}}}) = \left\{\varphi \in C^1([-r, 0]; X) : \ \varphi(0) \in D(A) \ \text{ and } \varphi(0)' \in \overline{D(A)}\right\} \\ \widetilde{\mathcal{A}_{\mathcal{U}}}\varphi = \varphi' + X_0(A\varphi(0) + L(\varphi) - \varphi(0)'). \end{cases} \tag{8}$$

Definition 1. We say a semigroup $(\mathcal{U}(t))_{t\geq 0}$ is hyperbolic if

$$\sigma(\mathcal{A}_{\mathcal{U}}) \cap i\mathbb{R} = \emptyset \tag{9}$$

For the sequel, we make the following assumption:
($\mathbf{H_1}$) $T_0(t)$ is compact on $\overline{D(A)}$ for every $t > 0$.

Theorem 2. ([2]) *Assume that* ($\boldsymbol{H_0}$) *and* ($\boldsymbol{H_1}$) *hold. Then the semigroup* $(\mathcal{U}(t))_{t\geq 0}$ *is compact for* $t > r$.

From the compactness of the semigroup $(\mathcal{U}(t))_{t\geq 0}$, we get the following result on the spectral decomposition of the phase space C_0.

Theorem 3. ([9]) *Assume that* ($\boldsymbol{H_1}$) *holds. If the semigroup* $(\mathcal{U}(t))_{t\geq 0}$ *is hyperbolic, then the space* C_0 *is decomposed as a direct sum*

$$C_0 = S \oplus U \tag{10}$$

of two $\mathcal{U}(t)$ *invariant closed subspaces* S *and* U *such that the restricted semigroup on* $\mathcal{U}$ *is a group and there exist positive constants* $\overline{M}$ *and* ω *such that*

$$|\mathcal{U}(t)\varphi| \leq \overline{M}e^{-\omega t}|\varphi| \ \ for\ \ t \geq 0 \ \ and \ \ \varphi \in S$$

$$|\mathcal{U}(t)\varphi| \leq \overline{M}e^{\omega t}|\varphi| \ \ for\ \ t \leq 0 \ \ and \ \ \varphi \in U,$$

where S *and* U *are called respectively the stable and unstable space,* Π^s *and* Π^u *denote respectively the projection operator on* S *and* U.

3 Almost Automorphic Functions and (μ, ν) Ergodic Functions

In this section, we recall some properties about (μ, ν)-pseudo almost automorphic functions. Let $BC(\mathbb{R}; X)$ be the space of all bounded and continuous function from $\mathbb{R}$ to X equipped with the uniform topology norm. We denote by $\mathcal{B}$ the Lebesgue σ-field of $\mathbb{R}$ and by $\mathcal{M}$ the set of all positive measures μ on $\mathcal{B}$ satisfying $\mu(\mathbb{R}) = +\infty$ and $\mu([a, b]) < \infty$, for all $a, b \in \mathbb{R}$ $(a \leq b)$.

Definition 2. A bounded continuous function $\phi : \mathbb{R} \to X$ is called almost automorphic if for each real sequence (s_m), there exists a subsequence (s_n) such that

$$g(t) = \lim_{n\to+\infty} \phi(t + s_n) \tag{11}$$

is well defined for each $t \in \mathbb{R}$ and

$$\lim_{n\to+\infty} g(t - s_n) = \phi(t) \tag{12}$$

for each $t \in \mathbb{R}$.

We denote by $AA(\mathbb{R}, X)$, the space of all such functions.

Proposition 1. ([13]) $AA(\mathbb{R}, X)$ *equipped with the sup norm is a Banach space.*

Definition 3. Let X_1 and X_2 be two Banach spaces. A bounded continuous function $\phi : \mathbb{R} \times X_1 \to X_2$ is called almost automorphic in $t \in \mathbb{R}$ uniformly for each x in X_1 if for every real sequence (s_m), there exists a subsequence (s_n) such that

$$g(t,x) = \lim_{n\to+\infty} \phi(t+s_n, x) \text{ in } X_2 \tag{13}$$

is well defined for each $t \in \mathbb{R}$ and each $x \in X_1$ and

$$\lim_{n\to+\infty} g(t-s_n, x) = \phi(t,x) \text{ in } X_2 \tag{14}$$

for each $t \in \mathbb{R}$ and for every $x \in X_1$.

Denote by $AA(\mathbb{R} \times X_1; X_2)$ the space of all such functions.

Definition 4. A bounded continuous function $\phi : \mathbb{R} \to X$ is called compact almost automorphic if for each real sequence (s_m), there exists a subsequence (s_n) such that

$$g(t) = \lim_{n\to+\infty} \phi(t+s_n) \text{ and } \lim_{n\to+\infty} g(t-s_n) = \phi(t) \tag{15}$$

uniformly on compact subsets of $\mathbb{R}$.
We denote by $AA_c(\mathbb{R}; X)$, the space of all such functions.

It is well known that $AA_c(\mathbb{R}; X)$ is closed subsets of $(BC(\mathbb{R}, X), |\cdot|_\infty)$.

Lemma 2. ([13]) *$AA_c(\mathbb{R}; X)$ equipped with the sup norm is a Banach space.*

Definition 5. Let X_1 and X_2 be two Banach spaces. A bounded continuous function $\phi : \mathbb{R} \times X_1 \to X_2$ is called compact almost automorphic in $t \in \mathbb{R}$ if every real sequence (s_m), there exists a subsequence (s_n) such that

$$g(t,x) = \lim_{n\to+\infty} \phi(t+s_n, x) \text{ and } \lim_{n\to+\infty} g(t-s_n, x) = \phi(t,x) \text{ in } X_2 \tag{16}$$

where the limits are uniform on compact subsets of $\mathbb{R}$ for each $x \in X_1$.
Denote by $AA_c(\mathbb{R} \times X_1; X_2)$ the space of all such functions.

In the sequel, we recall some preliminary results concerning the (μ,ν)-Pseudo almost automorphic functions. $\mathcal{E}(\mathbb{R}; X, \mu, \nu)$ stands for the space of functions

$$\mathcal{E}(\mathbb{R}; X, \mu, \nu) = \Big\{u \in BC(\mathbb{R}; X) : \lim_{\tau\to+\infty} \frac{1}{\nu([-\tau,\tau])} \int_{-\tau}^{+\tau} |u(t)| d\mu(t) = 0\Big\}. \tag{17}$$

To study delayed differential equations for which the history belong to $C([-r,0]; X)$, we need to introduce the space

$$\mathcal{E}(\mathbb{R}; X, \mu, \nu, r) = \Big\{u \in BC(\mathbb{R}; X) : \lim_{\tau\to+\infty} \frac{1}{\nu([-\tau,\tau])} \int_{-\tau}^{+\tau} \Big(\sup_{\theta\in[t-r,t]} |u(\theta)|\Big) d\mu(t) = 0\Big\}. \tag{18}$$

In addition to above-mentioned space, we consider the following spaces

$$\mathcal{E}(\mathbb{R}\times X_1, X_2, \mu, \nu) = \left\{u \in BC(\mathbb{R}\times X_1; X_2) : \lim_{\tau\to+\infty} \frac{1}{\nu([-\tau,\tau])}\int_{-\tau}^{+\tau} |u(t,x)|_{X_2} d\mu(t) = 0\right\}, \tag{19}$$

$$\mathcal{E}(\mathbb{R}\times X_1; X_2, \mu, \nu, r) = \left\{u \in BC(\mathbb{R}\times X_1; X_2) : \lim_{\tau\to+\infty} \frac{1}{\nu([-\tau,\tau])}\int_{-\tau}^{+\tau} \left(\sup_{\theta\in[t-r,t]} |u(\theta,x)|_{X_2}\right) d\mu(t) = 0\right\}, \tag{20}$$

where in both cases the limit (as $\tau \to +\infty$) is uniform in compact subset of X_1.

In view of previous definitions, it is clear that the spaces $\mathcal{E}(\mathbb{R}; X, \mu, \nu, r)$ and $\mathcal{E}(\mathbb{R}\times X_1; X_2, \mu, \nu, r)$ are continuously embedded in $\mathcal{E}(\mathbb{R}; X, \mu, \nu)$ and $\mathcal{E}(\mathbb{R}\times X_1, X_2, \mu, \nu)$, respectively.

Definition 6. We say that a continuous function f is ρ-weighted pseudo almost automorphic if $f = g + \phi$, where g is almost automorphic and ϕ is ergodic with respect to some weighted function ρ in the sense that

$$\lim_{\tau\to+\infty} \frac{1}{m(\rho,\tau)}\int_{-\tau}^{+\tau} |\phi(t)|\rho(t)dt = 0, \tag{21}$$

where $m(\rho,\tau) = \displaystyle\int_{-\tau}^{+\tau} \rho(t)dt$ and ρ is assumed to be positive and locally integrable.

On the other hand, one can observe that a ρ-weighted pseudo almost automorphic functions is μ-pseudo almost automorphic, where the measure μ is absolutely continuous with respect to the Lebesgue measure and its Radon-Nikodym derivative is ρ:

$$d\mu(t) = \rho(t)dt. \tag{22}$$

Example 1. [3] Let ρ be a nonnegative $\mathcal{B}$-measurable function. Denote by μ the positive measure defined by

$$\mu(A) = \int_A \rho(t)dt, \quad \text{for } A \in \mathcal{B}, \tag{23}$$

where dt denotes the Lebesgue measure on $\mathbb{R}$. The function ρ which occurs in Eq. 23 is called the Radon-Nikodym derivative of μ with respect to the Lebesgue measure on $\mathbb{R}$.

4 (μ,ν)-Stepanov-Like Pseudo Almost Periodic Functions

Definition 7. The Bochner transform $f^b(t,s)$, $t\in\mathbb{R}$, $s\in[0,1]$, of a function $f(t)$ on $\mathbb{R}$, with values in X, is defined by

$$f^b(t,s) = f(t+s). \tag{24}$$

Remark 1. If $f = h + \varphi$, then $f^b = h^b + \varphi^b$. Moreover, $(\lambda f)^b = \lambda f^b$ for each scalar λ.

Definition 8. The Bochner transform $F^b(t,s,u)$, $t \in \mathbb{R}$, $s \in [0,1]$, $u \in X$ of a function $F(t,u)$ on $\mathbb{R} \times X$, with values in X, is defined by

$$F^b(t,s,u) = F(t+s,u) \text{ for each } u \in X. \tag{25}$$

Definition 9. Let $p \in [1,+\infty[$. The space $BS^p(\mathbb{R},X)$ of all Stepanov bounded functions, with the exponent p, consists of all measurable functions F on $\mathbb{R}$ with values in X such that $f^b \in L^\infty(\mathbb{R}, L^p([0,1],X))$. This is a Banach space with the norm

$$\|f\|_{S^p} = \sup_{t\in\mathbb{R}} \Big(\int_t^{t+1} |f(s)|^p ds\Big)^{\frac{1}{p}}. \tag{26}$$

Definition 10. A bounded continuous function $\phi : \mathbb{R} \to X$ is called (μ,ν)-pseudo almost automorphic if $\phi = \phi_1 + \phi_2$ where $\phi_1 \in AA(\mathbb{R};X)$ and $\phi_2 \in \mathscr{E}(\mathbb{R};X,\mu,\nu)$.

We denote by $PAA(\mathbb{R};X,\mu,\nu)$, the space of all such functions.

Definition 11. Let X_1 and X_2 be two Banach spaces. A bounded continuous function $\phi : \mathbb{R} \times X_1 \to X_2$ is called uniformly (μ,ν)-pseudo almost automorphic if $\phi = \phi_1 + \phi_2$, where $\phi_1 \in AA(\mathbb{R} \times X_1; X_2)$ and $\phi_2 \in \mathscr{E}(\mathbb{R} \times X_1; X_2, \mu, \nu)$.

We denote by $PAA(\mathbb{R} \times X_1; X_2, \mu, \nu)$, the space of all such functions.
We now introduce some new spaces used in the sequel.

Definition 12. A function $f \in BS^p(\mathbb{R},X)$ is called (μ,ν)-S^p-pseudo-almost automorphic (or Stepanov-like pseudo-almost automorphic) if it can be expressed as $f = h+\varphi$, where $h^b \in AA(\mathbb{R}, L^p((0,1),X))$ and $\varphi^b \in \mathscr{E}(\mathbb{R}, L^p((0,1),X),\mu,\nu)$. The collection of such functions will be denoted by $PAAS^p(\mathbb{R},X,\mu,\nu)$.

In other words, a function $f \in L^p(\mathbb{R},X)$ is said to be S^p-pseudo-almost automorphic if its Bochner transform $f^b : \mathbb{R} \to L^p(0,1),X)$ is pseudo-almost periodic in the sense that there exist two functions $h,\varphi : \mathbb{R} \to X$ such that $f = h+\varphi$, where $h^b \in AA(\mathbb{R}, L^p((0,1),X))$ and $\varphi^b \in \mathscr{E}(\mathbb{R}, L^p((0,1),X),\mu,\nu)$, i.e., according to [11] (Definition 2.5 Page 2660.), for each real sequence (s_m), there exists a subsequence (s_n) and a function $g \in L^p_{loc}(\mathbb{R};X)$ such that

$$\lim_{n\to+\infty} \Big(\int_t^{t+1} |g(s) - h(s+s_n)|^p ds\Big)^{\frac{1}{p}} = 0 \tag{27}$$

and

$$\lim_{n\to+\infty} \Big(\int_t^{t+1} |g(s+s_n) - h(s_n)|^p ds\Big)^{\frac{1}{p}} = 0 \tag{28}$$

pointwise on $\mathbb{R}$ and

$$\lim_{\tau\to+\infty} \frac{1}{\nu[-\tau,\tau]} \int_{-\tau}^{\tau} \Big(\int_t^{t+1} |\varphi(s)|^p ds\Big)^{\frac{1}{p}} d\mu(t) = 0 \text{ for } t \in \mathbb{R}. \tag{29}$$

Definition 13. A function $f \in BS^p(\mathbb{R}, X)$ is called (μ,ν)-S^p-pseudo-almost automorphic of class r (or Stepanov-like pseudo-almost automorphic of class r) if it can be expressed as $f = h + \varphi$, where $h^b \in AA(\mathbb{R}, L^p((0,1),X))$ and $\varphi^b \in \mathscr{E}(\mathbb{R}, L^p((0,1),X), \mu, \nu, r)$ i.e.,

$$\lim_{\tau\to+\infty} \frac{1}{\nu[-\tau,\tau]} \int_{-\tau}^{\tau} \sup_{\theta\in[t-r,t]} \left(\int_{\theta}^{\theta+1} |\varphi(s)|^p ds \right)^{\frac{1}{p}} d\mu(t) = 0. \tag{30}$$

The collection of such functions will be denoted by $PAAS^p(\mathbb{R}, X, \mu, \nu, r)$.

Definition 14. A function $f : \mathbb{R} \times X_1 \to X_2$, $(t,x) \mapsto f(t,x)$ with $f(.,x) \in L^p(\mathbb{R}, X_2)$ is called (μ,ν)-S^p-pseudo-almost automorphic of class r (or Stepanov-like pseudo-almost automorphic of class r) if it can be expressed as $f = h + \varphi$, where $h^b \in AA(\mathbb{R}\times X_1, L^p((0,1),X_2))$ and $\varphi^b \in \mathscr{E}[(\mathbb{R}\times X_1, L^p((0,1),X_2), \mu, \nu, r]$ i.e.,

$$\lim_{\tau\to+\infty} \frac{1}{\nu[-\tau,\tau]} \int_{-\tau}^{\tau} \sup_{\theta\in[t-r,t]} \left(\int_{\theta}^{\theta+1} |\varphi(s,x)|_{X_2}^p ds \right)^{\frac{1}{p}} d\mu(t) = 0. \tag{31}$$

The collection of such functions will be denoted by $PAAS^p[\mathbb{R} \times X_1, L^p((0,1),X_2), \mu, \nu, r]$.

Definition 15. A bounded continuous function $\phi : \mathbb{R} \to X$ is called (μ,ν)-S^p-pseudo compact almost automorphic of class r if $\phi = \phi_1 + \phi_2$ where $\phi_1 \in AA_c(\mathbb{R}, L^p((0,1),X))$ and $\phi_2 \in \mathscr{E}(\mathbb{R}; L^p((0,1),X), \mu, \nu, r)$.

We denote by $PAA_cS^p(\mathbb{R}, L^p((0,1),X), \mu, \nu, r)$, the space of all such functions.

Definition 16. Let X_1 and X_2 be two Banach spaces. A bounded continuous function $\phi : \mathbb{R} \times X_1 \to X_2$ is called uniformly (μ,ν)-S^p-pseudo almost automorphic of class r (respectively uniformly pseudo compact almost automorphic of class r) if $\phi = \phi_1 + \phi_2$, where $\phi_1 \in AA(\mathbb{R} \times X_1, L^p((0,1),X_2))$ and $\phi_2 \in \mathscr{E}(\mathbb{R} \times X_1, L^p((0,1),X_2), \mu, \nu, r)$ (respectively if $\phi = \phi_1 + \phi_2$, where $\phi_1 \in AA_c(\mathbb{R} \times X_1, L^p((0,1),X_2))$ and $\phi_2 \in \mathscr{E}(\mathbb{R} \times X_1, L^p((0,1),X_2), \mu, \nu, r)$).

We denote by $PAAS^p(\mathbb{R}\times X_1, L^p((0,1),X_2), \mu, \nu, r)$ (respectively $PAA_cS^p(\mathbb{R}\times X_1, L^p((0,1),X_2), \mu, \nu, r)$) the space of all such functions.

5 Properties of μ-Stepanov-Like Pseudo Almost Automorphic Functions of Class r

From $\mu, \nu \in \mathcal{M}$, we formulate the following hypothesis.

($\mathbf{H_2}$) Let $\mu, \nu \in \mathcal{M}$ be such that $\limsup_{\tau\to+\infty} \frac{\mu([-\tau,\tau])}{\nu([-\tau,\tau])} = \alpha < \infty$.

Proposition 2. *$PAAS^p(\mathbb{R}; L^p((0,1),X), \mu, \nu, r)$ is a closed subspace of $BS^p(\mathbb{R}; X)$.*

Proof. Let $(x_n)_n$ be a sequence in $PAAS^p(\mathbb{R}; L^p((0,1),X),\mu,\nu,r)$ such that $\lim_{n\to\infty} x_n = x$ in $BC(\mathbb{R};X)$. For each n, let $x_n = y_n + z_n$ with $y_n^b \in AA(\mathbb{R}; L^p((0,1),X))$ and $z_n^b \in \mathscr{E}(\mathbb{R}; L^p((0,1),X),\mu,\nu,r)$. $(y_n)_n$ converges to some $y \in BS^p(\mathbb{R};X)$ and $(z_n)_n$ also converges to some $z \in BS^p(\mathbb{R};X)$.

In [11] Theorem 2.3, $y \in AA(\mathbb{R}; L^p((0,1),X))$. It remains to show that $z \in \mathscr{E}(\mathbb{R}; L^p((0,1),X),\mu,\nu,r)$ By the Minkowski's inequality, we have

$$\begin{aligned}
&\frac{1}{\nu[-\tau,\tau]}\int_{-\tau}^{+\tau}\sup_{\theta\in[t-r,t]}\Big(\int_\theta^{\theta+1}|z(s)|^p ds\Big)^{\frac{1}{p}} d\mu(t)\\
&= \frac{1}{\nu[-\tau,\tau]}\int_{-\tau}^{+\tau}\sup_{\theta\in[t-r,t]}\Big(\int_\theta^{\theta+1}|z(s)-z_n(s)+z_n(s)|^p ds\Big)^{\frac{1}{p}} d\mu(t)\\
&\leq \frac{1}{\nu[-\tau,\tau]}\int_{-\tau}^{+\tau}\sup_{\theta\in[t-r,t]}\Big[\Big(\int_\theta^{\theta+1}|z(s)-z_n(s)|^p ds\Big)^{\frac{1}{p}}+\Big(\int_\theta^{\theta+1}|z_n(s)|^p)ds\Big)^{\frac{1}{p}}\Big] d\mu(t)\\
&\leq \frac{1}{\nu[-\tau,\tau]}\int_{-\tau}^{+\tau}\sup_{t\in\mathbb{R}}\Big(\int_t^{t+1}|z(s)-z_n(s)|^p ds\Big)^{\frac{1}{p}} d\mu(t)\\
&\quad+\frac{1}{\nu[-\tau,\tau]}\int_{-\tau}^{+\tau}\sup_{\theta\in[t-r,t]}\Big(\int_\theta^{\theta+1}|z_n(s)|^p)ds\Big)^{\frac{1}{p}} d\mu(t)\\
&\leq \|z-z_n\|_{S^p}\times\frac{\mu([-\tau,\tau])}{\nu([-\tau,\tau])}+\frac{1}{\nu[-\tau,\tau]}\int_{-\tau}^{+\tau}\sup_{\theta\in[t-r,t]}\Big(\int_\theta^{\theta+1}|z_n(s)|^p)ds\Big)^{\frac{1}{p}} d\mu(t).
\end{aligned}$$

Then we get $z \in \mathcal{E}(\mathbb{R}; L^p((0,1),X),\mu,\nu,r)$, hence $x \in PAAS^p(\mathbb{R}; L^p((0,1),X),\mu,\nu,r)$.■

Proposition 3. *The space $PAAS^p(\mathbb{R}; L^p((0,1),X)\mu,\nu,r)$ endowed with the $\|\,.\,\|_{S^p}$ norm is a Banach space.*

Next result is a characterization of (μ,ν)-ergodic functions of class r.

Theorem 4. *Assume that $(\boldsymbol{H}_2)$ holds and let $\mu,\nu \in \mathcal{M}$ and I be a bounded interval (eventually $I=\emptyset$). Assume that $f \in BS^p(\mathbb{R},X)$. Then the following assertions are equivalent:*

(i) $f \in \mathcal{E}(\mathbb{R}; L^p((0,1),X),\mu,\nu,r)$.

(ii) $\displaystyle\lim_{\tau\to+\infty}\frac{1}{\nu([-\tau,\tau]\setminus I)}\int_{[-\tau,\tau]\setminus I}\sup_{\theta\in[t-r,t]}\Big(\int_\theta^{\theta+1}|f(s)|^p ds\Big)^{\frac{1}{p}} d\mu(t)=0.$

(iii) For any

$$\varepsilon>0,\ \lim_{\tau\to+\infty}\frac{\mu\Big(\Big\{t\in[-\tau,\tau]\setminus I:\ \sup_{\theta\in[t-r,t]}\Big(\int_\theta^{\theta+1}|f(s)|^p ds\Big)^{\frac{1}{p}}>\varepsilon\Big\}\Big)}{\nu([-\tau,\tau]\setminus I)}=0.$$

Proof. The proof is made like the proof of Theorem 2.13 in [3].

$(i) \Leftrightarrow (ii)$ Denote by $A = \nu(I)$, $B = \int_I \sup_{\theta \in [t-r,t]} \Big(\int_\theta^{\theta+1} |f(s)|^p ds \Big)^{\frac{1}{p}} d\mu(t)$. We have A and $B \in \mathbb{R}$, since the interval I is bounded and the function f is bounded and continuous. For $\tau > 0$ such that $I \subset [-\tau, \tau]$ and $\nu([-\tau, \tau] \setminus I) > 0$, we have

$$\frac{1}{\nu([-\tau,\tau] \setminus I)} \int_{[-\tau,\tau]\setminus I} \sup_{\theta \in [t-r,t]} \Big(\int_\theta^{\theta+1} |f(s)|^p ds \Big)^{\frac{1}{p}} d\mu(t)$$

$$= \frac{1}{\nu([-\tau,\tau]) - A} \Big[\int_{[-\tau,\tau]} \sup_{\theta \in [t-r,t]} \Big(\int_\theta^{\theta+1} |f(s)|^p ds \Big)^{\frac{1}{p}} d\mu(t) - B \Big]$$

$$= \frac{\nu([-\tau,\tau])}{\nu([-\tau,\tau]) - A} \Big[\frac{1}{\nu([-\tau,\tau])} \int_{[-\tau,\tau]} \sup_{\theta \in [t-r,t]} \Big(\int_\theta^{\theta+1} |f(s)|^p ds \Big)^{\frac{1}{p}} d\mu(t) - \frac{B}{\nu([-\tau,\tau])} \Big].$$

From above equalities and the fact that $\nu(\mathbb{R}) = +\infty$, we deduce that (ii) is equivalent to

$$\lim_{\tau \to +\infty} \frac{1}{\nu([-\tau,\tau])} \int_{-\tau}^{+\tau} \sup_{\theta \in [t-r,t]} \Big(\int_\theta^{\theta+1} |f(s)|^p ds \Big)^{\frac{1}{p}} d\mu(t) = 0,$$

that is (i).
$(iii) \Rightarrow (ii)$ Denote by A_τ^ε and B_τ^ε the following sets

$$A_\tau^\varepsilon = \Big\{ t \in [-\tau,\tau] \setminus I : \sup_{\theta \in [t-r,t]} \Big(\int_\theta^{\theta+1} |f(s)|^p ds \Big)^{\frac{1}{p}} > \varepsilon \Big\}$$

and

$$B_\tau^\varepsilon = \Big\{ t \in [-\tau,\tau] \setminus I) : \sup_{\theta \in [t-r,t]} \Big(\int_\theta^{\theta+1} |f(s)|^p ds \Big)^{\frac{1}{p}} \le \varepsilon \Big\}.$$

Assume that (iii) holds, that is

$$\lim_{\tau \to +\infty} \frac{\mu(A_\tau^\varepsilon)}{\nu([-\tau,\tau] \setminus I)} = 0. \tag{32}$$

From the equality

$$\int_{[-\tau,\tau]\setminus I} \sup_{\theta \in [t-r,t]} \Big(\int_\theta^{\theta+1} |f(s)|^p ds \Big)^{\frac{1}{p}} d\mu(t) = \int_{A_\tau^\varepsilon} \sup_{\theta \in [t-r,t]} \Big(\int_\theta^{\theta+1} |f(s)|^p ds \Big)^{\frac{1}{p}} d\mu(t)$$
$$+ \int_{B_\tau^\varepsilon} \sup_{\theta \in [t-r,t]} \Big(\int_\theta^{\theta+1} |f(s)|^p ds \Big)^{\frac{1}{p}} d\mu(t).$$

We deduce that for τ sufficiently large

$$\frac{1}{\nu([-\tau,\tau] \setminus I)} \int_{[-\tau,\tau]\setminus I} \sup_{\theta \in [t-r,t]} \Big(\int_\theta^{\theta+1} |f(s)|^p ds \Big)^{\frac{1}{p}} \mu(t) \le \|f\|_{S^p} \frac{\mu(A_\tau^\varepsilon)}{\nu([-\tau,\tau] \setminus I)} + \varepsilon \frac{\mu(B_\tau^\varepsilon)}{\nu([-\tau,\tau] \setminus I)},$$

By using ($\mathbf{H_2}$), it follows that

$$\text{for all } \varepsilon > 0, \quad \lim_{\tau \to +\infty} \frac{1}{\nu([-\tau,\tau]\setminus I)} \int_{-\tau}^{+\tau} \sup_{\theta \in [t-r,t]} \Big(\int_{\theta}^{\theta+1} |f(s)|^p ds \Big)^{\frac{1}{p}} d\mu(t) \le \alpha\varepsilon, \quad \text{for any } \varepsilon > 0,$$

consequently (ii) holds.
(ii) $\Rightarrow$ *(iii)* Assume that *(ii)* holds. From the following inequality

$$\int_{[-\tau,\tau]\setminus I} \sup_{\theta \in [t-r,t]} \Big(\int_{\theta}^{\theta+1} |f(s)|^p ds \Big)^{\frac{1}{p}} d\mu(t) \ge \int_{A_\tau^\varepsilon} \sup_{\theta \in [t-r,t]} \Big(\int_{\theta}^{\theta+1} |f(s)|^p ds \Big)^{\frac{1}{p}} d\mu(t)$$

$$\frac{1}{\nu([-\tau,\tau]\setminus I)} \int_{[-\tau,\tau]\setminus I} \sup_{\theta \in [t-r,t]} \Big(\int_{\theta}^{\theta+1} |f(s)|^p ds \Big)^{\frac{1}{p}} d\mu(t) \ge \varepsilon \frac{\mu(A_\tau^\varepsilon)}{\nu([-\tau,\tau]\setminus I)}$$

$$\frac{1}{\varepsilon\nu([-\tau,\tau]\setminus I)} \int_{[-\tau,\tau]\setminus I} \sup_{\theta \in [t-r,t]} \Big(\int_{\theta}^{\theta+1} |f(s)|^p ds \Big)^{\frac{1}{p}} d\mu(t) \ge \frac{\mu(A_\tau^\varepsilon)}{\nu([-\tau,\tau]\setminus I)},$$

for τ sufficiently large, we obtain Eq. 32, that is *(iii)*. ■

From $\mu \in \mathcal{M}$, we formulate the following hypotheses.

($\mathbf{H_3}$) For all a, b and $c \in \mathbb{R}$, such that $0 \le a < b \le c$, there exist δ_0 and $\alpha_0 > 0$ such that

$$|\delta| \ge \delta_0 \Rightarrow \mu(a+\delta, b+\delta) \le \alpha_0 \mu(\delta, c+\delta). \tag{33}$$

($\mathbf{H_4}$) For all $\tau \in \mathbb{R}$, there exist $\beta > 0$ and a bounded interval I such that

$$\mu(\{a+\tau : \ a \in A\} \le \beta\mu(A) \ \text{ when } A \in \mathscr{B} \text{ satisfies } A \cap I = \emptyset. \tag{34}$$

We have the following results due to [3]

Lemma 3. ([3]) *Hypothesis* ($\boldsymbol{H_4}$) *implies* ($\boldsymbol{H_3}$).

Proposition 4. ([6]) *$\mu, \nu \in \mathcal{M}$ satisfy ($\boldsymbol{H_3}$) and $f \in PAA(\mathbb{R}; X, \mu, \nu)$ be such that*

$$f = g + h \tag{35}$$

where $g \in AA(\mathbb{R}, X)$ and $h \in \mathscr{E}(\mathbb{R}, X, \mu, \nu)$. Then

$$\{g(t), \ t \in \mathbb{R}\} \subset \overline{\{f(t), \ t \in \mathbb{R}\}} \ \textit{(the closure of the range of f)}. \tag{36}$$

Corollary 1. ([6]) *Assume that ($\boldsymbol{H_3}$) holds. Then the decomposition of a (μ, ν)-pseudo almost automorphic function in the form $f = g + \phi$ where $g \in AA(\mathbb{R}; X)$ and $\phi \in \mathscr{E}(\mathbb{R}; X, \mu, \nu)$, is unique.*

The following corollary is a consequence of Corollary 1.

Corollary 2. *Let $\mu, \nu \in \mathcal{M}$. Assume ($\boldsymbol{H_3}$) holds. Then the decomposition of a (μ, ν)-pseudo-almost periodic function $\phi = \phi_1 + \phi_2$, where $\phi_1 \in AA(\mathbb{R}; X)$ and $\phi_2 \in \mathscr{E}(\mathbb{R}; X, \mu, \nu, r)$, is unique.*

Definition 17. Let $\mu_1, \mu_2 \in \mathcal{M}$. We say that μ_1 is equivalent to μ_2, denoting this as $\mu_1 \sim \mu_2$ if there exist constants α and $\beta > 0$ and a bounded interval I (eventually $I = \emptyset$) such that

$$\alpha\mu_1(A) \le \mu_2(A) \le \beta\mu_1(A), \ \text{ when } A \in \mathscr{B} \text{ satisfies } A \cap I = \emptyset. \tag{37}$$

From [3] $\sim$ is a binary equivalence relation on $\mathcal{M}$. the equivalence class of a given measure $\mu \in \mathcal{M}$ will then be denoted by

$$cl(\mu) = \{\varpi \in \mathcal{M} : \ \mu \sim \varpi\}. \tag{38}$$

Theorem 5. *Let $\mu_1, \mu_2, \nu_1, \nu_2 \in \mathcal{M}$. If $\mu_1 \sim \mu_2$ and $\nu_1 \sim \nu_2$, then $PAAS^p(\mathbb{R}; L^p((0,1), X), \mu_1, \nu_1, r) = PAPA^p(\mathbb{R}; L^p((0,1), X), \mu_2, \nu_2, r)$.*

Proof. Since $\mu_1 \sim \mu_2$ and $\nu_1 \sim \nu_2$ there exist some constants $\alpha_1, \alpha_2, \beta_1, \beta_2 > 0$ and a bounded interval I (eventually $I = \emptyset$) such that $\alpha_1\mu_1(A) \leq \mu_2(A) \leq \beta_1\mu_1(A)$ and $\alpha_2\nu_1(A) \leq \nu_2(A) \leq \beta_2\nu_1(A)$ for each $A \in \mathscr{B}$ satisfies $A \cap I = \emptyset$ i.e

$$\frac{1}{\beta_2\nu_1(A)} \leq \frac{1}{\nu_2(A)} \leq \frac{1}{\alpha_2\nu_1(A)}. \tag{39}$$

Since $\mu_1 \sim \mu_2$ and $\mathscr{B}$ is the Lebesgue σ-field, we obtain for τ sufficiently large, it follows that

$$\frac{\alpha_1\mu_1\Big(\Big\{t \in [-\tau,\tau] \setminus I : \sup\limits_{\theta\in[t-r,t]} \Big(\int_\theta^{\theta+1} |f(s)|^p ds\Big)^{\frac{1}{p}} > \varepsilon\Big\}\Big)}{\beta_2\nu_1([-\tau,\tau] \setminus I)}$$

$$\leq \frac{\mu_2\Big(\Big\{t \in [-\tau,\tau] \setminus I : \sup\limits_{\theta\in[t-r,t]} \Big(\int_\theta^{\theta+1} |f(s)|^p ds\Big)^{\frac{1}{p}} > \varepsilon\Big\}\Big)}{\nu_2([-\tau,\tau] \setminus I)}$$

$$\leq \frac{\beta_1\mu_1\Big(\Big\{t \in [-\tau,\tau] \setminus I : \sup\limits_{\theta\in[t-r,t]} \Big(\int_\theta^{\theta+1} |f(s)|^p ds\Big)^{\frac{1}{p}} > \varepsilon\Big\}\Big)}{\alpha_2\nu_1([-\tau,\tau] \setminus I)}$$

By using Theorem 4 we deduce that $\mathscr{E}(\mathbb{R}, X, \mu_1, \nu_1, r) = \mathscr{E}(\mathbb{R}, X, \mu_2, \nu_2, r)$. From the definition of a μ-pseudo almost periodic function, we deduce that $PAAS^p(\mathbb{R}; X, \mu_1, \nu_1, r) = PAAS^p(\mathbb{R}; X, \mu_2, \nu_2, r)$. ∎

Let $\mu, \nu \in \mathcal{M}$ we denote by

$$cl(\mu, \nu) = \{\varpi_1, \varpi_2 \in \mathcal{M} : \ \mu \sim \varpi_2 \text{ and } \ \nu \sim \varpi_2\}. \tag{40}$$

Proposition 5. ([6]) *Let $\mu, \nu \in \mathcal{M}$ satisfy ($\boldsymbol{H_4}$). Then $PAA(\mathbb{R}, X, \mu, \nu)$ is invariant by translation, that is $f \in PAA(\mathbb{R}, X, \mu, \nu)$ implies $f_\alpha \in PAP(\mathbb{R}, X, \mu, \nu)$ for all $\alpha \in \mathbb{R}$.*

Corollary 3. *Let $\mu \in \mathcal{M}$ satisfy ($\boldsymbol{H_3}$). Then $PAAS^p(\mathbb{R}, X, \mu, r)$ is invariant by translation, that is $f \in PAAS^p(\mathbb{R}, X, \mu, \nu, r)$ implies $f_\alpha \in PAAS^p(\mathbb{R}, X, \mu, \nu, r)$ for all $\alpha \in \mathbb{R}$.*

Proof. It suffices to prove that $\mathcal{E}(\mathbb{R}; L^p((0,1), X), \mu, \nu, r)$ is invariant by translation. Let $f \in \mathcal{E}(\mathbb{R}; L^p((0,1), X), \mu, \nu, r)$ and $F^t(\theta) = \sup\limits_{\theta\in[t-r,t]} \Big(\int_\theta^{\theta+1} |f(s)|^p ds\Big)^{\frac{1}{p}}$.

Then $F^t \in \mathcal{E}(\mathbb{R}, \mathbb{R}, \mu, \nu)$, but since $\mathcal{E}(\mathbb{R}, \mathbb{R}, \mu, \nu)$ is invariant by translation, it follows that

$$\begin{aligned}&\lim_{\tau\to+\infty} \frac{1}{\nu([-\tau,\tau])}\int_{-\tau}^{\tau} F^t(\theta+\alpha)d\mu(t)\\&= \lim_{\tau\to+\infty} \frac{1}{\nu([-\tau,\tau])}\int_{-\tau}^{\tau} \sup_{\theta\in[t-r,t]}\Big(\int_{\theta}^{\theta+1}|f(s+\alpha)|^p ds\Big)^{\frac{1}{p}} d\mu(t) = 0, \quad (41)\end{aligned}$$

which implies that $f(.+\alpha) \in PAAS^p(\mathbb{R}, X, \mu, \nu, r)$. ■

In what follows, we prove some preliminary results concerning the composition of μ-pseudo almost periodic functions of class r.

Theorem 6. *Let $\mu \in \mathcal{M}$, $\phi \in PAAS^p(\mathbb{R}\times X, L^p((0,1),X),\mu,\nu,r)$ and $h \in PAAS^p(\mathbb{R}, L^p((0,1),X),\mu,\nu,r)$. Assume that there exists a function $L_\phi : \mathbb{R} \to [0,+\infty[$ sastisfies*

$$|\phi^b(t,x_1)-\phi^b(t,x_2)| \le L_\phi(t)\|x_1-x_2\|_{L^p} \quad \text{for} \quad t\in\mathbb{R} \quad \text{and for} \quad x_1,x_2 \in L^p((0,1),X). \quad (42)$$

If

$$\begin{aligned}&\lim_{\tau\to+\infty}\frac{1}{2\tau}\int_{-\tau}^{+\tau}\Big(\sup_{\theta\in[t-r,t+1]} L_\phi^p(\theta)\Big)^{\frac{1}{p}}dt\\&<\infty \text{ and } \lim_{\tau\to+\infty}\frac{1}{2\tau}\int_{-\tau}^{+\tau}\Big(\sup_{\theta\in[t-r,t+1]} L_\phi^p(\theta)\Big)^{\frac{1}{p}}\Big(\int_\theta^{\theta+1}|\xi(s)|^p ds\Big)^{\frac{1}{p}}dt = 0 \quad (43)\end{aligned}$$

for each $\xi \in \mathcal{E}(\mathbb{R}, L^p((0,1),\mathbb{R}))$, then the function $t \to \phi(t,h(t))$ belongs to $PAAS^p(\mathbb{R}; L^p((0,1),X),\mu,\nu,r)$.

Proof. Assume that $\phi = \phi_1+\phi_2$, $h = h_1+h_2$ where $\phi_1^b \in AA(\mathbb{R}\times X, L^p((0,1),X))$, $\phi_2^b \in \mathcal{E}(\mathbb{R}\times X, L^p((0,1),X),\mu,\nu,r)$ and $h_1^b \in AA(\mathbb{R}; L^p((0,1),X))$, $h_2^b \in \mathcal{E}(\mathbb{R}; L^p((0,1),X),\mu,\nu,r)$. Consider the following decomposition

$$\phi^b(t,h^b(t)) = \phi_1^b(t,h_1^b(t)) + [\phi^b(t,h^b(t)) - \phi^b(t,h_1^b(t))] + \phi_2^b(t,h_1^b(t)). \quad (44)$$

From [16] $\phi_1^b(.,h_1^b(.)) \in AA(\mathbb{R}; L^p((0,1),X))$. It remains to prove that both $\phi^b(.,h(.))-\phi^b(.,h_1^b(.))$ and $\phi_2^b(.,h_1^b(.))$ belong to $\mathcal{E}(\mathbb{R}; L^p((0,1),X),\mu,\nu,r)$. Using inequality Eq. 42, it follows that

$$\begin{aligned}&\frac{\mu\Big(\Big\{t\in[-\tau,\tau] : \sup_{\theta\in[t-r,t]}\Big(\int_\theta^{\theta+1}|\phi(\theta,h(\theta))-\phi(\theta,h_1(\theta))|^p ds\Big)^{\frac{1}{p}} > \varepsilon\Big\}\Big)}{\nu([-\tau,\tau])}\\&\le \frac{\mu\Big(\Big\{t\in[-\tau,\tau] : \sup_{\theta\in[t-r,t]}\Big(\int_\theta^{\theta+1}(L_\phi(\theta)|h_2(\theta)|)^p ds\Big)^{\frac{1}{p}} > \varepsilon\Big\}\Big)}{\nu([-\tau,\tau])}\\&\le \frac{\mu\Big(\Big\{t\in[-\tau,\tau] : \Big(\sup_{\theta\in[t-r,t+1]} L_\phi(\theta)\Big)\Big(\sup_{\theta\in[t-r,t]}\Big(\int_\theta^{\theta+1}|h_2(\theta)|^p ds\Big)^{\frac{1}{p}}\Big) > \varepsilon\Big\}\Big)}{\nu([-\tau,\tau])}.\end{aligned}$$

Since h_2 is (μ,ν)-ergodic of class r, Theorem 4 and Eq. 43 yield that for the above-mentioned ε, we have

$$\lim_{\tau\to+\infty}\frac{\mu\Big(\Big\{t\in[-\tau,\tau]:\Big(\sup_{\theta\in[t-r,t+1]}L_\phi(\theta)\Big)\Big(\sup_{\theta\in[t-r,t]}\Big(\int_\theta^{\theta+1}|h_2(\theta)|^p ds\Big)^{\frac{1}{p}}\Big)>\varepsilon\Big\}\Big)}{\nu([-\tau,\tau])}=0,$$

and then we obtain

$$\lim_{\tau\to+\infty}\frac{\mu\Big(\Big\{t\in[-\tau,\tau]:\sup_{\theta\in[t-r,t]}\Big(\int_\theta^{\theta+1}|\phi(\theta,h(\theta))-\phi(\theta,h_1(\theta))|^p ds\Big)^{\frac{1}{p}}>\varepsilon\Big\}\Big)}{\nu([-\tau,\tau])}=0, \tag{45}$$

By Theorem 4, Eq. 45 shows that $t\mapsto\phi(t,h(t))-\phi(t,h_1(t))$ belongs to $\mathcal{E}(\mathbb{R};L^p((0,1),X),\mu,\nu,r)$.

Since ϕ_2^b is uniformly continuous on the compact set $K=\overline{\{h_1^b(t):t\in\mathbb{R}\}}$ with respect to the second variable x, we deduce that for given $\varepsilon>0$, there exists $\delta>0$ such that, for all $t\in\mathbb{R}$, ξ_1 and $\xi_2\in K$, one has

$$\|\xi_1-\xi_2\|_{L^p}\le\delta\Rightarrow|\phi_2^b(t,\xi_1(t))-\phi_2^b(t,\xi_2(t))|\le\varepsilon. \tag{46}$$

Therefore, there exist $n(\varepsilon)$ and $\{z_i\}_{i=1}^{n(\varepsilon)}\subset K$, such that

$$K\subset\bigcup_{i=1}^{n(\varepsilon)}B_\delta(z_i,\delta) \tag{47}$$

and then by Minkowski's inequality we have

$$\begin{aligned}\Big(\int_t^{t+1}|\phi_2(t,h_1(t))|^p ds\Big)^{\frac{1}{p}}&\le\Big(\int_t^{t+1}|\phi_2(t,h_1(t))-\phi_2(t,z_i)|^p ds\Big)^{\frac{1}{p}}+\Big(\int_t^{t+1}|\phi_2(t,z_i))|^p ds\Big)^{\frac{1}{p}}\\&\le\varepsilon+\sum_1^{n(\varepsilon)}\Big(\int_t^{t+1}|\phi_2(t,z_i)|^p ds\Big)^{\frac{1}{p}}\end{aligned}$$

Since

$$\forall i\in\{1,...,n(\varepsilon)\},\quad\lim_{\tau\to+\infty}\frac{1}{\nu([-\tau,\tau])}\int_{-\tau}^{\tau}\sup_{\theta\in[t-r,t]}\Big(\int_\theta^{\theta+1}|\phi_2(\theta,z_i)|^p ds\Big)^{\frac{1}{p}}d\mu(t)=0, \tag{48}$$

we deduce that

$$\forall\varepsilon>0,\quad\limsup_{\tau\to+\infty}\frac{1}{\nu([-\tau,\tau])}\int_{-\tau}^{\tau}\sup_{\theta\in[t-r,t]}\Big(\int_\theta^{\theta+1}|\phi_2(\theta,h_1(\theta))|^p ds\Big)^{\frac{1}{p}}d\mu(t)\le\varepsilon, \tag{49}$$

that implies

$$\forall \varepsilon > 0, \quad \lim_{\tau \to +\infty} \frac{1}{\nu([-\tau, \tau])} \int_{-\tau}^{\tau} \sup_{\theta \in [t-r,t]} \left(\int_{\theta}^{\theta+1} |\phi_2(\theta, h_1(\theta))|^p ds \right)^{\frac{1}{p}} d\mu(t) = 0. \tag{50}$$

Consequently $t \mapsto \phi_2(t, h(t))$ belongs to $\mathcal{E}(\mathbb{R}; L^p((0,1), X), \mu, \nu, r)$. ■

For $\mu \in \mathcal{M}$ and $\alpha \in \mathbb{R}$, we denote μ_α the positive measure on $(\mathbb{R}, \mathscr{B})$ defined by

$$\mu_\alpha(A) = \mu([a + \alpha : \ a \in A]) \tag{51}$$

Lemma 4. ([3]) *Let $\mu \in \mathcal{M}$ satisfy $(\boldsymbol{H_3})$. Then the measures μ and μ_α are equivalent for all $\alpha \in \mathbb{R}$.*

Lemma 5. ([3]) $(\boldsymbol{H_3})$ *implies*

$$\text{for all } \sigma > 0 \ \limsup_{\tau \to +\infty} \frac{\mu([-\tau - \sigma, \tau + \sigma])}{\mu([-\tau, \tau])} < +\infty. \tag{52}$$

We have the following result.

Theorem 7. *Let $u \in PAA_cS^p(\mathbb{R}; L^p((0,1), X), \mu, \nu, r)$, then the function $t \to u_t$ belongs to $PAA_cS^p(C([-r,0], L^p((0,1), X)), \mu, \nu, r)$.*

Proof. Assume that $u = g + h$ where $g^p \in AA(\mathbb{R}; L^p((0,1), X))$ and $h^p \in \mathcal{E}(\mathbb{R}; L^p((0,1), X), \mu, \nu, r)$. We can see that $u_t = g_t + h_t$. We want to show that $g_t^p \in AA_c(\mathbb{R}; L^p((0,1), X))$ and $\varphi_t^p \in \mathcal{E}(\mathbb{R}; L^p((0,1), X), \mu, \nu, r)$.

Firstly for a given sequence $(s_m)_{m \in \mathbb{N}}$ of real numbers, fix a subsequence $(s_n)_{n \in \mathbb{N}}$ and $v \in BS^p(\mathbb{R}; X)$ such that $g(s + s_n) \to v(s)$ uniformly on compact subsets of $\mathbb{R}$. Let $K \subset \mathbb{R}$ be an arbitrary compact and $L > 0$ such that $K \subset [-L, L]$. For $\varepsilon > 0$, fix $N_{\varepsilon,L} \in \mathbb{N}$ such that $\|g(s+s_n) - v(s)\|_{S^p} \leq \varepsilon$ for $s \in [-L, L]$ whenever $n \geq N_{\varepsilon,L}$. For $t \in K$ and $n \geq N_{\varepsilon,L}$, we have

$$\begin{aligned} \|g_{t+s_n} - v_t\|_{S^p} &\leq \sup_{\theta \in [-L,L]} \|g(\theta + s_n) - v(\theta)\|_{S^p} \\ &\leq \varepsilon. \end{aligned}$$

In view of above, g_{t+s_n} converges to v_t uniformly on K. Similary, one can prove that v_{t-s_n} converges to u_t uniformly on K. Thus, the function $s \to g_s$ belongs to $AA_c(L^p((0,1), C))$.

Let us denote by

$$M_\alpha(\tau) = \frac{1}{\nu_\alpha([-\tau, \tau])} \int_{-\tau}^{+\tau} \left(\sup_{\theta \in [t-r,t]} \left(\int_{\theta}^{\theta+1} |h(\theta)| \right)^{\frac{1}{p}} d\mu_\alpha(t), \tag{53}$$

where μ_α and ν_α are the positive measure defined by Eq. 51. By using Lemma 4, it follows that μ_α and μ are equivalent, then by using Theorem 5 we have $\mathscr{E}(\mathbb{R}; L^p((0,1),X),\mu_\alpha,\nu_\alpha,r) = \mathscr{E}(\mathbb{R}; L^p((0,1),X),\mu,\nu,r)$, therefore $h \in \mathscr{E}(\mathbb{R}; L^p((0,1),X),\mu_\alpha,\nu_\alpha,r)$, that is

$$\lim_{\tau\to+\infty} M_\alpha(\tau) = 0, \quad \text{for all } \alpha \in \mathbb{R}. \tag{54}$$

On the other hand, for $r > 0$ and $\xi \in [-r,0]$ we have

$$\begin{aligned}
&\frac{1}{\nu([-\tau,\tau])}\int_{-\tau}^{+\tau} \sup_{\theta\in[t-r,t]} \Big(\int_\theta^{\theta+1} \Big(|h_s(\xi)|\Big)^p ds\Big)^{\frac{1}{p}} d\mu(t)\\
&= \frac{1}{\nu([-\tau,\tau])}\int_{-\tau}^{+\tau} \sup_{\theta\in[t-r,t]} \Big(\int_\theta^{\theta+1} \Big(|h(s+\xi)|\Big)^p ds\Big)^{\frac{1}{p}} d\mu(t)\\
&= \frac{1}{\nu([-\tau,\tau])}\int_{-\tau}^{+\tau} \sup_{\theta\in[t-r,t]} \Big(\int_{\theta+\xi}^{\theta+1+\xi} \Big(|h(s)|\Big)^p ds\Big)^{\frac{1}{p}} d\mu(t)\\
&\le \frac{1}{\nu([-\tau,\tau])}\int_{-\tau}^{+\tau} \sup_{\theta\in[t-r,t]} \Big(\sup_{\xi\in[-r,0]}\int_{\theta+\xi}^{\theta+1+\xi} |h(s)|^p ds\Big)^{\frac{1}{p}} d\mu(t)\\
&\le \frac{1}{\nu([-\tau,\tau])}\int_{-\tau}^{+\tau} \sup_{\theta\in[t-2r,t]} \Big(\int_\theta^{\theta+1} |h(s)|^p ds\Big)^{\frac{1}{p}} d\mu(t)\\
&\le \frac{1}{\nu([-\tau,\tau])}\int_{-\tau}^{+\tau} \sup_{\theta\in[t-2r,t-r]} \Big(\int_\theta^{\theta+1} |h(s)|^p ds\Big)^{\frac{1}{p}} d\mu(t)\\
&\quad + \frac{1}{\nu([-\tau,\tau])}\int_{-\tau}^{+\tau} \sup_{\theta\in[t-r,t]} \Big(\int_\theta^{\theta+1} |h(s)|^p ds\Big)^{\frac{1}{p}} d\mu(t)\\
&\le \frac{1}{\nu([-\tau,\tau])}\int_{-\tau-r}^{+\tau+r} \sup_{\theta\in[t-r,t]} \Big(\int_\theta^{\theta+1} |h(s)|^p ds\Big)^{\frac{1}{p}} d\mu(t+r)\\
&\quad + \frac{1}{\nu([-\tau,\tau])}\int_{-\tau}^{+\tau} \sup_{\theta\in[t-r,t]} \Big(\int_\theta^{\theta+1} |h(s)|^p ds\Big)^{\frac{1}{p}} d\mu(t)\\
&\le \Big[\frac{\nu([-\tau-r,\tau+r])}{\nu([-\tau,\tau])}\Big] \times \frac{1}{\nu([-\tau-r,\tau+r]}\int_{-\tau-r}^{+\tau+r} \sup_{\theta\in[t-r,t]} \Big(\int_\theta^{\theta+1} |h(s)|^p ds\Big)^{p} d\mu(t+r)\\
&\quad + \frac{1}{\nu([-\tau,\tau])}\int_{-\tau}^{+\tau} \sup_{\theta\in[t-r,t]} \Big(\int_\theta^{\theta+1} |h(s)|^p ds\Big)^{p} d\mu(t).
\end{aligned}$$

Consequently

$$\begin{aligned}
\frac{1}{\nu([-\tau,\tau])}\int_{-\tau}^{+\tau} \Big(\sup_{\theta\in[t-r,t]}\Big[\sup_{\xi\in[-r,0]} |h(\theta+\xi)|\Big]\Big) d\mu(t) &\le \Big[\frac{\nu([-\tau-r,\tau+r])}{\nu([-\tau,\tau])}\Big] \times M_r(\tau+r)\\
&+ \frac{1}{\nu([-\tau,\tau])}\int_{-\tau}^{+\tau} \Big(\sup_{\theta\in[t-r,t]} |h(\theta)|\Big) d\mu(t),
\end{aligned}$$

which shows using Lemmas 4 and 5 that h_t belongs to $PAAS^p(C([-r,0], L^p((0,1),X)),\mu,\nu,r)$. Thus, we obtain the desired result. ■

6 Weighted Stepanov Like Pseudo Almost Automorphic Solutions of Class r

Theorem 8. ([8]) *Assume that* $(\boldsymbol{H_0})$, $(\boldsymbol{H_1})$ *and* $(\boldsymbol{H_2})$ *hold and the semigroup* $(\mathcal{U}(t))_{t\geq 0}$ *is hyperbolic. If* $f \in BC(\mathbb{R};X)$, *then there exists a unique bounded solution* u *of Eq. 2 on* $\mathbb{R}$, *given by*

$$u_t = \lim_{\lambda\to+\infty}\int_{-\infty}^{t}\mathcal{U}^s(t-s)\Pi^s(\widetilde{B}_\lambda X_0 f(s))ds + \lim_{\lambda\to+\infty}\int_{+\infty}^{t}\mathcal{U}^u(t-s)\Pi^u(\widetilde{B}_\lambda X_0 f(s))ds \ \text{for}\ t\in\mathbb{R}, \tag{55}$$

where Π^s *and* Π^u *are the projections of* C *onto the stable and unstable subspaces, respectively and* $\widetilde{B}_\lambda = \lambda(\lambda I - \widetilde{\mathcal{A}}_{\mathcal{U}})^{-1}$ *for* $\lambda > \widetilde{\omega}$..

Theorem 9. *Let* $g \in \mathcal{E}(\mathbb{R}; L^p((0,1),X),\mu,\nu,r)$ *and* Γ *be the mapping defined by*

$$\begin{aligned}\Gamma g(t) = &\lim_{\lambda\to+\infty}\int_{-\infty}^{t}\mathcal{U}^s(t-s)\Pi^s(\widetilde{B}_\lambda X_0 g(s))ds\\ &+\lim_{\lambda\to+\infty}\int_{+\infty}^{t}\mathcal{U}^u(t-s)\Pi^u(\widetilde{B}_\lambda X_0 g(s))ds \ \text{for}\ t\in\mathbb{R}.\end{aligned} \tag{56}$$

Then $\Gamma g \in \mathcal{E}(\mathbb{R}; X,\mu,\nu,r)$.

Proof. For each $n = 1, 2, 3, ...,$ set

$$\begin{aligned}X_n(t) = &\lim_{\lambda\to+\infty}\int_{t-n}^{t-n+1}\mathcal{U}^s(t-s)\Pi^s(\widetilde{B}_\lambda X_0 g(s))ds\\ &-\lim_{\lambda\to+\infty}\int_{t+n-1}^{t+n}\mathcal{U}^u(t-s)\Pi^u(\widetilde{B}_\lambda X_0 g(s))ds \ \text{ for }\ t\in\mathbb{R}.\end{aligned} \tag{57}$$

We have

$$|X_n(t)| \leq \overline{M}\widetilde{M}|\Pi^s|\int_{t-n}^{t-n+1}e^{-\omega(t-s)}|g(s)|ds + \overline{M}\widetilde{M}|\Pi^u|\int_{t+n-1}^{t+n}e^{-\omega(t-s)}|g(s)|ds. \tag{58}$$

Set

$$K = \max(\overline{M}\widetilde{M}|\Pi^s|, \overline{M}\widetilde{M}|\Pi^u|). \tag{59}$$

Let q such that $\dfrac{1}{q}+\dfrac{1}{p} = 1$. Using the Hölder's inequality, we obtain

$$\begin{aligned}|X_n(t)| \leq\ & K\Big(\int_{t-n}^{t-n+1}e^{-q\omega(t-s)}ds\Big)^{\frac{1}{q}}\Big(\int_{t-n}^{t-n+1}|g(s)|^p ds\Big)^{\frac{1}{p}}\\ &+K\Big(\int_{t+n-1}^{t+n}e^{q\omega(t-s)}ds\Big)^{\frac{1}{q}}\Big(\int_{t+n-1}^{t+n}|g(s)|^p ds\Big)^{\frac{1}{p}}\\ \leq\ & \frac{K}{\sqrt[q]{q\omega}}(e^{-q\omega(n-1)} - e^{-q\omega n})^{\frac{1}{q}}\Big(\int_{t-n}^{t-n+1}|g(s)|^p ds\Big)^{\frac{1}{p}}\\ &+\frac{K}{\sqrt[q]{q\omega}}(e^{q\omega(1-n)} - e^{-q\omega n})^{\frac{1}{q}}\Big(\int_{t+n-1}^{t+n}|g(s)|^p ds\Big)^{\frac{1}{p}}\end{aligned}$$

$$\leq \frac{Ke^{-\omega n}}{\sqrt[q]{q\omega}}(e^{q\omega}-1)^{\frac{1}{q}}\Big[\Big(\int_{t-n}^{t-n+1}|g(s)|^p ds\Big)^{\frac{1}{p}}+\Big(\int_{t+n-1}^{t+n}|g(s)|^p ds\Big)^{\frac{1}{p}}\Big]$$
$$\leq \frac{Ke^{-\omega n}}{\sqrt[q]{q\omega}}(e^{q\omega}+1)^{\frac{1}{q}}\Big[\Big(\int_{t-n}^{t-n+1}|g(s)|^p ds\Big)^{\frac{1}{p}}+\Big(\int_{t+n-1}^{t+n}|g(s)|^p ds\Big)^{\frac{1}{p}}\Big].$$

Since the series

$$\frac{K}{\sqrt[q]{q\omega}}(e^{q\omega}+1)^{\frac{1}{q}}\times\sum_{n=1}^{+\infty}e^{-q\omega n}=\frac{K}{\sqrt[q]{q\omega}}(e^{q\omega}+1)^{\frac{1}{q}}\times\frac{e^{-q\omega}}{1-e^{-q\omega}} \tag{60}$$

is convergent, it follows from the Weierstrass test that the sequence of functions $\sum_{n=1}^{N}X_n(t)$ is uniformly convergent on $\mathbb{R}$.

Since $g\in\mathcal{E}(\mathbb{R};L^p((0,1),X),\mu,\nu,r)$ and

$$|X_n(t)|\leq C_q(K,\omega)\Big[\Big(\int_{t-n}^{t-n+1}|g(s)|^p ds\Big)^{\frac{1}{p}}+\Big(\int_{t+n-1}^{t+n}|g(s)|^p ds\Big)^{\frac{1}{p}}\Big], \tag{61}$$

where $C_q(K,\omega)=\frac{K}{\sqrt[q]{q\omega}}(e^{q\omega}+1)^{\frac{1}{p}}\times\sum_{n=1}^{+\infty}e^{-q\omega n}$, we conclude that $X_n\in\mathcal{E}(\mathbb{R},X,\mu,\nu,r)$. Thus $\sum_{n=1}^{N}X_n(t)\in\mathcal{E}(\mathbb{R},X,\mu,\nu,r)$ and its uniform limit belongs $\mathcal{E}(\mathbb{R},X,\mu,\nu,r)$ by Theorem 2. Observing that

$$\Gamma g(t)=\sum_{n=1}^{+\infty}X_n(t), \tag{62}$$

it follows that $\Gamma g(t)\in\mathcal{E}(\mathbb{R},X,\mu,\nu,r)$. ∎

Theorem 10. *Let* $g\in\mathcal{E}(\mathbb{R};L^p((0,1),X),\mu,\nu,r)$, *then* $\Gamma g\in\mathcal{E}(\mathbb{R};L^p((0,1),X),\mu,\nu,r)$.

Proof. For each $n=1,2,3,...$, let be X_n defined as previously, we have

$$|X_n(t)|^p\leq C_q^p(K,\omega)\Big[\Big(\int_{t-n}^{t-n+1}|g(s)|^p ds\Big)^{\frac{1}{p}}+\Big(\int_{t+n-1}^{t+n}|g(s)|^p ds\Big)^{\frac{1}{p}}\Big]^p. \tag{63}$$

Using the Minkowski's inequality, we obtain

$$\Big(\int_\theta^{\theta+1}|X_n(s)|^p ds\Big)^{\frac{1}{p}}\leq C_q(K,\omega)\Big[\Big(\int_\theta^{\theta+1}\Big[\Big(\int_{s-n}^{s-n+1}|g(\xi)|^p d\xi\Big)^{\frac{1}{p}}ds\Big]^p\Big)^{\frac{1}{p}}$$
$$+\Big(\int_\theta^{\theta+1}\Big[\Big(\int_{s+n-1}^{s+n}|g(\xi)|^p d\xi\Big)^{\frac{1}{p}}\Big]^p\Big)^{\frac{1}{p}}ds\Big]$$

$$\leq C_q(K,\omega)\Big[\Big(\sup_{s\in[\theta,\theta+1]}\int_{s-n}^{s-n+1}|g(\xi)|^p d\xi\Big)^{\frac{1}{p}}$$

$$+\Big(\sup_{s\in[\theta,\theta+1]}\int_{s+n-1}^{s+n}|g(\xi)|^p d\xi\Big)^{\frac{1}{p}}\Big]$$

$$\leq C_q(K,\omega)\Big[\Big(\int_{\theta-n}^{\theta-n+2}|g(s)|^p ds\Big)^{\frac{1}{p}}+\Big(\int_{\theta+n-1}^{\theta+n+1}|g(s)|^p ds\Big)^{\frac{1}{p}}\Big]$$

$$\leq C_q(K,\omega)\Big[\Big(\int_{\theta-n}^{\theta-n+1}|g(s)|^p ds\Big)^{\frac{1}{p}}+\Big(\int_{\theta-n+1}^{\theta-n+2}|g(s)|^p ds\Big)^{\frac{1}{p}}$$

$$+\Big(\int_{\theta+n-1}^{\theta+n}|g(s)|^p ds\Big)^{\frac{1}{p}}+\Big(\int_{\theta+n}^{\theta+n+1}|g(s)|^p ds\Big)^{\frac{1}{p}}\Big].$$

It follows that

$$\frac{1}{\nu([-\tau,\tau])}\int_{-\tau}^{+\tau}\sup_{\theta\in[t-r,t]}\Big(\int_{\theta}^{\theta+1}|X_n(s)|^p ds\Big)^{\frac{1}{p}}d\mu(t)$$

$$\leq C_q(K,\omega)\Big[\frac{1}{\nu([-\tau,\tau])}\int_{-\tau}^{+\tau}\sup_{\theta\in[t-r,t]}\Big(\int_{\theta-n}^{\theta-n+1}|g(s)|^p ds\Big)^{\frac{1}{p}}d\mu(t)$$

$$+\frac{1}{\nu([-\tau,\tau])}\int_{-\tau}^{+\tau}\sup_{\theta\in[t-r,t]}\Big(\int_{\theta-n+1}^{\theta-n+2}|g(s)|^p ds\Big)^{\frac{1}{p}}d\mu(t)$$

$$+\frac{1}{\nu([-\tau,\tau])}\int_{-\tau}^{+\tau}\sup_{\theta\in[t-r,t]}\Big(\int_{\theta+n-1}^{\theta+n}|g(s)|^p ds\Big)^{\frac{1}{p}}d\mu(t)+$$

$$+\frac{1}{\mu([-\tau,\tau])}\int_{-\tau}^{+\tau}\sup_{\theta\in[t-r,t]}\Big(\int_{\theta+n}^{\theta+n+1}|g(s)|^p ds\Big)^{\frac{1}{p}}d\mu(t)\Big].$$

We conclude that $X_n \in \mathcal{E}(\mathbb{R}; L^p((0,1),X),\mu,\nu,r)$. Thus $\sum_{n=1}^{N}X_n(t) \in \mathcal{E}(\mathbb{R}; L^p((0,1),X),\mu,\nu,r)$ and its uniform limit belongs $\mathcal{E}(\mathbb{R}; L^p((0,1),X),\mu,\nu,r)$ by Lemma 2. Observing that

$$\Gamma g(t)=\sum_{n=1}^{+\infty}X_n(t), \tag{64}$$

it follows that $\Gamma g(t)\in\mathcal{E}(\mathbb{R}; L^p((0,1),X),\mu,\nu,r)$. ■

Theorem 11. *Let $h\in AA_c(\mathbb{R}; L^p((0,1),X))$, then $\Gamma h\in AA_c(\mathbb{R}; L^p((0,1),X))$.*

Proof. For a given sequence $(s_m)_{m\in\mathbb{N}}$ of real numbers, fix a subsequence $(s_n)_{n\in\mathbb{N}}$ and $v\in BS^p(\mathbb{R};X)$ such that $h(t+s_n)$ converges to $v(t)$ and $v(t-s_n)$ converges to $h(t)$ uniformly on compact subsets of $\mathbb{R}$. We get the following estimates

$$(\Gamma v)(s) - (\Gamma h)(s+s_n) = \lim_{\lambda\to+\infty}\int_{-\infty}^{s} \mathcal{U}^s(s-\theta)\Pi^s(\widetilde{B}_\lambda X_0[v(\theta)-h(\theta+s_n)]d\theta$$

$$+\lim_{\lambda\to+\infty}\int_{+\infty}^{s} \mathcal{U}^u(s-\theta)\Pi^u(\widetilde{B}_\lambda X_0[v(\theta)-h(\theta+s_n)]d\theta$$

$$|(\Gamma h)(s+\xi) - (\Gamma h)(s)| = K\int_{-\infty}^{s} e^{-\omega(s-\theta)}|v(\theta)-h(\theta+s_n)|d\theta$$

$$+K\int_{+\infty}^{s} e^{\omega(s-\theta)}|v(\theta)-h(\theta+s_n)|d\theta.$$

where

$$K = \max(\overline{M}\widetilde{M}|\Pi^s|, \overline{M}\widetilde{M}|\Pi^u|). \tag{65}$$

For each $k = 1, 2, 3, ...,$ set

$$X_k(s) = K\int_{s-k}^{s-k+1} e^{-\omega(s-\theta)}|v(\theta)-h(\theta+s_n)|d\theta + K\int_{s+k-1}^{s+k} e^{\omega(s-\theta)}|v(\theta)-h(\theta+s_n)|d\theta. \tag{66}$$

Using the Hölder's inequality, we obtain

$$X_k(s) \le \frac{Ke^{-q\omega k}}{\sqrt[q]{q\omega}}(e^{q\omega}+1)^{\frac{1}{p}}\Big[\Big(\int_{s-k}^{s-k+1}|v(\theta)-h(\theta+s_n)|^p d\theta\Big)^{\frac{1}{p}}$$

$$+\Big(\int_{s+k-1}^{s+k}|v(\theta)-h(\theta+s_n)|^p d\theta\Big)^{\frac{1}{p}}\Big],$$

and by the Minkowski's inequality, we have

$$\Big(\int_t^{t+1}|X_k(s)|^p ds\Big)^{\frac{1}{p}} \le \frac{Ke^{-q\omega k}}{\sqrt[q]{q\omega}}(e^{q\omega}+1)^{\frac{1}{p}}\Big[\Big(\int_t^{t+1}\Big[\Big(\int_{s-k}^{s-k+1}|v(\theta)-h(\theta+s_n)|^p d\theta\Big)^{\frac{1}{p}}\Big]^p\Big)^{\frac{1}{p}}$$

$$+\Big(\int_t^{t+1}\Big[\Big(\int_{s+k-1}^{s+k}|v(\theta)-h(\theta+s_n)|^p d\theta\Big)^{\frac{1}{p}}\Big]^p\Big)^{\frac{1}{p}}\Big]$$

$$\le \frac{Ke^{-q\omega k}}{\sqrt[q]{q\omega}}(e^{q\omega}+1)^{\frac{1}{p}}\Big[\Big(\sup_{s\in[t,t+1]}\int_{s-k}^{s-k+1}|v(\theta)-h(\theta+s_n)|^p d\theta\Big)^{\frac{1}{p}}$$

$$+\Big(\sup_{s\in[t,t+1]}\int_{s+k-1}^{s+k}|v(\theta)-h(\theta+s_n)|^p d\theta\Big)^{\frac{1}{p}}\Big]$$

$$\le \frac{2Ke^{-q\omega k}}{\sqrt[q]{q\omega}}(e^{q\omega}+1)^{\frac{1}{p}}\sup_{t\in\mathbb{R}}\Big(\int_t^{t+1}|v(\theta)-h(\theta+s_n)|^p ds\Big)^{\frac{1}{p}}.$$

Since

$$\sum_{k=1}^{+\infty} X_k(s) = K\int_{-\infty}^{s} e^{-\omega(s-\theta)}|v(\theta)-h(\theta+s_n)|d\theta + K\int_{+\infty}^{s} e^{\omega(s-\theta)}|v(\theta)-h(\theta+s_n)|d\theta, \tag{67}$$

it follows that

$$\Big(\int_t^{t+1} |(\Gamma v)(s) - (\Gamma h)(s+s_n)|^p ds\Big)^{\frac{1}{p}} \leq 2C_q(K,\omega) \sup_{t\in\mathbb{R}} \Big(\int_t^{t+1} |v(\theta) - h(\theta+s_n)|^p ds\Big)^{\frac{1}{p}}. \tag{68}$$

Fix $L > 0$ and $N_\varepsilon \in \mathbb{N}$ such that $K \subset [\frac{-L}{2}, \frac{L}{2}]$ with

$$|h(s+s_n) - v(s)| \leq \varepsilon \text{ for } n \geq N_\varepsilon \text{ and } s \in [-L, L].$$

Then, for each $t \in K$, ones has

$$\Big(\int_t^{t+1} |(\Gamma v)(s) - (\Gamma h)(s+s_n)|^p ds\Big)^{\frac{1}{p}} \leq 2C_q(K,\omega)\varepsilon, \tag{69}$$

which proves that the convergence is uniform on K, by the fact that the last estimate is independent of $t \in K$. Proceeding as previously, one can similarly prove that

$$\lim_{n\to+\infty} \Big(\int_t^{t+1} |(\Gamma v)(s-s_n) - (\Gamma h)(s)|^p ds\Big)^{\frac{1}{p}} = 0, \tag{70}$$

which implies that $\Gamma h \in AA_c(\mathbb{R}; L^p((0,1), X))$. ■

For the existence of pseudo almost automorphic solution, we make the following assumption.

($\mathbf{H_5}$) $f : \mathbb{R} \to X$ is $cl(\mu,\nu)$-S^p-pseudo almost automorphic of class r.

Theorem 12. *Assume* ($\boldsymbol{H_0}$), ($\boldsymbol{H_1}$), ($\boldsymbol{H_2}$), ($\boldsymbol{H_3}$) *and* ($\boldsymbol{H_5}$) *hold. Then Eq. 1 has a unique* $cl(\mu,\nu)$*-*S^p*-pseudo almost automorphic solution of class* r.

Proof. Since f is S^p-pseudo almost automorphic function, f has a decomposition $f = f_1 + f_2$ where $f_1^b \in AA(\mathbb{R}; L^p((0,1), X_1))$ and $f_2^b \in \mathcal{E}(\mathbb{R}; L^p((0,1), X_1), \mu, \nu, r)$. Using Theorems 8, 10 and 11, we get the desired result. ■

Our next objective is to show the existence of pseudo almost automorphic solutions of class r for the following problem

$$u'(t) = Au(t) + L(u_t) + f(t, u_t) \text{ for } t \in \mathbb{R} \tag{71}$$

where $f : \mathbb{R} \times C \to X$ is a continuous.

For the sequel, we make the following assertion.

($\mathbf{H_6}$) The instable space $U \equiv \{0\}$.

($\mathbf{H_7}$) $f : \mathbb{R} \times C \to X$ is uniformly pseudo compact almost automorphic such that there exists a function $L_f \in L^p(\mathbb{R}, \mathbb{R}^+)$ such that

$$|f(t,\varphi_1) - f(t,\varphi_2)| \leq L_f(t)\|\varphi_1 - \varphi_2\|_{S^p}, \quad \text{for all } t \in \mathbb{R}, \ \varphi_1, \varphi_2 \in BS^p([-r,0]; X)) \tag{72}$$

where L_f satisfies conditions of Theorem 6.

Theorem 13. *Assume* $(\boldsymbol{H}_0)$, $(\boldsymbol{H}_1)$, $(\boldsymbol{H}_2)$, $(\boldsymbol{H}_3)$, $(\boldsymbol{H}_6)$ *and* $(\boldsymbol{H}_7)$ *hold. Then Eq. 71 has a unique* $cl(\mu,\nu)$-S^p*-pseudo compact almost automorphic mild solution of class* r.

Proof. Let x be a function in $PAA_c(\mathbb{R}; L^p((0,1),X),\mu,\nu,r)$, from Theorem 7 the function $t \to x_t$ belongs to $PAA_c(\mathbb{R}; L^p((0,1),C),\mu,\nu,r)$. Hence Theorem 6 implies that the function $g(.) := f(.,x_.)$ is in $PAA_c(\mathbb{R}; L^p((0,1),X),\mu,\nu,r)$. Since the unstable space $U \equiv \{0\}$, then $\Pi^u \equiv 0$. Consider now the mapping

$$\mathcal{H} : PAA_c(\mathbb{R}; L^p((0,1),X),\mu,\nu,r) \to PAA_c(\mathbb{R}; L^p((0,1),X),\mu,\nu,r) \tag{73}$$

defined for $t \in \mathbb{R}$ by

$$(\mathcal{H}x)(t) = \Big[\lim_{\lambda\to+\infty}\int_{-\infty}^{t} \mathcal{U}^s(t-s)\Pi^s(\widetilde{B}_\lambda X_0 f(s,x_s))ds\Big](0). \tag{74}$$

From Theorems 8, 10 and 11, we can infer that $\mathcal{H}$ maps $PAA_c(\mathbb{R}; L^p((0,1),X),\mu,\nu,r)$ into $PAA_c(\mathbb{R}; L^p((0,1),X),\mu,\nu,r)$. It suffices now to show that the operator $\mathcal{H}$ has a unique fixed point in $PAA_c(\mathbb{R}; L^p((0,1),X),\mu,\nu,r)$.

Case 1: $L_f \in L^1(\mathbb{R})$.
Let $x_1, x_2 \in PAA_c(\mathbb{R}; L^p((0,1),X),\mu,\nu,r)$. Then we have

$$\begin{aligned}
|\mathcal{H}x_1(t) - \mathcal{H}x_2(t)| &\le \Big|\lim_{\lambda\to+\infty}\int_{-\infty}^{t} \mathcal{U}^s(t-s)\Pi^s(\widetilde{B}_\lambda X_0(f(s,x_{1s}) - f(s,x_{2s})))ds\Big| \\
&\le \overline{M}\widetilde{M}|\Pi^s|\ \|x_1 - x_2\|_{S^p}\int_{-\infty}^{t} e^{-\omega(t-s)}L_f(s)ds \\
&\le \overline{M}\widetilde{M}|\Pi^s|\ \|x_1 - x_2\|_{S^p}\int_{-\infty}^{t} L_f(s)ds.
\end{aligned}$$

It follows that

$$\begin{aligned}
|\mathcal{H}^2x_1(t) - \mathcal{H}^2x_2(t)| &\le \Big|\lim_{\lambda\to+\infty}\int_{-\infty}^{t} \mathcal{U}^s(t-s)\Pi^s(\widetilde{B}_\lambda X_0(f(s,\mathcal{H}x_{1t}) - f(s,\mathcal{H}x_{2t}))ds\Big| \\
&\le (\overline{M}\widetilde{M}|\Pi^s|)^2\|x_1 - x_2\|_{S^p}\int_{-\infty}^{t} L_f(s)\int_{-\infty}^{s} L_f(\delta)d\delta ds \\
&\le \frac{(\overline{M}\widetilde{M}|\Pi^s|)^2}{2}\Big(\int_{-\infty}^{t} L_f(s)ds\Big)^2\|x_1 - x_2\|_{S^p}.
\end{aligned}$$

Induction on n in the same way, gives

$$|\mathcal{H}^n x_1 - \mathcal{H}^n x_2| \le \frac{(\overline{M}\widetilde{M}|\Pi^s|)^n}{n!}\Big(\int_{-\infty}^{t} L_f(s)ds\Big)^n\|x_1 - x_2\|_{S^p}. \tag{75}$$

Therefore

$$\|\mathcal{H}^n x_1 - \mathcal{H}^n x_2\|_{S^p} \le \frac{(\overline{M}\widetilde{M}|\Pi^s|\ |L_f|_{L^1(\mathbb{R})})^n}{n!}\|x_1 - x_2\|_{S^p}. \tag{76}$$

Let n_0 be such that $\dfrac{(\overline{M}\widetilde{M}|\Pi^s|\ |L_f|_{L^1(\mathbb{R})})^{n_0}}{n_0!} < 1$. By Banach fix point Theorem, $\mathcal{H}$ has a unique point fixed and this fixed point satisfies the integral equation

$$u_t = \lim_{\lambda\to+\infty}\int_{-\infty}^{t} \mathcal{U}^s(t-s)\Pi^s(\widetilde{B}_\lambda X_0 f(s,u_s))ds. \tag{77}$$

Case 2: $L_f \in L^p(\mathbb{R})$; $(1 < p < \infty)$.

First, put

$$\mu(t) = \int_{-\infty}^{t} (L_f(s))^p ds. \tag{78}$$

Then we define an equivalent norm over $PAAS^p(\mathbb{R}, X)$ as follows,

$$|f|_c = \sup_{t\in\mathbb{R}} e^{-c\mu(t)}\Big(\int_t^{t+1} |f(s)|^p\Big)^{\frac{1}{p}}, \tag{79}$$

where c is a fixed positive number to be precised later. Using the Hölder inequality we have

$$\begin{aligned}
|\ \mathcal{H}x_1(t) - \mathcal{H}x_2(t)| &\le \Big|\lim_{\lambda\to+\infty}\int_{-\infty}^{t} \mathcal{U}^s(t-s)\Pi^s(\widetilde{B}_\lambda X_0(f(s,x_{1s}) - f(s,x_{2s})))ds\Big| \\
&\le \overline{M}\widetilde{M}|\Pi^s|\int_{-\infty}^{t} e^{-\omega(t-s)}L_f(s)\|x_{1s} - x_{2s}\|_{S^p}ds \\
&\le \overline{M}\widetilde{M}|\Pi^s|\int_{-\infty}^{t} e^{-\omega(t-s)}e^{-c\mu(s)}e^{c\mu(s)}L_f(s)\|x_{1s} - x_{2s}\|_{S^p}ds \\
&\le \overline{M}\widetilde{M}|\Pi^s|\int_{-\infty}^{t} \Big(e^{-\omega(t-s)}e^{c\mu(s)}L_f(s)\Big)\sup_{s\in\mathbb{R}}\Big(e^{-c\mu(s)}\|x_{1s} - x_{2s}\|_{S^p}\Big)ds \\
&\le \overline{M}\widetilde{M}|\Pi^s|\int_{-\infty}^{t} \Big(e^{-\omega(t-s)}e^{c\mu(s)}L_f(s)ds\Big)|x_1 - x_2|_c \\
&\le \overline{M}\widetilde{M}|\Pi^s|\Big(\int_{-\infty}^{t} e^{pc\mu(s)}(L_f(s))^p ds\Big)^{\frac{1}{p}}\Big(\int_{-\infty}^{t} e^{-\omega q(t-s)}ds\Big)^{\frac{1}{q}}|x_1 - x_2|_c
\end{aligned}$$

$$\begin{aligned}
&\le \overline{M}\widetilde{M}|\Pi^s|\Big(\int_{-\infty}^{t} e^{pc\mu(s)}\mu'(s)ds\Big)^{\frac{1}{p}}\Big(\int_{-\infty}^{t} e^{-\omega q(t-s)}ds\Big)^{\frac{1}{q}}|x_1 - x_2|_c \\
&\le \overline{M}\widetilde{M}|\Pi^s|\Big(\frac{1}{(pc)^{\frac{1}{p}}} \times \frac{1}{(\omega q)^{\frac{1}{q}}}\Big)e^{c\mu(t)}|x_1 - x_2|_c.
\end{aligned}$$

Consequently

$$|\mathcal{H}x_1(t) - \mathcal{H}x_2(t)|_c \le \frac{\overline{M}\widetilde{M}|\Pi^s|}{(pc)^{\frac{1}{p}} \times (\omega q)^{\frac{1}{q}}}|x_1 - x_2|_c.$$

Fix $c > 0$ so large, then the function $c \mapsto \dfrac{1}{(pc)^{\frac{1}{p}}}$ converges to 0 when c tends to $+\infty$. It follows that for $c > 0$ so large enough we have $\dfrac{\overline{M}\widetilde{M}|\Pi^s|}{(pc)^{\frac{1}{p}} \times (\omega q)^{\frac{1}{q}}} < 1$.

Thus $\mathcal{H}$ is a contractive mapping. Using the same argument as in Theorem 3.3 of [14], We conclude that Eq. 71, has one and only one $cl(\mu,\nu)$-S^p-pseudo almost automorphic solution of class r which ends the proof. ■

Corollary 4. *Assume* $(\boldsymbol{H}_0)$, $(\boldsymbol{H}_1)$, $(\boldsymbol{H}_2)$, $(\boldsymbol{H}_3)$ *and* $(\boldsymbol{H}_7)$ *hold and* f *is Lipschitzian with respect the second argument. If*

$$Lip(f) < \frac{\omega}{\overline{M}\widetilde{M}|\Pi^s|}, \tag{80}$$

then Eq. 71 has a unique $cl(\mu)$*-pseudo almost automorphic solution of class* r*, where* $Lip(f)$ *is the Lipschitz constant of* f*.*

Proof. Let us pose $k = Lip(f)$, we have

$$\begin{aligned}
|\mathcal{H}x_1(t) - \mathcal{H}x_2(t)| &\leq \left| \lim_{\lambda\to+\infty} \int_{-\infty}^{t} \mathcal{U}^s(t-s)\Pi^s(\widetilde{B}_\lambda X_0(f(s,x_{1s}) - f(s,x_{2s})))ds \right| \\
&\leq |\Pi^s|\overline{M}\widetilde{M}k\|x_1 - x_2\|_{S^p}\left(\int_{-\infty}^{t} e^{-\omega(t-s)}\right) \\
&\leq \frac{|\Pi^s|\overline{M}\widetilde{M}k\|x_1 - x_2\|_{S^p}}{\omega}.
\end{aligned}$$

Consequently $\mathcal{H}$ is a strict contraction if $k < \dfrac{\omega}{\overline{M}\widetilde{M}|\Pi^s|}$. ■

7 Application

For illustration, we propose to study the existence of solutions for the following model

$$\begin{cases}
\dfrac{\partial}{\partial t}z(t,x) = \dfrac{\partial^2}{\partial x^2}z(t,x) + \displaystyle\int_{-r}^{0} G(\theta)z(t+\theta,x))d\theta + \sin\left(\frac{1}{2+\cos t + \cos\sqrt{2}t}\right) + g(t) \\
\qquad\qquad + \displaystyle\int_{-r}^{t} e^{\omega(-t+\theta)}h(\theta, z(t+\theta,x))d\theta \text{ for } t \in \mathbb{R} \text{ and } x \in [0,\pi] \\
\\
z(t,0) = z(t,\pi) = 0 \text{ for } t \in \mathbb{R},
\end{cases} \tag{81}$$

where $r \geq 0$, $G : [-r,0] \to \mathbb{R}$ is a continuous function, $h : [-r,0] \times \mathbb{R} \to \mathbb{R}$ is continuous and lipschitzian with respect to the second argument, ω is a positive real number and $g : \mathbb{R} \times [0,\pi] \to \mathbb{R}$ is a bounded continuous function defined by

$$g(t) = \begin{cases} 0 \text{ for } t \leq 0 \\ \\ te^{-t} \text{ for } t \geq 0. \end{cases} \tag{82}$$

For example, we can take $G(\theta) = \dfrac{\theta^2 - 1}{(\theta^2+1)^2}$ for and $\theta \in [-r,0]$ and $h(\theta,x) = \theta^2 + \sin\left(\dfrac{x}{2}\right)$ for $(\theta,x) \in [-r,0] \times \mathbb{R}$. We can see that $G : [-r,0] \to \mathbb{R}$ is a continuous

function and $|h(\theta,x_1)-h(\theta,x_2)| \leq \frac{1}{2}|x_1-x_2|$, which implies $h:[-r,0]\times\mathbb{R}\to\mathbb{R}$ is continuous and lipschitzian with respect to the second argument.
To rewrite Eq. 81 in the abstract form, we introduce the space $X = C_0([0,\pi];\mathbb{R})$ of continuous function from $[0,\pi]$ to $\mathbb{R}^+$ equipped with the uniform norm topology. Let $A: D(A)\to X$ be defined by

$$\begin{cases} D(A) = \{y \in X\cap C^2([0,\pi],\mathbb{R}) :\ y'' \in X\} \\ Ay = y''. \end{cases} \tag{83}$$

Then A satisfied the Hille-Yosida condition in X. Moreover the part A_0 of A in $\overline{D(A)}$ is the generator of strongly continuous compact semigroup $(T_0(t))_{t\geq 0}$ on $\overline{D(A)}$. It follows that $(\mathbf{H_0})$ and $(\mathbf{H_1})$ are satisfied.

We define $f:\mathbb{R}\times C\to X$ and $L: C\to X$ as follows

$$f(t,\varphi)(x) = \sin\Big(\frac{1}{2+\cos t+\cos\sqrt{2}t}\Big) + g(t) + \int_{-r}^{t} e^{\omega(-t+\theta)} h(\theta,\varphi(\theta)(x))d\theta \ \text{ for } x\in[0,\pi] \text{ and } t\in\mathbb{R},$$

$$L(\varphi)(x) = \int_{-r}^{0} G(\theta)\varphi(\theta)(x))d\theta \ \text{ for } \ -r\leq\theta\leq 0 \text{ and } x\in[0,\pi].$$

Let us pose $v(t) = z(t,x)$. Then Eq. 81 takes the following abstract form

$$v'(t) = Av(t) + L(v_t) + f(t,v_t) \text{ for } t\in\mathbb{R}. \tag{84}$$

Consider the measures μ and ν where its Radon-Nikodym derivative are respectively $\rho_1,\rho_2:\mathbb{R}\to\mathbb{R}$ defined by

$$\rho_1(t) = \begin{cases} 1 \ \text{ for } \ t>0 \\ \\ e^t \ \text{ for } \ t\leq 0. \end{cases} \tag{85}$$

and

$$\rho_2(t) = |t| \ \text{ for } \ t\in\mathbb{R} \tag{86}$$

i.e $d\mu(t) = \rho_1(t)dt$ and $d\nu(t) = \rho_2(t)dt$ where dt denotes the Lebesgue measure on $\mathbb{R}$ and

$$\mu(A) = \int_A \rho_1(t)dt \ \text{ for } \ \nu(A) = \int_A \rho_2(t)dt \ \text{ for } \ A\in\mathscr{B}. \tag{87}$$

From [3] $\mu,\nu \in \mathcal{M}$, μ,ν satisfy Hypothesis $(\mathbf{H_4})$ and from [7], $\sin\Big(\frac{1}{2+\cos t+\cos\sqrt{2}t}\Big)$ is compact almost automorphic..

We have

$$\limsup_{\tau\to+\infty}\frac{\mu([-\tau,\tau])}{\nu([-\tau,\tau])} = \limsup_{\tau\to+\infty}\frac{\displaystyle\int_{-\tau}^{0} e^t dt + \int_0^{\tau} dt}{2\displaystyle\int_0^{\tau} tdt} = \limsup_{\tau\to+\infty}\frac{1-e^{-\tau}+\tau}{\tau^2} = 0 < \infty, \tag{88}$$

which implies that $(\mathbf{H_2})$ is satisfied.

Let $p \geq 1$, since r is given then we have

$$\begin{aligned}
&\frac{1}{\nu([-\tau,\tau])}\int_{-\tau}^{+\tau}\sup_{\theta\in[t-r,t]}\Big(\int_{\theta}^{\theta+1}|g(s)|^p ds\Big)^{\frac{1}{p}}d\mu(t)\\
&\leq \frac{1}{\mu([-\tau,\tau])}\int_{0}^{+\tau}\sup_{\theta\in[t-r,t]}\Big(\int_{\theta}^{\theta+1}|g(s)|^p ds\Big)^{\frac{1}{p}}dt\\
&\leq \frac{1}{\nu([-\tau,\tau])}\int_{0}^{+\tau}\sup_{\theta\in[t-r,t]}\Big(\int_{t-r}^{t+1}s^p e^{-ps}ds\Big)^{\frac{1}{p}}dt\\
&\leq \frac{1}{\nu([-\tau,\tau])}\int_{0}^{+\tau}\Big(\int_{t-r}^{t+1}s^p e^{-s}ds\Big)^{\frac{1}{p}}dt\\
&\leq \frac{1}{\nu([-\tau,\tau])}\int_{0}^{+\tau}\Big[\Big(\int_{t-r}^{t+1}s^{p^2}\Big)^{\frac{1}{p}}\Big(\int_{t-r}^{t+1}e^{-qs}ds\Big)^{\frac{1}{q}}\Big]^{\frac{1}{p}}dt\\
&\leq \frac{1}{\mu([-\tau,\tau])}\int_{0}^{+\tau}\Big[[(t+1)^{p^2+1}]^{\frac{1}{p}}e^{-(t-r)}\Big]^{\frac{1}{p}}dt\\
&\leq \frac{1}{\nu([-\tau,\tau])}\int_{0}^{+\infty}\Big[(t+1)^{p+1}e^{-(t-r)}\Big]^{\frac{1}{p}}dt\\
&\leq \frac{1}{\nu([-\tau,\tau])}\int_{0}^{+\infty}(t+1)^2 e^{-\frac{(t-r)}{p}}dt\\
&\leq \frac{p(4p+1)e^{\frac{r}{p}}}{\tau^2}.
\end{aligned}$$

Consequently

$$\lim_{\tau\to+\infty}\frac{1}{\nu([-\tau,\tau])}\int_{-\tau}^{+\tau}\sup_{\theta\in[t-r,t]}\Big(\int_{\theta}^{\theta+1}|g(s)|^p ds\Big)^{\frac{1}{p}}dt = 0. \tag{89}$$

It follows that $g \in \mathcal{E}(\mathbb{R}; L^p((0,1),X),\mu,\nu,r)$, consequently, f is uniformly μ-S^p-pseudo almost automorphic of class r. Consequently, f is uniformly (μ,ν)-S^p-pseudo almost automorphic of class r. Moreover, L is a bounded linear operator from C to X.

For every $\varphi_1, \varphi_2 \in BS^p(\mathbb{R};X))$ and $t \geq 0$, we have

$$|f(t,\varphi_1) - f(t,\varphi_2)| \leq \frac{1}{2}\int_{-\infty}^{t} e^{\omega(-t+\theta)}|\varphi_1(\theta)(x) - \varphi_2(\theta)(x)|d\theta.$$

For each $n = 1,2,3,...$, set

$$X_n(t) = \frac{1}{2}\int_{t-n}^{t-n+1} e^{\omega(-t+\theta)}|\varphi_1(\theta)(x) - \varphi_2(\theta)(x)|d\theta. \tag{90}$$

Let q such that $\dfrac{1}{q}+\dfrac{1}{p}=1$. Using the Hölder's inequality, we obtain

$$\begin{aligned} X_n(t) &\leq \frac{1}{2}\Big(\int_{t-n}^{t-n+1} e^{-q\omega(t-\theta)}d\theta\Big)^{\frac{1}{q}}\Big(\int_{t-n}^{t-n+1}|\varphi_1(\theta)(x)-\varphi_2(\theta)(x)|^p d\theta\Big)^{\frac{1}{p}} \\ &\leq \frac{1}{2\sqrt[q]{q\omega}}(e^{-q\omega(n-1)}-e^{-q\omega n})^{\frac{1}{q}}\Big(\int_{t-n}^{t-n+1}|\varphi_1(\theta)(x)-\varphi_2(\theta)(x)|^p d\theta\Big)^{\frac{1}{p}} \\ &\leq \frac{e^{-\omega n}}{2\sqrt[q]{q\omega}}(e^{q\omega}+1)^{\frac{1}{q}}\Big(\int_{t-n}^{t-n+1}|\varphi_1(\theta)(x)-\varphi_2(\theta)(x)|^p d\theta\Big)^{\frac{1}{p}}. \end{aligned}$$

Since the series

$$\frac{(e^{q\omega}+1)^{\frac{1}{q}}}{2\sqrt[q]{q\omega}}\times\sum_{n=1}^{+\infty}e^{-q\omega n} \tag{91}$$

is convergent, it follows from the Weierstrass test that the sequence of functions $\sum_{n=1}^{N}X_n(t)$ is uniformly convergent on $\mathbb{R}$. Thus we have

$$\begin{aligned} |f(t,\varphi_1)-f(t,\varphi_2)| &\leq \frac{C_q(\omega)}{2}\sup_{t\in\mathbb{R}}\Big(\int_t^{t+1}|\varphi_1(\theta)(x)-\varphi_2(\theta)(x)|^p d\theta\Big)^{\frac{1}{p}} \\ &\leq \frac{C_q(\omega)}{2}\sup_{0\leq x\leq\pi}\|\varphi_1(x)-\varphi_2(x)\|_{S^p}. \end{aligned}$$

where $C_q(\omega)=\dfrac{(e^{q\omega}+1)^{\frac{1}{p}}}{\sqrt[q]{q\omega}}\times\displaystyle\sum_{n=1}^{+\infty}e^{-q\omega n}=\dfrac{(e^{q\omega}+1)^{\frac{1}{p}}}{\sqrt[q]{q\omega}}\times\dfrac{e^{-q\omega}}{1-e^{-q\omega}}$. Consequently, we conclude that f is Lipschitz continuous and $cl(\mu,\nu)$-S^p-pseudo almost periodic of class r.

Lemma 6. ([8,15]) *If* $\int_{-r}^{0}|G(\theta)|d\theta<1$, *then the semigroup* $(\mathcal{U}(t))_{t\geq 0}$ *is hyperbolic and the instable space* $U\equiv\{0\}$.

We can see that

$$\int_{-r}^{0}|G(\theta)|d\theta=\int_{-r}^{0}\Big|\frac{\theta^2-1}{(\theta^2+1)^2}\Big|d\theta=\Big[\frac{\theta}{\theta^2+1}\Big]_{-r}^{0}=\frac{r}{r^2+1}<1 \text{ if } r<1$$

and

$$\int_{-r}^{0}|G(\theta)|d\theta=\int_{-r}^{0}\Big|\frac{\theta^2-1}{(\theta^2+1)^2}\Big|d\theta=\int_{-r}^{-1}\frac{\theta^2-1}{(\theta^2+1)^2}d\theta+\int_{-1}^{0}\frac{-\theta^2+1}{(\theta^2+1)^2}d\theta=1-\frac{r}{r^2+1}<1 \text{ if } r\geq 1,$$

then $(\mathbf{H_6})$ holds. Consequently by Corollary 4 we deduce the following result.

Corollary 5. *Under the above assumptions, if $Lip(h) = \frac{1}{2} < \frac{1}{C_q(\omega)}$, i.e $C_q(\omega) < 2$, then Eq. 84 has a unique $cl(\mu,\nu)$-S^p-pseudo almost automorphic solution v of class r.*

Acknowledgement. The authors would like to thank the referees for their careful reading of this article. Their valuable suggestions and critical remarks made numerous improvements throughout this article.

References

1. Adimy, M., Elazzouzi, A., Ezzimbi, K.: Bohr-Neugebauer type theorem for some partial neutral functional differential equations. Nonlinear Anal. Theory Methods Appl. **66**(5), 1145–1160 (2007)
2. Adimy, M., Ezzinbi, K., Laklach, M.: Spectral decomposition for partial neutral functional differential equations. Can. Appl. Math. Q. **1**, 1–34 (2001)
3. Blot, J., Cieutat, P., Ezzinbi, K.: New approach for weighted pseudo almost periodic functions under the light of measure theory, basic results and applications. Appl. Anal. **92**(3), 493–526 (2013)
4. Bochner, S.: Continuous mappings of almost automorphic and almost automorphic functions. Proc. Natl. Sci. USA **52**, 907–910 (1964)
5. Bochner, S.: A New Approach to Almost-Periodicity. Proc. Natl. Acad. Sci. USA **48**, 2039–2043 (1962)
6. Diagana, T., Ezzinbi, K., Miraoui, M.: Pseudo-almost periodic and pseudo-almost automorphic solutions to some evolution equations involving theoretical measure theory. CUBO A Math. J. **16**(02), 01–31 (2014)
7. Ezzinbi, K., Fatajou, S., N'Guérékata, G.M.: Pseudo almost automorphic solutions for dissipative differential equations in Banach spaces. J. Math. Anal. Appl. **351**(2), 765–772 (2009)
8. Ezzinbi, K., Fatajou, S., N'Guérékata, G.M.: C^n-almost automorphic solutions for partial neutral functional differential equations. Appl. Anal. **86**(9), 1127–1146 (2007)
9. Ezzinbi, K., N'Guérékata, G.M.: Almost automorphic solutions for some partial functional differential equations. Appl. Math. Lett. **328**, 344–358 (2007)
10. N'Guérékata, G.M.: Spectral Theory of Bounded Functions and Applications to Evolution Equations. Nova Science Publisher, New York (2017)
11. N'Guérékata, G.M., Pankov, A.: Stepanov-like almost automorphic functions and monotone evolution equations. Nonlinear Anal. **68**, 2658–2667 (2008)
12. N'Guérékata, G.M.: Topics in Almost Automorphy. Springer, New York (2005)
13. N'Guérékata, G.M.: Almost Automorphic and Almost Periodic Functions. Kluwer Academic Publishers, New York (2001)
14. Xiao, T.J., Liang, J., Zhang, J.: Pseudo almost automorphic solutions to semilinear differential equations in Banach spaces. Semigroup Forum **76**, 518–524 (2008)
15. Zabsonré, I., Touré, H.: Pseudo almost periodic and pseudo almost automorphic solutions of class r under the light of measure theory. Afr. Diaspora J. Math. **19**(1), 58–86 (2016)
16. Zhang, R., Chang, Y.K., N'Guérékata, G.M.: New composition theorems of Stepanov-like almost automorphic functions and applications to nonautonomous evolution equations. Nonlinear Anal. RWA **13**, 2866–2879 (2012)

Printed by Printforce, the Netherlands